Environmental Sustainability
Sustainable Development Goals and Human Rights

Editors

Sônia Regina da Cal Seixas
Center for Environmental Studies and Research (NEPAM)
University of Campinas, UNICAMP
São Paulo, Brazil

João Luiz de Moraes Hoefel
Center for Sustainability and Cultural Studies
UNIFAAT University Center (NESC/CEPE/UNIFAAT)
Atibaia, São Paulo, Brazil

CRC Press
Taylor & Francis Group
Boca Raton London New York

CRC Press is an imprint of the
Taylor & Francis Group, an **informa** business

A SCIENCE PUBLISHERS BOOK

First edition published 2021
by CRC Press
6000 Broken Sound Parkway NW, Suite 300, Boca Raton, FL 33487-2742

and by CRC Press
2 Park Square, Milton Park, Abingdon, Oxon, OX14 4RN

Library of Congress Cataloging-in-Publication Data

Names: Seixas, Sonia Regina da Cal, 1956- editor. | Hoefel, João Luiz de
  Moraes, 1955- editor.
Title: Environmental sustainability : sustainable development goals and
  human rights / editors, Sonia Regina da Cal Seixas, Center for
  Environmental Studies and Research (NEPAM), University of Campinas,
  UNICAMP, São Paulo, Brazil, João Luiz de Moraes Hoefel, Center for
  Sustainability and Cultural Studies, UNIFAAT University Center
  (NESC/CEPE/UNIFAAT), Atibaia, São Paulo, Brazil.
Description: First edition. | Boca Raton : CRC Press, Taylor & Francis
  Group, 2022. | Series: AAP research notes on optimization and
  decision-making theories | Includes bibliographical references and
  index.
Identifiers: LCCN 2021018826 | ISBN 9780367861698 (hardcover)
Subjects: LCSH: Sustainable development. | Human rights. |
  Environmentalism.
Classification: LCC HC79.E5 E59198 2022 | DDC 338.9/27--dc23
LC record available at https://lccn.loc.gov/2021018826

ISBN: 978-0-367-86169-8 (hbk)
ISBN: 978-1-032-04627-3 (pbk)
ISBN: 978-1-003-02016-5 (ebk)

Typeset in Palatino Roman
by Innovative Processors

Foreword

This is a rich book, rich in insights, reflections and agenda for the present and the future.

In fact, this is a book about social transformation, about building a sustainable society from a risk one.

Building a sustainable society means protecting and preserving wilderness and diversity. In this way, it rejects the capitalist view of a world divided between environment and society.

This book explores the process of a sustainable society, and it focuses mainly on problems of climate change at a variety of scales and how it may affect the health status of millions of people. Those with low adaptive capacity are the poorest and belong to the most vulnerable groups, mainly based on the prediction of climate change because these could alter the spatial distribution of some infectious disease vectors.

It seeks to foster a dialogue between the factors that directly define the health of populations (education, medical care, prevention and public health infrastructure, and sustainable development) and solidarity between people and their governments will consolidate a society that prioritizes human rights and preservation for the defence of the dignity of life and not exclusively of economic systems.

In short, open and read the book Environmental Sustainability – Sustainable Development Goals and Human Rights that is directly associated with the present moment and the critical approaches that each of the authors tried to offer in the different chapters of this work.

The book is organized into four sections: Sustainability and Human Rights, Natural Resources and Sustainability, Environmental Risks and Sustainability, and Mobility and Sustainability.

This means that it provides an almost exhaustive guide to the significant overview of climate change's sustainability science and social science.

I hope that this book will serve to interest the reader in sustainability issues in general and in the kind of lifestyles and political actions that flow from the adoption of a sustainable orientation towards the world.

Sustainability issues represent the particular approach to a more democratic world, which has inspired the authors.

I want to invite you to read this book, and I hope it will inspire you.

Leila da Costa Ferreira
Full Professor
Campinas State University

Contents

Presentation

Sônia Regina da Cal Seixas[1] and João Luiz de Moraes Hoefel[2]

[1] UNICAMP, Brazil
[2] UNIFAAT, Brazil

In mid-2019, we started designing this book and it was approved in October of the same year. Thus, sharing with the different fellow authors and collaborators of this project, we joyfully began this beautiful project that has just been published.

However, little did we know that in December 2019, the world would start a long and arduous challenge, established by the Covid-19 pandemic, declared by the UN on March 11, 2020. Thus, it is impossible to not bring up the context of the production and realization of this book, according to the meaning it brings, and the bond it has with our work. It is also worth noting that the pandemic alone did not establish the serious civilization crisis that we have been facing since the middle of this decade alone. The growing project to consolidate neoliberal capitalism has collapsed world economic systems, undoing multilateral environmental preservation agreements and causing unprecedented environmental changes, which lead to environmental contamination, deforestation, climate change, increasing poverty and inequality combined with the loss of rights and the maintenance of the preservation of the dignity of life that together only make an extreme event, significant like this, put us in a position of excessive social vulnerability and impotence in face of the current socio-environmental reality. And also of a deep shaking in the national and international institutions. Allied to this, we are also faced with great forces and networks that promote society disinformation and disbelief about science.

It is curious to note that since 1988, when it was created by the UN to evaluate data and provide reliable scientific evidence for climate negotiations, the Intergovernmental Panel on Climate Change (IPCC) has been warning about the great impacts that humanity could experience in face of extreme events caused by climate change, understood here as the cruel consequence of this (in)sustainable development model and of a consumption pattern that consumes us as humanity.

In 2007, the researchers responsible for that year's report highlighted the important health problems that are directly related to climate change as well as indicating that there is strong evidence that exposure to climate change may affect the health status of millions of people, especially those with low adaptive capacity, that is the poorest and belonging to the most vulnerable groups, mainly based on the prediction of climate change because these could alter the spatial distribution of some infectious disease vectors (IPCC, 2007: 10-11).

Thus, scientists warned us 13 years ago that viruses that are pathogenic to humans (e.g., ebola, influenza, type H5N1, and avian influenza among some) occur naturally on the planet and live within a natural cycle where wild animals are their hosts, but when human actions alter this cycle, the virus undergoes genetic mutations and recombinations and they can bind to organisms that do not adapt to it, like us human beings, which is exactly the challenge we are facing at this moment (Buss 2020, Rabello and Oliveira 2020, Seixas and Ferreira 2020).

Covid-19 is a new respiratory and cardiovascular disease, which has become a serious pandemic, spreading across the planet at unprecedented speed. The first few cases were found in the city of Wuhan, China, in early December 2019, and at the end of December, the Chinese Center for Disease Control and Prevention after evaluations sent a notification to the World Health Organization (WHO) since at that time it managed to exclude causes, such as influenza, avian influenza, SARS-CoV, and MERS-CoV among the possible ones. On January 7, 2020, the pathogen was identified as a new coronavirus (Allaerts 2020, Zhang 2020), spreading across Asia, Europe, and the Americas. It arrived in Brazil on February 26, and by September 2020 it had tragically offered alarming data, such as around 150,000 deaths and a total of 4,748,882 confirmed cases (Seixas and Ferreira 2020).

What we believe as important to highlight in the presentation of our book, which coincides strongly with the moment we are living in, is that the learning and the challenges that must be present at this moment are related to an important set of alerts: (1) the Covid-19 pandemic is the great challenge we are facing as humanity; (2) it is necessary that we develop learning strategies about measures to contain the spread of the disease, effectiveness, and support capacity of health systems and socioeconomic measures to serve the most vulnerable populations; (3) we need to strengthen the strategic set of development of any country, namely, science, researchers, public university, and research institutes; and lastly (4) it is essential to train populations on the severity of extreme events and climate change. These alerts, impose us, at the same time that we must create the conditions for adaptation and mitigation through the investments of nation-states and their institutions, the formation of citizens and new generations for environmental risks, and the significance of climate change for the planet.

In this way, the factors that directly define the health of populations (education, medical care, prevention and public health infrastructure, and sustainable development), coupled with solidarity between peoples and their governments will consolidate a society that prioritizes human rights and preservation for the defense of the dignity of life and not exclusively in economic systems. Thus, the challenge of bringing the reader, the book *Environmental Sustainability – Sustainable Development Goals and Human Rights* is directly associated at the present moment and in the critical approaches that each of the authors tried to offer in the different chapters of this work.

The book is organized into four sections, namely, Sustainability and Human Rights, Natural Resources and Sustainability, Environmental Risks and Sustainability, and Mobility and Sustainability.

In the first section, Sustainability and Human Rights, are the following chapters: "The Future is Here – Universities, Human rights, and Sustainability" by Néri de Barros Almeida (UNICAMP). This essay moves in many directions in the intention of arguing that facing the great challenges of the advancement of human dignity on our planet depends on being prepared for a systemic approach and that universities may prove to be strategic allies in this purpose.

"Ecofeminism and Sustainability" by Micheli Machado (UNIFAAT). This chapter seeks to analyze and investigate the contribution of ecofeminism to sustainability, considering in this context the relationship of gender issues with environmental degradation and conservation with the current development model as well as with the violence suffered by women and the environment. It was noted that there are still many challenges to be overcome concerning gender and sustainability issues and that ecofeminism has a fundamental role in overcoming these challenges.

"Gender Equity, Sustainability and Human Rights: Considerations About Gender and Race in the COVID-19 Pandemic Scenario" by Sônia Regina da Cal Seixas, João Luiz de Moraes Hoefel, Amasa Ferreira Carvalho, Luana Aparecida Ribeiro Javoni, Gianlucca Consoli and Waldo Emerson de Souza Nascimento. This chapter investigates the relationship between sustainability and gender equity based on human rights, its aspects of the crisis caused by the COVID-19 pandemic, and its impact on the population of greatest socio-environmental vulnerability. It also brought reflections on the current situation of gender equality in the Brazilian and global perspective and included ways of implementing concrete actions to achieve the Sustainable Development Objective 05 (SDG05) of Agenda 2030.

In the second session, Natural Resources and Sustainability, are the following chapters: "Challenges and Opportunities for Water Security in São Paulo Metropolitan Region, Brazil" by Luana Dandara Barreto Torres and Gabriela Farias Asmus. This chapter reports the latest discussions and political deeds on water security regarding the Metropolitan Region of São Paulo (MRSP), considering that it is the largest metropolitan area in Latin America and the most densely occupied and economically developed area in Brazil. It is located within one of the largest watersheds of the continent – the Paraná water basin. The authors expect that this review will help further develop and integrate scientific knowledge on the social, environmental, and economic fronts of water management in the MRSP as well as serve as a tool for sustainable political decision-making in the region.

"A New Relationship With Nature: The Paradigm Shifts between Society and Environment and Its Consequences" by Ana Paula Leal Pinheiro Cruz and Sônia Regina da Cal Seixas. This chapter analyzes the connection between sustainability, climate change, and human rights and discusses the expansion of concepts and concerns about current environmental issues. It highlights that it is necessary to once again signify values and rights attributed to nature to safeguard and preserve human life. For this, it is guided by the possibilities presented by the Constitution of Ecuador. Part of the reflections on the appropriation and use of common goods is related to mining activity actions. The recent episodes of rupture of tailings dams in Minas Gerais stand out, leading the discussion to the dynamics around the mineral extraction practices that have prevailed in Latin America for centuries.

"Sustainability and Water Quality Recovery" by Almerinda Antonia Barbosa Fadini and Pedro Sérgio Fadini. The main objective of this chapter is to deal with sustainability and water quality recovery for which the following scope was defined: to present a brief overview of water resources in Brazil and the world, strengthen the right of everyone to access drinking water, demonstrate the relevance of integrated and participative management emphasizing Brazilian scenario, and expose some innovative technologies of environmental sanitation.

"Environmental Perception of Students at General Secondary School Located in Niassa Special Reserve" by Francisco Gonçalves Nhachungue, Sônia Regina da Cal Seixas and Benjamim Olinda Bandeira. This work deals with the environmental perceptions of students on June 16, General Secondary School located in the central block of Niassa Special Reserve, in Mozambique. The general objective of the study is to evaluate students' environmental perception in relation to their insertion in an area of biodiversity conservation. The study sought specifically to characterize students' socio-demographic variables, identify subjects at secondary school education that deal with environmental aspects, analyze the relationship between students' environmental knowledge and their insertion in the area of biodiversity conservation, and capture students' perceptions about socio-environmental conflicts.

In the third session, Environmental Risks and Sustainability, are the following chapters:"Mining and Sustainability" by Maria José Mesquita, Rosana Icassatti Corazza, Maria Cristina O. Souza, Guilherme Nascimento Gomes, Isabela Noronha, and Dione Macedo. Building on a bibliographic and documental survey and supported by two brief bibliometric exercises, this chapter outlines urgent issues to be addressed and improved by the mining industry, provides an insight into the literature on the subject as an emergent, interdisciplinary, and still fragmented field of research, and proposes a set of recommendations to spur further debate and strategies toward a more sustainable horizon for the mining sector.

"CRIAB Project – Conflicts, Risks, and Impacts Associated with Dams: Looking for Sustainability and Human Rights" by Jefferson de Lima Picanço, João Frederico da Costa Azevedo Meyer, José Mario Martinez, and Claudia Regina Castellanos Pfeiffer. This chapter deals with one association, CRIAB (the acronym in Portuguese for research and action group on conflicts, risks, and environmental impacts associated with dams), constituted immediately after the rupture of Vale's 01 dam in the Córrego do Feijão Mine in Brumadinho, State of Minas Gerais. This group, still in the process of formation, intends to make an effective contribution mainly in what regards the affected populations directly and immediately in mid and long-term consequences, involving both scientific knowledge and solidarity, as well as relevant contributions in terms of policies and strategies for emergencies, contingencies, prevention actions, and computational simulations of different scenarios.

"Habitability and Health: A Relationship with Energy Efficiency" by Andrea Lobato-Cordero and Sônia Regina da Cal Seixas. This chapter discusses the role of the indoor environment of a house in terms of energy, health, and temperature. Further, it proposes that evaluating thermal comfort is an incomplete metric if it is not contrasted with its effects on the health of the people that live in the house. If not properly designed and built, a house could have a negative impact on its inhabitants contrary to its purpose of protecting them. Habitability and a healthy environment could result from an energy-efficient house, but a proper methodology is needed to identify the factors that link the health and sustainability goals.

"Data or Misconceptions? Understanding the Role of Economic Expertise in the Development of Sustainable Marine Aquaculture in Santa Catarina, Brazil" by Thomas G. Safford and Marcus Polette. This study assesses how socialized beliefs among stakeholders about the likely economic contributions from mariculture affected the development of the industry. Next, it examines to what extent government officials, scientists, and growers prioritized and valued accessing economic expertise and business-related data to support science-based mariculture planning. Finally, it applies insights from the sociological study of institutions to inform our analysis of the social forces shaping consideration of threats to mariculture commerce and economic sustainability and assesses its effectiveness in helping achieve UNSDGs 14, 12, and 8.

In the fourth session, Mobility and Sustainability, are the following chapters: "Mobility and Sustainability: Individual and Collective Rights" by Ennio Peres da Silva. Mobility and sustainability are two concepts linked by numerous parameters, like technical, economic, social, and cultural. This multidimensionality makes approaching this relationship quite complex, being it usual to look at them as segments, taking limited analyzes among some of these parameters. Here, in this chapter, the energy aspect of mobility (fuel expenditure) and the respective emissions of pollutants and greenhouse gases, which impact its sustainability, will be considered.

"Estimating Vehicular CO_2 Emissions on Highways Using a 'Bottom-up' Method" by Estevão Brasil Ruas Vernalha and Sônia Regina da Cal Seixas. This chapter considers the transport sector and evaluates the methodological possibilities for estimation of vehicular CO_2 emissions on a major highway in Brazil – D. Pedro. I Highway (SP-65). These emissions are considered the main source for GHGs concentration increasing in the atmosphere and are thus associated with climate change.

"Urban Collective Mobility: Global Challenges Towards Sustainability" by Daniela Godoy Falco, Leonardo Mattoso Sacilotto and Carla Kazue Nakao Cavaliero. This chapter discusses challenges toward sustainability that emerge from the demand for urban collective mobility and its value chain. Many aspects and direct and indirect impacts of the urban collective mobility value chain are presented and distributed amid the SDG, whether delimited by the energy production sector, i.e. oil and gas, biofuels and electricity, by vehicles production sectors i.e. batteries, and by public transportation services sectors. The goals and challenges are assigned in the five 'Ps' that shaped SDG.

"Proposals for Sustainable Urban Mobility Solutions for South America" by Alyson da Luz Pereira Rodrigues. This chapter discusses urban mobility in Argentina, Brazil, and Chile, which are the most economically advanced and populous countries in South America. To this purpose, a general description of the current situation of urban mobility in each country was made, considering how the transport structures affect energy consumption and GHG emissions in addition to public urban mobility policies. Sustainable urban mobility projects have been proposed, socioeconomic aspects, legal structures, and individual policies of each country.

The book concludes with *Final Considerations* by Sônia Seixas and João Luiz Hoefel.

References

Allaerts, W. 2020. How Could This Happen? Acta Biotheor. https://doi.org/10.1007/s10441-020-09382-z

Buss, P.M. 2020. De pandemias, desenvolvimento e multilateralismo. Seção Opinião. Rio de Janeiro: Fiocruz/Agência Fiocruz de Notícias, 03 abr. 2020. 7 p. Agencia Fiocruz de Notícias, 15 de abril de 2020, vide: https://agencia.fiocruz.br.

IPCC. 2007. Intergovernmental Panel on Climate Change. Mudança do Clima Impactos, Adaptação e Vulnerabilidade Contribuição do Grupo de Trabalho II ao Quarto Relatório de Avaliação do Painel Intergovernamental sobre Mudança do Clima Sumário para os Formuladores de Políticas, Bruxelas, p. 10-11

Rabello, A.M. e D.B. Oliveira. 2020. Impactos ambientais antrópicos e o surgimento de pandemias.Unifesspa: Painel Reflexão em tempos de crise. 26 mai. Disponível em: <https://acoescovid19.unifesspa.edu.br/images/conteudo/Impactos_ambientais_antrópicos_e_o_surgimento_de_pandemias_Ananza_e_Danielly.pdf> Acesso em 08 jul. 2020.

Seixas, S.R.C. e Leila C. Ferreira. 2020. Mudanças climáticas e Covid 19: perspectivas futuras para enfrentamentos de eventos extremos. Jornal da UNICAMP (JU), 05 de outubro. Disponível em https://www.unicamp.br/unicamp/ju/artigos/ambiente-e-sociedade/mudancas-climaticas-e-covid-19-perspectivas-futuras-para

Zhang, Y. 2020. The Epidemiological characteristics of an outbreak of 2019 Novel Coronavirus Diseases (COVID-19) – China, 2020. Chinese Centre for Disease Control and Prevention – CCDC Weekly, 2(8): 113-122.

Part I
Sustainability and Human Rights

The Future is Here – Universities, Human Rights, and Sustainability

Néri de Barros Almeida

Executive director of Human Rights State University of Campinas/Unicamp, Brazil

"In a time in which those who detain power are seduced by the most limited logic dictated by the interests of privileged groups, speaking about development as a coming together with the creative genius of our culture may seem like a mere escape into utopia. But the utopic is often the fruit of our perception of secret dimensions of reality, a blooming of contained energies that anticipates the broadening of a horizon of possibilities open to society. Required avant-garde action constitutes one of the noblest tasks to be fulfilled by intellectual workers in times of crisis. It is up to them to deepen the perception of social reality to prevent stains of irrationality that feed political adventures from spreading; it is up to them to project light onto the attics of history, where the crimes committed by those who abuse power are hidden; it is up to them to listen to and translate the anxieties as aspirations of social forces that are yet to find their own means of expression." FURTADO, Celso. "Quando o futuro chegar" InSACHS, Ygnacy, WILHEIM, Jorge, e PINHEIRO, Paulo Sérgio (Org.). *Brasil: um século de transformações*. São Paulo: Cia das Letras, 2009, 4ª. Ed., p. 425.

The Covid-19 pandemic, which started last December, has led to an 'outbreak' of global communication. Alongside the interest in the virus and how to survive it, other demands connected to human rights have gained force. The context has enabled manifestations in favor of greater regulation of communication systems in order to guarantee access to reliable data. Similarly, our full-time connection to communication devices has given local movements more agility, as in the case of antiracist protests that appeared after the murder of George Floyd, the African-American man killed by members of the Minneapolis police (May 25, 2020). The information that the pandemic was a result of an imbalance caused by human interference in a forest environment gave new breath and helped to make the environmental agenda more popular: the destruction of forests, global warming, preservation and promotion of biodiversity, conservation of sources for generating

---

Email: neri@unicamp.br

hydric power, alternative energies, residue management, impacts of extensive cattle raising, etc. In the middle of all these uncertainties, discussions about a planetary future became more frequent. It is still not possible to say that the awareness of the dimensions of this environmental crisis has reached a point that is enough to press for the necessary mitigating actions. It is true, however, that the relationship between human rights and environmental rights has become clearer to a larger number of people.

This essay moves in many directions with the intention of arguing that facing the great challenges of the advancement of human dignity in our planet depends on being prepared for a systemic approach and that universities may prove to be strategic allies in this purpose. There are, however, challenges for this to become possible. One of them is to make society recover its trust in the university, a trust that has been shaken by an obscurantist, revisionist, and denialist culture that has taken control of some conservative means around the globe. Another challenge is to make sure the universities themselves take an interest in accomplishing the changes that will put the knowledge produced by them in the forefront of actions in favor of respecting the rights of present and future generations.

1. The Tragedy of Technology

Daily life with technological objects and means (laptops, tablets, smartphones, internet of things, etc.) does not give the just measure of the technological demands involved in their production. We live with technology, but we do not know its secret life: the haunting machinery involved in the exploitation and transportation of ores, the multiple manners of soil destruction, the contamination of water and air, the disorganization and destruction of communities, the decisions following pacts between entrepreneurs north and south of the Equator that hovers over our heads and determines the present and the future. Because of our difficulty to perceive the prevailing technological model in all its scope, it may be interesting to discuss technology through its register in global culture – equally produced by technological gadgets.

Science fiction is the contemporary narrative genre by excellence. In general, it explores future worlds and despite its name, it worries less about science and more about its offspring technology. Science fiction's most striking purpose is to explore questions such as 'what if we were able to?'. Its questions are fascinating from the point of view of fantasy, but we cannot ignore its intention of power and control. In science fiction, we find technological solutions that are out of our reach today (interstellar traveling, cure of deadly diseases, etc.). Through this path, science fiction—or better say futuristic fiction or technological fiction—records one of the distortions that marks the contemporary perception of the world: we collectively believe that current technological choices are leading us to the desired future with no violence, inequality, or physical suffering. The mistake lies in the fact that we have no power over some of technology's ends, one of the most strategic domains of contemporary life. Until now, most of us have been more object than the subject in technological choices.

This situation has an impact on science fiction, dividing it into two branches. One in which the future is the scenery of a fantastic adventure whose subject is new. The other in which the future is the result of a precise analysis of the objective subject of the present and that, as a rule, ends up in a dystopia or tragedy. In this second model, we can see science fiction interested in assessing the risks that come from the way we think about and perceive technology.

Science fiction is counted among the minor genres of art. But from time to time, a work that can overcome this condition appears. This is what happened with Stanley Kubrick's (1928-1999) *2001. A Space Odyssey* and what some managed to achieve in literature, like Aldous Huxley (1894-1963) and George Orwell (1903-1950) did with *Brave New World* (1932) and *Nineteen Eighty-Four* (1949), respectively. A similar achievement was also obtained more recently by American author George Saunders (1958).

Entitled *Tenth of December* (2013)—date in which the Universal Declaration of Human Rights was approved by the United Nations General Assembly in 1948— Saunders's collection of short stories invites us into its futuristic universe from the point of view of human rights. Although George Saunders does not explicitly confirm the connection between the story that gives the collection its title and his wish to place its reading under the sign of human rights, reading his book from this perspective is fascinating.

It is, in fact, pertinent to discuss human rights through futuristic-technological fiction since all attacks and threats to human dignity over the last 130 years have been in some way connected to technology understood as a form of exploitation. The most emblematic case is that of petroleum, due to the variety and impact of its technological demands: the extremely wide chain of products and consumption it creates, the social inequalities it produces, and the violence and international instability it causes, eventually supporting the war technology industry. Our historical-social memory has only a limited answer to these facts. Its logic, informed by hegemonic culture, presents the current technological and productive system as a result of human ingenuity's natural course of evolution in the face of necessity. In a contradictory way, the more our condition as objects of the technological system of production and consumption deepens, the more the efficient subtraction of our time keeps us from seeing this system as a producer of social inequality and environmental injustice as well as a liberty suppressing agent. The possibility of assessing the individual and collective risks involved in our existence is placed completely outside the tangible human sphere nowadays. Culture is the structure in which this system is legitimized on a large scale. Changing this is essential in order for technology to be re-thought from the perspective of its commitment to life dignity.

The middle class, as the opinion former and reproducer of hegemonic culture, believes that its consumption capacity ensures its place away from the margins and exclusion. However, precarious work has been climbing the social ladder. Income concentration has also allowed consumers to become so sophisticated and has mobilized such a high level of resources that intermediary economic ranges between the middle and the high classes are disappearing. As the high-luxury market becomes broader and more diverse due to the concentration of wealth, we arrive at a situation similar to that lived in the Middle Ages, where luxury commerce had a relevant weight compared to whole segments of non-superfluous goods. Add to this the mechanization of work, we have a scenario where the progressive disappearance of dignified work and employability tends to spread to all professions, no longer being restricted to layers of professionals with no qualification. Even though precarious work will not necessarily represent a reduction of income for everyone, the fact is that this precariousness is something new and for which the middle class is not prepared.

The fantasy that exploitation (of the environment and workers) will generate growth and that this will at some point ensure an equal and satisfactory standard of life for everyone has submitted us to new forms of exploitation. It has also led us to live, without choice or full awareness of the risks, with the waste products from this growth, one of them being exactly the systematic disappearance of dignified work. For at least 130 years, humanity has been dragged by the economic model that promises freedom but enslaves us to the imperatives of a technological productive model. It is easy to understand today why the technological world that Huxley describes in *Brave New World* is founded on two very distinct human groups. On one side we have different castes of humans whose conditions, rights, and fates come from a genetic intervention. Living in cities, they spend their days at work, and at night they are 'free' to consume technological services and entertainment. On the other side, those excluded from genetic selection, who live in a wild area limited by electrified fences and are immersed in absolute paucity, religious obscurantism, and hypocritical moralism.

Going back to George Saunders, his short stories are placed in a time that is identical to ours in almost every aspect. We understand the characters' feelings, values, and ways of living and moving, which is what most efficiently constitutes a narrative's temporal landscape. The dissimilarity between our world and the one described by Saunders lies in its technical-technological reach, which emerges in small details such as a multi ethnic group of living dolls that adorn the small garden of a middle-class house and are pinned to a sort of technological clothesline. Observed up close, these small details present a future that is based on values from our current culture. At the same time that his stories show happy families, where all care is present, this care is always mediated by things. The result is a bizarre mixture between anguish and childish satisfaction. Just like us, Saunders's characters are their orphans, living wishes that are projected onto their souls and that drag them towards unhappiness that cannot be confessed when you enjoy the privileges of superfluous consumption. Therefore, his characters consume other lives and are programmed to consume their existences in the name of false aspirations and false benefits. Saunders's characters—partially cruel, partially ignorant, and partially alienated— live surrounded by technological devices that have no real connection to any desire born from within. Inversely, this apparatus can be mobilized to convince them of another humanity, as in the case of the young prisoner submitted to an experiment that intends to demonstrate the impossibility of altruistic empathy.

Good science fiction writers show us that human time and the time of technology are very different. Human ability to perceive and understand the world is not enough to follow technological change, which is much faster than the standards of our perception. This makes us unable to develop a perception of risk that is appropriate in light of the development of hegemonic technology. Contrary to the stereotype that has been explored by science fiction, we will not be dominated by machines endowed with consciousness because they do not need one to dominate us. The annihilation they will perpetrate is slower and more subtle and is applied to the very basis of our existence.

Unfortunately, everything indicates that the current risks of climatic and environmental emergency will only be noticed and faced properly when the human scale of perception is sensitive to them. And by then, it may be too late. Another possibility is to bring about a deep cultural change capable of accelerating this awareness.

2. The Tragedy of Science

The modernity that came into light in Europe at the end of the fifteenth century is, in essence, a global phenomenon. Its globalizing vocation is associated with a culture of conquest and exploitation based on a supposed superiority of its truth and its purposes. This culture was not new when modernity started. In the Middle Ages, the Crusades— which began in the eleventh century and were still present in the minds of the sixteenth century—installed a long-lasting logic of external conquest and exploration, first through sacks and then through commerce. The ideological basis of these actions was the defense of the Christian community and proselytism. However, when scientific thought changes the thinking standards of knowledge and control of the natural world at the beginning of Modernity, it joins religion to justify the right of the western world to conquer and it supports the new forms of exploitation. The problem, obviously, is not intrinsic to science. It comes from the cultural and ideological foundations that led science to obliterate or impose itself over all other forms of knowledge. This medieval commitment to science has been put into question and scientists must be willing to discuss it. It is, without a doubt, a great challenge, but it may also represent a freer science that is more aware of its purposes.

It is undeniable that in relation to pre-modern knowledge, scientific culture has brought us huge accomplishments. The reach of western medicine is undoubtedly the best example. But every gesture of creation by science corresponds to steps of destruction. Healthcare is still a good example. Medicine leaflets make it clear that the use of synthetic products may represent some harm to health and that all the support to healing the body can have an environmental impact. Western science's commitment to a hegemonic project of domination and exploitation blocks the legitimacy of traditional medicine even when it proves to be efficient. The resistance to acupuncture and phytotherapy are good examples of this posture. The hegemony intended by western medicine excludes from our imagination ways of knowing the world and ways of life that are potentially non-destructive to human life or the environment.

We know today that entire parts of the Amazon Rainforest are not constituted by primary forest but are in fact the result of thousands of years of handling by original people, for food and medicinal use. In this experience, we can find an efficient form of science and technology. However, we have difficulties in granting value to this because it does not answer the demands of a system where everything is liable to become an object of exploitation, i.e. must become a commodity and generate the concentration of wealth and power. An example that challenges this standard comes from bioeconomics, which aims to establish links between science and traditional knowledge. In bioeconomics, we can find a mixed approach of forest economics in which universities and high technology are allies to traditional knowledge, promoting the preservation of the woods, the way of life of the communities involved, and generating local wealth. This is the goal of the project entitled "Amazonia 4.0", coordinated by Brazilian scientist Carlos Nobre, and of the Amazon Center of Indigenous Medicine, coordinated by indigenous anthropologist João Paulo Barreto among others.

There is no doubt that western science has produced technology of great impact on human beings' quality of life. However, these accomplishments have never been made universally accessible. The Covid-19 pandemic has shown us the magnitude of planetary vulnerabilities related to access to healthcare, despite the existing medical techniques. Workers' living conditions at the epicenter of the industrial revolution have improved greatly since the eighteenth century. They no longer sleep on the streets or get drunk to be able to bear the exhaustive work journeys, but social development in these regions—universal education, housing, and some public benefits—has created chronic and never before seen destitution in peripheral zones. Brazil is a good example. Since its assimilation to European influence, Brazil has performed a subaltern role in the world economy. Slavery and exploitation of natural resources structured an elitist culture that never allowed the emergence of an organized job market or access to land. The destruction of the Amazon Forest for extensive cattle raising purposes is one of the byproducts of this world system. When one hectare of lettuce produces wealth that corresponds to 24 hectares of bovine raising, as is the case today it becomes clear that the goal of the destruction of the forest is to reproduce an exploitation system. This system presents itself as a network that overwhelms populations, territories, and nations around the planet. It is not exactly about how much money you make, but that the way this money is made contributes to reproducing a preexisting form of domination. The efficacy of the old logic of the conqueror, where environmental destruction and social exclusion go hand-in-hand becomes clear.

The result of the model in which development and exploitation are associated was the global transformation of poverty into destitution. The hijacking of workers' land and time made it impossible for them to accomplish any kind of work for themselves. In this way, workers became hostages to the world market of essential goods. This overview shows us that something happens in the transfiguration of the discoveries of science and technology. It is in this passage that universities can have an important role.

In *Machines Like Me and People Like You* (2019), Ian McEwan (1948) develops a retrograde futuristic fiction that takes place in an alternative early 1980s. His romance comes from the premise of what would have happened if Alan Turing (1912-1954) had not died precociously and technology of information and artificial intelligence had developed a few decades earlier. The result is a world where technology has been made abundant both for what is favorable and unfavorable to life. Do we know a world like this? The questions posed by this scenario are: where do the supposed results of this tension between benefit and harm in the current technological model point to, and is it possible to guide them toward solutions that are favorable to life on a planetary scale?

It is obvious that technology is not responsible for the choices that led to its development as hostile to the universalization of respect for human dignity. However, science, in some way, has helped this facet by transferring to technology its status of a neutral and democratic domain when, in fact, the decisions that guide technology are restricted to a few. Government policies insist that supporting these decisions is a way of creating—or at least saving—jobs. But what we see in practice is that these results are not being reached and, quite contrary, that the periphery of chronic destitution has increased even in rich countries. Consequently, since the 1970s, the financialization of the economy and a vertiginous investment in technology based on the idea of exploitation have banned human needs from the horizon of the planet's political-economic priorities.

The pandemic has laid bare this model that defends the growing deregulation and the retreat of State intervention in the guarantee of human rights (healthcare, work, education, the standard of living, nationality, security, political rights, freedom of expression, and justice). In showing the limits of thinking about society from a market perspective, the current crisis has shaken the legitimacy of neoliberalism. In a country like Brazil, until the pandemic hit, it was unthinkable for many members of the ruling elite to consider income recovery policies that are under discussion today. At a planetary level, the deceleration of production and commerce imposed by quarantines also showed that the limitations for change in economic patterns are not technical but political.

Despite the shock of reality represented by the pandemic, we are still far from collectively understanding our situation in the world. Most people resist acknowledging and extracting all the consequences of two facts: on the one hand, that the environmental crisis is the critical problem of our time, notably the climatic emergency that might make the continuation of human life on Earth impossible and, on the way to this, create an enormous civilizational crisis. On the other hand, that it is necessary to develop an awareness of the systemic character of this crisis and that in order to face it, we will have to critique the basis of hegemonic culture contained in the triad conquest-exploitation-development.

When the politics of science and technology is submitted to models of exploitative development that destroy the planet, then we need to admit that scientists do not have control over this politics. Therefore, their political "neutrality" and their resistance to criticizing the technological standard have been inefficient in guaranteeing their credibility and influence. In Brazil, where public universities are responsible for 95% of the country's research, this has not prevented attacks and cuts to funding that seriously threatens their existence. Maybe a political position—not necessarily partisan or ideological—is what science needs to strengthen its position in society. This position—beyond the current commitment in favor of the material conditions for the development of science and technology—can find solid ground in an unnegotiable commitment of science to the promotion of the dignity of life.

Universities preserve a repository of critical knowledge that is essential to scientific and educational construction and the politics of change that the planet needs. In order for science and technology to overcome their relationship with the culture of exploitation and enter a cultural regime based on human and environmental rights, concrete actions are needed. Interdisciplinary research groups are an achievement, but we need to go beyond the interdisciplinarity that is internal to these groups and develop the conditions for exchange between groups and areas. Universities must become leaders in the spreading of ideas, principles, concepts, and values that come from what they attest about the planet's environmental imbalance. It is important that, from the current prognostics, they establish minimum local and global goals and form a network of dissemination and discussion of these goals.

The slowness of universities to adhere to the United Nation's Declaration of Climate Emergency or to establish ambitious commitments in relation to mitigating the damage they cause to the environment is significant. The rhythm and depth of this adhesion are not compatible with social and environmental demands. The greatest harm caused by this delay refers to the formation of students and new researchers concerned with environmental challenges. The goals of the United Nation's 2030 Agenda for Sustainable Development are an important political-pedagogical tool, but their impacts are still insufficient in higher-level education and research institutions.

This situation has hindered a systemic approach to a systemic problem. Close collaboration between exact, biological, and human sciences is the only way to face a crisis that is simultaneously social and environmental and that is based on an aging, albeit resistant political culture. The environmental situation is of such gravity and rooted in so many structural elements in the world we live in that it should be systematically considered by different research scenarios. But that is not what happens. One example, among many others, may be found in population studies that understand population growth merely from the sociological perspective (advances in women's rights to reproductive healthcare, countries' capacities to incorporate immigrants, etc.) and their impact on economic growth (understood as the economy of exploitation, exclusion, consumption, and generation of waste). If we arrive at around 8.8 or 11 billion inhabitants by 2100, this possibility cannot be assessed only in light of the dominating economic criteria. It is imperative to consider the social and environmental impacts. From this point of view, these numbers do not point to the success of certain economies but rather to planetary collapse. If we look at the projection of population growth from an environmental point of view, this suggests urgent actions in favor of environmentally sustainable economic solutions are needed. The question is not whether the adequacy of population percentage will ensure the growth of productive activities in some parts of the planet and change geopolitics but rather how much social exclusion and environmental destruction this situation will create. Once human life is relevant to human beings themselves, the environmental crisis must be considered in all projections of research results in which it is pertinent.

A hegemonic standard of science still makes the acknowledgment of the potential and the discussion around other technologies (traditional and modern) slow. The model of development as economic growth has already proven that it has no intention of delivering what it promised: universal access to the comforts of the middle class. However, the strength of this model renders any other form of alternative technology invisible and delegitimizes any other way of life, especially those where ideas of wealth and value do not coincide with the exploitative standard. Therefore, in order to have the change we need and have the participation of universities in these changes, this cultural inability needs to be overcome. For this to happen, it is essential and urgent to bring exact and biological sciences (capable of producing new technology) and human sciences (capable of criticizing the limits of our culture and thought systems and identifying human risks in technological processes) closer. One example is how ecosystemic services and bioeconomics have been highly valued by researchers who intend to counteract exploitative interests. This movement does not happen necessarily outside the logics of the productive *status quo* though. Production with social-environmental preservation must be associated with CO_2 emission targets applied to the whole circuit where bioeconomics products are disseminated. This takes planning that goes beyond the natural sciences to include political sciences, economics, law, and international relations.

Once the above-mentioned challenges are overcome, universities can enjoy an extremely important position in fighting the environmental crisis due to their role as agents in global partnerships. The Sustainable Development Goals demand a well-structured, strong network of local and global actions from which we are still very far at the moment. In light of the politics of weakening the United Nations by the United States and the growth of the extreme right in many parts of the planet, the main members of this network today are local political organizations (especially in cities), companies, and non-governmental agencies. Universities must occupy a highlighted place in this scenario as representatives of knowledge committed to social-environmental sustainability.

Universities have been present in the debate regarding sustainable development through specific research groups. They need to get involved with this agenda in a massive way so as to foment a global involvement of these institutions. Universities must be seen as strategic institutions in the resolution of large world problems, first, due to the network that they have already built through knowledge and cooperation agreements; second, through the aggregation of different knowledge that these spaces harbor in a unique way; and third due to their social capillarity, be it through the diversity of their students or the professionals it delivers to society periodically.

Because of the environmental crisis, its effects (on water and food safety, land and air quality, defrosting, raising of sea levels in densely populated areas, human exploitation, and social violence), and demands (control and decrease of greenhouse emissions, particularly CO_2 and methane, clean and sustainable energy, management of waste products, employability, and social order) represent great challenges to human rights, universities' responsibilities in safeguarding human dignity become clear. Universities offer an environment where problems and solutions can be analyzed following a global logic. Their commitment to the demands of the environmental crisis through the education of students and through methods, themes, research questions, and external commitments that strengthen ethics where the dignity of all forms of life is a first-rate value constitute the ingredients to safeguard and advance human rights in the new and decisive decade that opens before us.

3. The Tragedy of Memory

The expression 'environmental crisis' is inexact. It would be better to follow Bruno Latour and speak of "an ongoing, irreversible ecological mutation"[1]. The progressive environmental unbalance is already obvious to anyone attentive to daily life and the news. Prolonged droughts, unbearable heat waves in regions known for their temperate climates, changes to the regime of seasons (shorter winters and ever longer summers with clear effects on the life cycle of the plant and animal life), smaller differences of temperature between day and night, tropical storms in temperate zones, gigantic forest fires in both hemispheres, generalized contamination of the environment by plastic and fertilizers, defrosting in the poles and snowy peaks, increase of sea level advancing toward densely populated regions, infestation by locusts, the disappearance of species of bees and other pollinizers, etc. The most urgent issue today is to know whether we can—and want to—stop the advances of these catastrophic changes.

Brazil has the largest biodiversity on the planet and whatever happens in its territory will have an impact on everyone's future. At the moment, the prognostics are not good at all. If on the one hand countries react to the deforestation of the Amazon and Brazil's non-compliance to environmental goals by cutting funds destined to the Amazon Fund (almost 60 million dollars from Germany and Norway) make threats to the ratification of the free trade deal between the European Union and Mercosul, and companies track Brazilian products more carefully triggering the reaction of local businessmen[2].On the other hand, large mining companies increase their presence in Brazil, particularly in the Amazon territory without significant resistance from their countries of origin. On top of that, Brazil insists on not guaranteeing environmental protection[3].

Currently, 85% of the Brazilian population lives in an urban environment, and 57% of this total are concentrated in 6% of the cities. This 85% take up the equivalent of 0.63% of the country's territory. These percentages point to the importance of local communities in defining the fate of biomes. In a country with this profile, the right to land as part of social-environmental preservation is extremely important. Running in the opposite direction is the concentration of land, including the ones occurring through the invasion of public and indigenous lands which continues to happen

---

[1] http://www.bruno-latour.fr/sites/default/files/downloads/P-202-AOC-ENGLISH_1.pdf (consulted on August 3, 2020).

[2] https://veja.abril.com.br/economia/pressoes-e-ameacas-estrangeiras-poem-em-risco-o-agronegocio-do-brasil/(consulted on July 30, 2020)

[3] https://g1.globo.com/natureza/noticia/mineradoras-canadenses-souberam-de-extincao-de-reserva-na-amazonia-5-meses-antes-do-anuncio-oficial.ghtml; https://amazonia.org.br/2020/03/batalha-judicial-tenta-impedir-mineradora-canadense-belo-sun-de-explorar-ouro-em-terras-indigenas-no-brasil/; https://outraspalavras.net/outrasmidias/como-a-belo-sun-abocanhou-o-ouro-amazonico/(links consulted on July 29, 2020) and DIELE-VIEGAS, Luisa Maria; PEREIRA, Eder Johnson de Area Leão; ROCHA, Carlos Frederico Duarte. "The new Brazilian gold rush: Is Amazonia at risk?", *Forest Policy and Economics*, 119, October 2020, Available in: https://www.sciencedirect.com/science/article/abs/pii/S1389934120303750?via%3Dihub(consulted on August 1, 2020)

for the benefit of mineral exploitation, cattle raising, and the production of soy for export. Legislative, executive and judiciary powers need to ensure the human rights of local populations in order to prevent social-environmental catastrophes entrenched in exploitation. The results of the installation of the Belo Monte hydroelectric plant in the state of Pará of ore mining in Mariana and Brumadinho in the state of Minas Gerais, and of the 'Day of Fire' in the Amazon (August 10, 2019) are examples that attracted large-scale media attention but that represents the logic that governs everyday life and has led our country to a state of chronic and ever-growing social-environmental disaster. Brazil is only an expression of a world system though. The impacts of its current environmental politics must be seen as a warning to the urgent need to make efficient global decisions that aim at the protection of biomes and populations and the promotion of a productive system that makes sense from the perspective of the imperatives of the dignity of life.

Changing the cultural system to which the exploitation system is connected necessarily goes through the deconstruction of the founding narrative of a hegemonic regime of memory. Since the fifteenth century, the institutional production of a memory that bypasses human and environmental tragedy has been the constant side kick to conquest and exploitation culture. Two general consequences come from this 'forgetting'. On the one hand, the inability to perceive the damages, risks, and failures of this system always seen as necessary, civilizing, and the only one possible. From this point of view, human destiny is indissociable in submitting to the demands connected to the techniques of exploitation. On the other hand, even those who can see the disaster—usually those who can afford a high-quality education—are unable to imagine or wish for a different fate and continue to bet on a predatory adventure. The whimsical plans of the wealthy to escape this cataclysm are all over the media. Thus, the planet's gigantic wealth—extracted through the financial system and by reverting public resources—is applied to private projects that offer 'solutions' that at the same time ignore and deepen the problem. These projects include everything from water and mineral exploitation in asteroids, the creation of colonies on Mars, and the construction of ultra-technological bunkers in places that are candidates to being the final frontiers on Earth able to hold human life[4]. We can oppose these elites' interests in cheap dystopic science fiction projections that do not make us think and are limited to worst-case scenarios which cannot make us personally accountable with the possibilities that concrete situations indicate. The first one comes from the scientifically recognized fact that environmental disaster is caused by human intervention.

The second one is related to the possibility of overcoming our induced blindness by betting on production and cooperation models that are truly alternative in a regimen

---

[4] https://brasil.elpais.com/internacional/2020-08-03/bilionarios-se-preparam-para-o-fim-da-civilizacao.html, https://brasil.elpais.com/ciencia/2020-07-30/eua-iniciam-a-missao-para-encontrar-sinais-de-vida-em-marte.html, https://brasil.elpais.com/ciencia/2020-07-23/china-se-lanca-a-conquista-de-marte.html, https://brasil.elpais.com/ciencia/2020-07-17/emirados-arabes-entram-para-a-corrida-rumo-a-marte.html, https://brasil.elpais.com/brasil/2017/09/29/ciencia/1506674185_145347.html; https://brasil.elpais.com/brasil/2018/06/29/ciencia/1530264726_949579.html (consulted on August 6, 2020)

of memory that is open to that which the myth of western superiority has kept hidden. The value of marginal experience and techniques must be recognized by those who, in some way, are committed to natural sciences and the technical basis of the current production model. From this perspective, one of the most strategic categories is undoubtedly that of original peoples due to their wisdom of extraordinary resilience and of the fact that their shadowed history can bear witness to the nature of the West. The level of respect that is granted to original peoples over the next decisive ten years will be an indication of the path we will follow as a species.

The West needs to reach maturity. But one can only reach this stage of life when 'being with the others' is understood and practiced with respect. Concretely, this means that attention to the rights of original peoples and vulnerable populations must enter our culture truly and decisively. Guaranteeing ways of life, values, language, education, and health is at the heart of the defense of the environment and, consequently, at the heart of human rights all over the world. Therefore, advancing in the protection of the environment and of human rights depends on a social awareness favorable to land demarcations and heavy investments in projects—elaborated in partnership with the communities—to foment revolution in education, healthcare, sustainable global economic development, and, finally, high investments in inspection and juridical education so that we can overcome the impunity of those who violate human and environmental rights.

Although it is not easy to reach, the development of a human rights culture has the potential to become the accelerating factor for the necessary changes to stop the environmental collapse. This movement depends on an organized society, the circulation of information about this collapse, the interest for the development of efficient solutions, the courage to live with very difficult news, and even greater courage to acknowledge the difference between right and privilege. As Bruno Latour has proposed, the pandemic has shown us that we can decelerate production without collapse[5]. But poverty also needs to be stopped. The poorest pay when there are production and consumption, but they also pay when these decrease as the pandemic has demonstrated[6]. From the point of view of the conditions of existence of a growing mass of excluded and vulnerable people, in and out of cities, it is easy to realize the relationship between human rights and sustainable development in its primary and ethical meaning of a right to dignified life to present and future generations and not as economic growth at all costs as the expression is frequently used.

The present situation in which we see exclusion expanding over habitually vulnerable social groups but also over new social segments makes it clear that the rights granted for centuries to the dominating classes in the name of economic rationality would produce a future where the majority would enjoy a minimum

---

[5] http://www.bruno-latour.fr/sites/default/files/downloads/P-202-AOC-ENGLISH_1.pdf (consulted on August 3, 2020)

[6] Oxfam's report on the increase of income concentration in Latin America and the Caribbean can be accessed on the link https://www.oxfam.org.br/noticias/bilionarios-da-america-latina-e-do-caribe-aumentaram-fortuna-em-us-482-bilhoes-durante-a-pandemia-enquanto-maioria-da-populacao-perdeu-emprego-e-renda/ (consulted on August 3, 2020)

standard of comfort and material safety are nothing but a piece of the machinery of a supra-state system of political domination. The analysis of its basic fundamental structures shows us that what this economic rationality professes will never be reached because it is not part of its ends. This history is, obviously, not a linear or homogenous one. The welfare state that was established in some parts of the world is proof of that, and we must admit that its cost (land expropriation, environmental degradation, and social destitution) was transferred to the planet's peripheral regions. In protecting financial exploitation and the exploitation of nature, these concessions allow for human and environmental exploitation to advance over all of the environments on Earth.

The denial of the right to land (in favor of mineral exploitation and industrial agriculture and cattle raising), the expropriation of work (through its continuous precariousness that little by little excludes workers both as labor force and as consumers), and the delegitimizing of political will (with actions that deconstruct rights earned by society over decades, or even centuries, of fights) walk together and, at this moment, present themselves as challenges to the continuity of human life on Earth. We are, undoubtedly, far from the sixteenth century, but the primitive narrative of the right to conquest, followed by exclusion and subsequent destruction/ exploitation continues to operate. In a system that produces such violent power inequality, that prejudice, and racism—the strongest instrument for transforming human beings into objects of exclusion, exploitation, and extermination—remain as underlying motivations for the narrative of progress. The exploitation of humankind and nature are the same.

As it is presented nowadays, the narrative that structures our perception of the facts is no longer sufficient to offer a collective understanding of the moment we live in. Evidence from the present clashes against a historical narrative that affirms human rights over nature and culture as a rational attitude. We can highlight two moments of this construction. The first one takes place in the fifth century BCE when people of the Greek language launched a model of artificial memory codified by writing. We inherited two important elements from this model: first, the understanding of human experience through fundamentally sociological and political criteria. In this conception, human beings appear connected to an exclusive, moral, and rational system that is distinct from the natural world. Knowledge of the distant past of nature and humankind was integrated into different fields of knowledge: physics and nature. The second element was the codification of the relationship between peoples in terms of contrast and conflict.

After going through several forms of appropriation over time, this tradition was retrieved with new urges between the fourteenth and nineteenth centuries. The nineteenth-century added a linear and progressive view to this narrative that was slowly created by the meeting between Christian temporal perspective and the acceleration of the processes for understanding and controlling the physical world. Medieval eschatological pessimism gave way to trust in eternal progress. The presumption of superiority of this productive-technical-scientific system to progressively solve humanity's problems and lead it to happiness makes it difficult to admit the relevance of its risks and damages.

Over the past 200 years, we have learned a lot about our biological history and the million and a half years of our pre-history. However, this knowledge is incorporated in a very imperfect way to the awareness of our natural condition. In general, historians have never made a real effort to integrate pre-history into historical memory in a relevant way and even less to the natural history of humankind. Cities and states that arose over eight thousand years ago have always raised more interest in the inhabitants of cities and states today than forests, rivers, seas, bonfires, hunts, and 'rudimentary' forms of language. Modern historians have, of course, always relied on their biologist colleagues to present human beings in their natural world. But working with their backs to each other, the awareness of our biological vulnerability has never actually entered our sociological memory. Similarly, we have advanced very little since the concept of biodiversity in the 1980s confronted us with the delicate and vast interdependence between the various forms of life (visible and invisible and known and unknown). Regardless of what we think of ourselves, our historical existence has always depended on this balance.

Overall, this memorial tradition has had three larger impacts on the perception that human beings have of themselves. First, it has made us uninterested in our natural condition. Second, it has made us unable to notice our failure and the tyrannical way in which we treat nature and our species. Finally, it has kept us from seeing in the long duration of our history that begins much earlier than the construction of the first temple or palace – the importance that selfless cooperation has in human survival. Because of this and of the hegemonic historical memory developed in the nineteenth century to describe the evolutionary trajectory of exploitation culture, it is easier to affirm competition, conflict, and supplanting as the only effective solutions in moments of extreme collective difficulty.

The evolutionary view of history has had the effect of leaving us without a pedagogy of memory since everything that is seen as better and more important in history is always inscribed in the present and the future. This way we end up without alternatives to the current model of development and society. That is why accelerating our responses to the environmental crisis depends on a change in historical culture that is able to legitimize the change in the matrix and finality of technology. A politics of memory that deals with the past not in the sense of overcoming it, but as an imaginative alternative space that allows us to defeat our insensibility to the possibilities are critically necessary. We are living a paralysis that comes not from the death of utopias, but from the way we make use of historical memory. Starting from the very immaturity in the face of the death of utopias, as if human history had started at the crossroads left by revolutionary processes that exploded at the end of the eighteenth century. We have given little attention to the fact that along with that of utopias, we have been living the death of capitalism as well. If we take long in realizing that we must envision change and a way to manage it, it will be a fatal mistake. The past cannot be seen, obviously, as a source of answers, but it can help us to unveil the mirage in which we see ourselves as mere consumers and producers of waste products.

Final Thoughts

In 2020, we have been confronted with our global fate. We stand at a crossroads, and it is still not possible to say for sure what our future will be like. It is certain, at this moment, that the powers of destruction are at an advantage and among them, our inability to assess risks. In the face of governments that reject the commitment to general wellbeing, the responsibilities of knowledge communities increase. In order to take over these responsibilities, the nature scientists and researchers of techniques must keep three facts in their eyesight. First, that the epic form of historical narrative, where the scientific and technological conquest of the world corresponds to a victory of civilization actually hides a trajectory of the destruction of biological, cultural, and intellectual diversity. Second, because of this, we are unable to recognize efficacy and value in existing alternative models of society and technology or in those that are still being proposed. Thus, we do not realize that the reason we believe there is no way out is that our regimen of memory points us to a single path. Finally, that our inability to fully assume that the hegemonic technological model robs us of our existence in its most elementary and biological dimension comes from a memory that incapacitates our self-perception as part of the natural world.

Our memory of modernity states that thinking is a deep instance where freedom remains impenetrable. Facts, however, show us that thinking has also become a commodity and, as such, a servile object. If it is worth retrieving the sense of freedom and relevance of science, we must face our condition and as Melville's (1819-1891) "Bartleby" says "prefer not to"[7].

---

[7] MELVILLE, Herman. *Bartleby, the Scrivener.*

Ecofeminism and Sustainability

Micheli Kowalczuk Machado

Center for Sustainability and Cultural Studies, UNIFAAT University Center
(NESC/CEPE/UNIFAAT), Estrada Municipal Juca Sanches 1050, CEP 12954-070,
Atibaia, São Paulo, Brazil

1. Introduction

According to data from the United Nations, 80% of the participants in ecological
activism are women (Camargo and Fontoura 2014), a reality that may be related to
the ecofeminist movement which has been developing since the 1970s as a school of
thought that has guided environmental movements and feminists in various parts of
the world, seeking to make an interconnection between the domination of nature and
the domination of women (Siliprandi 2000, Carmo et al. 2016).

For Gaard and Gruen (1993), the analysis of environmental problems often
helps to understand how and why women's oppression is linked to the unjustified
domination or exploitation of the environment. The authors cite as an example that
especially poor women in rural areas of less developed countries and who are heads
of households suffer disproportionate damage caused by environmental problems,
such as deforestation, water pollution and toxin contamination. Understanding this
issue helps to understand how the life and condition of women are connected to
contemporary environmental problems (Gaard and Gruen 1993).

Feminist environmentalism begins with the perception of similarities and
connections between forms and instances of human oppression, including the
oppression of women and the degradation of nature. A central position that underlies
ecofeminism is the belief that values, notions of reality and social practices are related
and that forms of oppression and domination, although historically and culturally
distinct, are interconnected and entangled (Ruether 1996).

In this context, the search for sustainability is related to the minimization
or resolution of environmental problems experienced by today's society that
have generated significant impacts on the environment and whose reflexes affect
different social groups in different ways and with different intensity. Gimenes et
al. (2019) point out that sustainability seeks a balance between social, economic

Email: michelimkm@gmail.com

and environmental factors and from a social point of view, women's empowerment becomes a key element for the transformation of the current system, which in addition to generating environmental impacts in different scales affect the quality of life of human beings and is also marked by inequalities, vulnerability, exclusion and discrimination concerning various social groups, especially women.

Women are essential in the practice of actions for the promotion of gender equality, aiming at fairer working relationships as well as strategies that aim at the full development of the female gender. This emancipatory performance is part of the discussions related to the perspective of sustainability and the ideological and behavioral change of society. The condition for modern society to achieving a more sustainable way of thinking about its development is related to the vision of equity in social relations between genders, developed by humanity (Couto and Wivaldo 2017).

For Lamim-Guedes and Inocêncio (2018), the debate on sustainability must contemplate more than the management of natural resources, including issues related to the part of the populations that suffer from socio-environmental injustices due to financial, racial, ethnic or gender reasons. Thus, the relationship between gender and the search for sustainability must be discussed in a way that encompasses quality of life and equal access to health, education, employment and political representation, going beyond aspects of resource management and business maintenance. The central issue is the need for urgent broad and unrestricted recognition that women are an essential part of the search for sustainability and the resolution of several problems, such as climate change, food insecurity and nature protection.

Given the above, this chapter aims to analyze and investigate the contribution of ecofeminism to sustainability, considering in this context the relationship of gender issues with environmental degradation and conservation with the current development model as well as with the violence suffered by women and the environment.

2. Women and the Environment

Currently, it is not possible to think about environmental issues and sustainability without considering a complex vision concerning the concept of the environment itself. In addition to biological factors and the recursive concept, it is necessary to include human beings in this context, as well as all the social, economic, historical, political and philosophical configurations and interactions inherent to the relationship between humans and nature.

In general, terms, when talking about the environment, nature is often considered external to human beings. However, because of the current search for environmentally healthier alternatives for human development, it is essential to consider human society as part of the environment (Foladori 2002).

Castro and Abramovay (2005) mention that in general the approach to the environment and sustainable development concerning environmental degradation involves, for example overexploitation and inadequate management of natural resources as well as deforestation and contamination of water, soil and air. However, a purely biological perspective, which does not consider the relationship that men and women and their different forms of the organization establish with their surroundings, has been the conservation discourse. "We refer to the relationships that human beings establish among themselves and with other beings in nature, through simpler, or elaborate, or even contradictory creations, as in the context of the wider society" (Castro and Abramovay 2005). So, environmental issues are eminently social (Leff 2002).

The current conservationist thinking influenced the prepositions for environmental problems, limiting itself to presenting superficial solutions that did not consider their causes which encompass economic, social and political problems (Pilon 2005, Pelicioni and Philippi Jr 2005). To conserve natural resources, it is necessary to think about whom, how and why to carry out this action, otherwise what is actually being carried out are assumptions or generalizations of little value for understanding reality or intervention in reality (Castro and Abramovay 2005).

It should also be noted that in society, the different human groups also develop different forms of relationship with nature as they are ordered, hierarchized, differentiated and occupy a certain position. However, when it comes to environmental issues, social analyzes linked to this context are often underestimated (Castro and Abramovay 2005).

Addressing issues related to the environment is about understanding and relating natural processes linked to human activity, so it is not about knowing social and natural processes in isolation. It is necessary to understand how humanity's use of resources today can interfere in natural processes that affect the quality of life of human beings and also to analyze possible alternative ways for negative impacts to be avoided or minimized (Castro and Abramovay 2005).

In this context, it is important to mention that the current development model affects men and women daily and that it is neither sustainable nor egalitarian. Thus, to improve the condition of human beings in society, it is important to review the power structures in which they are immersed (Castro and Abramovay 2005). A view that considers the issue of gender is essential within the environmental sphere to promote a critical look at the social origins of environmental problems that affect in a non-homogeneous way, different groups and human communities, particularly women (Lamim-Guedes and Inocêncio 2018).

Tornquist et al. (2010) mention that the "social inequality that affects poor women causes them to become the first and most affected by the ecological crisis and deterioration of the environment. Some examples of this are the increase in the workload of women in the face of climate change to satisfy the basic needs of the family in face of water scarcity, firewood, animal fodder, among others. Especially in rural areas, women have felt the effects of water and air contamination, the increasing exposure of them and their families to chemicals, or also in contemporary migration processes, some of them related to climate change" (Tornquist et al. 2010).

Jacobi et al. (2015) recall that the relationship between women and nature is not recent and that the reflections that establish the symbolism of the feminine and its proximity to nature are deeply present throughout the history of humanity. However, for the authors, it is essential to recognize that the social relations that pre-establish specific responsibilities for women due to gender relations are reflected in the way women interact with the environment. Such relationships differ according to the economic and social class in which women find themselves and in this way are socially constructed reflecting on the tasks that they have in the domestic and public domain.

Given all the environmental issues, it is evident that all social actors and actresses have to be contemplated to arrive at a new development model, considering their voices in this process. In this perspective, it is essential not only to incorporate women into analyzes of existing practical needs but to consider the gender perspective for the establishment of more just and equitable social policies (Castro and Abramovay 2005).

Often women have assumed the management and support of natural resources that are part of the daily lives of community groups, villages and the most excluded segments in different parts of the world (Jacobi et al. 2015, UN Women 2012, Mouro 2017). However, the participation of women in the conservation of the environment goes beyond this reality, considering that the role of women in society occurs in a multifaceted way, a reality that also includes the public space and not only in practices that guarantee the reproduction of social life in domestic spaces. Through various forms of production on a local scale related to the environment, women have contributed to the support of their families and their communities, configuring a dynamic production and their participation in the production chain (Jacobi et al. 2015, Castro and Abramovay 2005, Siliprandi 2000, Carmo et al. 2016). Raimi et al. (2019) also mention that women around the world play a fundamental role in protecting biological diversity.

According to Martínez-Alier (2007), the fight in defense of environmental wealth preservation and for communities' survival and their livelihoods throughout history has always had the participation of women, especially in the poorest countries. Research has shown that women are more aware and involved in preserving the environment. As an example, Ballew et al. (2018) present the results of a survey conducted by Yale University in which it was found that although a similar proportion of men and women think that human actions cause global warming, women have perceptions of higher risk of how global warming will harm them personally and will harm people in the US, animals and future generations. Besides, compared to men, a greater proportion of women are concerned about global warming and support climate change mitigation policies.

On the other hand, according to the UN (2020) women generally face greater risks and greater burdens from the impacts of climate change in situations of poverty and the majority of the world's poor are women. The unequal participation of women in decision-making processes and the labor markets increases inequalities and often prevents them from contributing fully to climate-related planning, policy-making and implementation.

Given the above, it is observed that for development to be truly sustainable, it is necessary to include a gender focus centered on the experience and incorporation of actions by men and women in policies and programs, promoting concrete proposals on how to ensure more effective women participation in this process. "A public policy or program approach from a gender perspective, involving men and women, would indicate with more specificity the different use of resources, based on the relationships that establish men and women among themselves, between groups, in the community and society in general, and with nature in particular" (Castro and Abramovay 2005).

Considering the search for sustainability, the authors draw attention to the fact that a gender approach does not refer only to the measures used to incorporate women in development processes. In fact, this approach questions the objective and content of development, demonstrating the need to seek new policies that contribute to changing existing inequality structures and sustainable environment use (Castro and Abramovay 2005).

According to Sachs (2000), the concept of development is multidimensional. It cannot be synonymous with economic growth, considering that there may be growth, but growth that generates great social and ecological costs is growth that leads to bad development. In this context, the criteria for development must take into account social, ecological and economic factors. Specifically, from a social point of view, the objective should be to enact the well-being of all based on the ethical principle and social justice and solidarity.

Although discussions and reflections on sustainability and sustainable development have advanced significantly, including concerning social issues and especially gender issues, in practice the current development model is extremely predatory, unsustainable and has been a source of oppression on women. In this scenario, even with the recognition that women have a fundamental role in environmental conservation and that they, on the other hand, have their lives directly affected by various environmental impacts. What is observed according to Jacobi et al. (2015) is that despite the discourse showing the need for a greater role for women, in daily practice women are still very much absent from decision-making processes about environmental policies. Thus, although the international community recognizes that without the full participation of women, it will not be possible to move consistently and steadily towards a more sustainable society, the incorporation of women in the formulation, planning and execution of environmental policies remains very slow.

"Sustainable development is not possible if female population continues to have incomplete rights and opportunities. In short, achieving sustainable development and gender equality potentially involves issues that need to be discussed widely and openly among different social groups. In these negotiations, the social dimensions of sustainability must be integrated, with formal state actions and other actors responsible for this active construction, including equal access to quality education, economic resources and political participation, fair jobs and wages, leadership, and decision-making at all levels for everyone—women, girls, men, and boys" (Seixas and Hoefel 2020).

Multilateral organizations see the lack of gender equity and equality as an obstacle to a more sustainable society based on principles that guarantee socio-environmental justice, recovery of fragile ecosystems, protection of the environment and food security (Jacobi et al. 2015).

According to Barcellos (2013) today, the era of development, linked to other events such as globalization and neoliberalism, has posed many challenges, including the strengthening of forms of sexual subordination, which contribute to deepening gender inequalities. On the other hand, impacts of development projects have led women from all over the world to engage in struggles of the most varied types, seeking to defend the preservation of their ecosystems, the physical integrity of their families and respect for their cultures. They have also worked in networks in an attempt to strengthen themselves politically. For the author, "at the same time, they have been developing concrete experiences of environmental recovery, in an attempt to achieve autonomy and decent livelihood conditions for themselves, their family and community. Because of countless social, environmental, economic and political initiatives, women have occupied more public spaces, seeking to give visibility to their struggles. The new spaces are arenas for their political action in search for social changes and justice" (Barcellos 2013).

In this context, ecofeminism is characterized as a determining factor that seeks to bring a voice to women and nature that have historically lived an experience of oppression and domination. This approach has collaborated significantly to enable gender issues and their relationship to sustainability to be present on international agendas and in the design and implementation of public environmental policies.

3. Ecofeminism

From the feminist movements of the 1970s, ecofeminist thinking appeared for the first time, at this point already influenced by the pacifist, antimilitarist and anti-nuclear movements that broke out across Europe and the United States in the 1960s and which gave rise to the environmental movements of the actuality (Siliprandi 2000, Valencia 2008, Carmo et al. 2016, Muller and Olorunju 2020).

Ecofeminism, like other modern progressive movements, played a role in the social changes that took place in the 1960s and 1970s. "Texts such as Rachel Carson's Silent Spring (1962), Rosemary Radford Ruether's New Woman/New Earth (1975), Mary Daly's Gyn/Ecology (1978), Susan Griffin's Woman and Nature: The Roaring Inside Her (1978), Elizabeth Dodson Gray's Green Paradise Lost (1979), and Carolyn Merchant's The Death of Nature (1980) " laid the groundwork for what would become the feminist approaching related to ecology and environment of the 1980s (Gaard and Gruen 1993).

It is also worth mentioning in this context the Indian researcher and activist Vandana Shiva, who in addition to theoretical and ideological discussions made an analysis in 1988 of how violence against women and nature in India and other third world countries originated on a material basis. For her, there is a relationship between the forms of domination over the people of third world countries and the destruction of nature, a relationship that has as its main consequence the destruction of conditions that should guarantee the very survival of women (Shiva 1995, Siliprandi 2000).

> "Com la destruccicón de los bosques, el agua y la tierra, estamos pierdendo los sistemas em que se apoia la vida. Esta destrucción se está llevando a cabo en nombre de "desarrolho" y el progresso, pero debe haver algo muy equivocado em um concepto de progresso que amenaza la própria supervivencia. La violencia hacia la naturaleza, que parece inherente al al modelo de desarrollo dominante, se asocia también con la violencia hacia las mujeres que dependen de la naturaleza para obter el sustento para ellas, sus famílias y sus sociedades" (Shiva 1995).

Given this scenario consisting of studies, research and analysis on the relationship of women situation and environmental issues, ecofeminist studies, especially from the 1980s onwards, have strengthened themselves as a field of knowledge and have proposed a world reading based on the feminine approach for all forms of life and diverse ecosystems and nature in its entirety. The female concern for the environment is justified by the recognition that there is an intersection between the system of unlimited exploitation of different ecosystems in favor of human interests and the system of oppression that women have historically experienced (Kuhnen 2017).

For the author, based on this way of accessing the world, it is clear that the same system that is responsible for the exploitation of animal life, environmental degradation and the progressive elimination of natural environmental areas oppresses women, blacks and the poor. "Therefore, if feminism brings together a set of conceptions united by the recognition of injustice, inherent to systems of exploitation and oppression, combating inequalities of power, resulting from a patriarchal order, which still prevails in modern world, then, as a theory and social movement I could not ignore the destruction of the environment and the exploration of the life of non-human animals" (Kuhnen 2017).

Siliprandi (2000) presents in a simplified way the principles of ecofeminist thinking in the following questions:

Table 1: Principles of Ecofeminist Thinking

From an economic point of view	There is a convergence between the way hegemonic Western thought sees women and nature, that is the domination of women and the exploitation of nature are two sides of the same coin of using 'natural resources' without cost in the service of accumulation
For ecofeminism	Western thought identifies, from a political point of view, women with nature and men with culture, with culture (in Western thought) being superior to nature; culture is a way of 'dominating' nature; hence the view (of ecofeminism) that women would have a special interest in ending the domination of nature because a society without exploitation of nature would be a condition for the liberation of women
Scientific and technological policies	What has guided modern economic development are policies that reinforce this view, not being 'gender-neutral' or environmental guided. The very way of researching history has followed these principles. It, therefore, has not shown how women were excluded from the world of 'scientific' knowledge nor how their worldview (of integration with nature) was being subjugated by the idea of domination.

Source: Elaborated by the author from Siliprandi (2002)

In addition to these principles, ecofeminism and public debate between the egalitarian movement and feminism of difference that occurred when celebrating the International Year of Women (1975) contribute to understanding and analysis of this subject. The egalitarian tradition, which internally presents different approaches, was against the inequalities of power structured around sexual differences seeking the universality of human dignity and fought for the expansion of civil rights, the entry of women into the public world and their autonomy from the point of view economic, social, political, etc. Difference feminism, on the other hand, criticizes this view, considering that the public world, as it is structured, reflects a masculine view of being and that women would have other contributions to make for a new form of structuring society that incorporates a wealth of the female universe, instead of devaluing it. In this perspective, it understands that the woman represents another way of being different from the masculine vision (Siliprandi 2000). According to Puleo (2012), the difference feminism states that men and women express opposing essences; women would be characterized by a non-aggressive and egalitarian eroticism and maternal qualities that predispose to pacifism and the preservation of nature, while men would be characterized for carrying out competitive and destructive activities.

Regarding the feminism of difference, this 'biologicism' has aroused strong criticism within feminism and is accused among other factors of demonizing men (Siliprandi 2000, Carmo et al. 2016, Garcia 1992, Puleo, 2012). This line can be seen as a form of essentialism based on the notion of an immutable and irreducible feminine essence. This type of approach goes against another, opposite, which demonstrates that the concepts of gender, culture and nature are historically and socially constructed and vary according to time and within cultures and periods (Agarwal 2004, Garcia 1992).

According to Siliprandi (2000), it is necessary to remember that within ecofeminism many currents involve those with a more anarchist ('radical') traditional socialists; those more liberal who favor institutional actions, in parliament, etc., and the spiritualist and even esoteric strands, which "understand how necessary is to rescue the 'magical' practices of knowledge of the reality that women have exercised since ancient times, as ways to reconstruct a female identity that has been lost over time" (Siliprandi 2000).

As a result of the discussions about the essence of ecofeminism as well as the criticisms arising from this process, in the context of ecofeminism characterization, it is important to highlight two other approaches, the post-colonial ecofeminist aspect and the constructivist aspect.

According to Carmo et al. (2016) and Puleo (2012) in the mid-1980s, Vandana Shiva inaugurates the post-colonial ecofeminist strand, still linked to the mystical tendencies of the first ecofeminism but rejecting the demonization of men. With the publication of Staying Alive, a work translated into Spanish with the title 'Abrazar la Vida', a new phenomenon occurs: feminist theory reaches the north from the south.

In that publication, Shiva (1995) presents a critical analysis of development, ecology, women, science, nature and gender. The author relates the forms of domination over people living in the third world countries through development programs that were oriented with the destruction of nature, and its main consequence was the destruction of conditions for the very survival of women due to the extinction of sources of food, water, biodiversity, etc. This development was reduced to being the continuation of the colonization process, a model based on the exploitation and exclusion of women (western and non-western), on the exploration and degradation of nature and the exploration and gradual destruction of other cultures. "Economic growth stole resources from those who needed it most; only instead of colonial powers, the exploiters were the new national elites" (Shiva 1995). For Shiva (1995) this "bad development" would be characterized by modern patriarchal postulates of homogeneity, domination and centralization. In this context, the disappearance of biodiversity and cultural diversity are closely linked processes.

Puleo (2012) emphasizes that because of Shiva's contribution, it was possible to know of the existence of successful women's movements to resist bad development. One of the first movements was by Chipko's women who, based on Gandhi's principles of creative nonviolence and in the name of the feminine principle of nature in Indian cosmology, managed to prevent total deforestation in the Himalayas by taking turns in policing the area and tying themselves to the trees whenever someone attempted cutting the trees down. These women confronted their husbands, who were willing to sell communal forests. Chipko women gained group awareness and later continued to fight domestic violence and political participation.

In the publication "Manifiesto para una Democracia de la Tierra. Justicia, sostenibilidad y paz", Shiva (2006) also presents the Plachimada women's movement that obtained a judicial sentence that recognized the community's rights over water against the devastating exploitation of multinationals. "Movements with the power struggle waged by the tribal women of a small village called Plachimada in the Indian state of Kerala against one of the largest companies in the world, Coca Cola, is a central example of the emerging Democracy of the Earth" (Shiva 2016). Democracy of the Earth seeks to celebrate diversity and tries to balance rights with responsibilities. In this perspective, it is necessary to create economies that protect life on Earth, supply basic needs and promote economic security for all through an active democracythat is inclusive. Thus, the economy, politics and society move from negative systems, which benefit only a few in the short term, to positive systems, which guarantee the fundamental right to life for all species. They also create a new paradigm for global governance while empowering local communities (Shiva 2006). "Earth's democracy is not just a concept, and it is made from multiple practices and diverse people who claim their common goods and spaces, their resources and means of production, their freedom, their dignity, their identity and their peace" (Shiva 2006).

Shiva's contributions to ecofeminism are undeniable, given that it focuses on the need for an active movement on the part of women to be heard, to participate in decision-making bodies to counter the current development model. According to Siliprandi (2000), from an ecological point of view, Shiva brings a vision of questioning the productivist paradigm of development and defense of biodiversity. Its positions also bring a strong focus on the reality of the third world as it questions the relations between the countries that dominate contemporary science and those that suffer more closely its consequences.

However, despite his important analysis regarding the consequences of poor development on women's living conditions, Shiva's explanations from the theoretical point of view on the causes of the separation between men, women and nature, which occurred in contemporary thought, put their considerations in the field of essentialism. Shiva's thinking was inserted in the criticisms of essentialism due to passages in her work that indicate an ontological relationship between women and nature through the "feminine principle" (Siliprandi 2000, Puleo, 2012, Carmo et al. 2016, Agarwal 2004). According to Puleo (2012), the best-known criticism is that of Bina Agarwal which is based on constructivist positions, theories that do not appeal to the essence of the relationship between woman and nature but to cultural constructions that would generate different identities. A trained economist, originally from India, Agarwal criticizes the theory that attributes the protective activity of the nature of women in his country to the feminine principle of his cosmology.

Agarwal (2004) suggests an alternative proposal for what she calls feminist environmentalism, whose structure brings a relationship between women and men with nature rooted in their material reality and their specific forms of interaction with the environment. For the author, this feminist reasoning with which ecofeminism is built is problematic, since:

> "Firstly, it postulates 'woman' as a unitary category and does not differentiate women according to class, race, ethnicity, among other factors. Thus, it ignores other forms of domination other than those of gender that also critically influence the position of women. Second, it locates the domination of women and nature almost exclusively in the field of ideology, ignoring the material sources of that domination [...]. Thirdly, even in the field of ideological creations, it says little about the social, economic and political structures within which these creations are produced and transformed [...]. Fourth, feminist reasoning does not take into account the relationship that women have with nature as opposed to the relationship that others or themselves can conceive. Fifth, the currents of ecofeminism that attribute the connection between women and nature to the biological can be considered as linked to a form of essentialism (a notion of the unchanging and irreducible female 'essence'. This formulation disappears in the face of an ample evidence that the concepts of nature, culture, gender, etc., were constructed historically and socially and vary from one culture to another, within the same culture and from one era to another" (Agarwal 2004).

According to Valencia (2008), Agarwal highlights that the processes of environmental degradation and appropriation of natural resources by a fewhave specific implications for class, gender and geographic location. At the same time, it considers that the connection between women and the environment is determined by a structure that includes several aspects, including gender, class (caste/race), production organization, reproduction and income distribution.

In this context, the constructivist ecofeminist thought has as one of its exponents, Bina Agarwal, considering that its theories do not refer to essences but to cultural constructions that generate different identities. Constructivist theory thinks holistically and in terms of community interaction and priority based on the material reality in which women are. Thus, it is not the effective or cognitive characteristics of women, but their interaction with the environment which favors their ecological awareness (Carmo et al.2016). Agarwal (2004) suggests that environmentalists should question and transform representations of the relationship between nature and people, in addition to current methods of appropriating natural resources for the benefit of a few.

Currently, new currents of thought, although they also assume a constructivist vision, seek to develop an ecofeminist theory according to feminist coordinates from their own cultural and vital contexts, as proposed by Puleo (2012). Called critical ecofeminism, this position refers to the need to recognize and affirm but also to critically review the illustrated legacy of criticism of prejudice and equality and empowerment of women. Ecofeminism offers an alternative to the crisis of values in today's consumerist and individualist society. The contributions of two critical reflections—feminism and environmentalism—promote the opportunity to confront the sexism of patriarchal society at the same time that it is possible to discover and denounce the androcentric subtext of domination of nature (Puleo 2012).

Because of the various reflections on ecofeminism presented, it is worth mentioning, as proposed by Valencia (2008), that addressing gender and environment issues represent a theoretical challenge, leading to the need to examine in detail the issues related to redistribution and development. This issue should no longer be seen under the neutrality of science, given that it is a dynamic and complex process that requires a plurality of perspectives. Ecofeminism, despite being criticized as essentialist, is a mandatory reference for analyzing proposals. That has to do with the relationship between environment and gender.

Ecofeminism can bring several innovative contributions as a current of thought that seeks to incorporate women's views into discussions about environmental issues as it draws attention to aspects that are not usually considered in development policies, such as the implications that certain economic activities have on the living and working conditions of women as well as on other segments of the population (Siliprandi 2000). "By giving importance to what was not 'economically relevant', such as local culture, quality of life, the values of the target populations of these policies (which go unnoticed in official statistics), it helps to question development visions based solely on in criteria such as income, production and productivity" (Siliprandi 2000).

For Puleo (2012), feminism should not be closed to the new concerns and sensitivities of women, and environmentalism is one of these concerns. When considering that feminism should raise utopian horizons in the etymological sense of 'utopia' (or-topos, what has not yet happened but may have), it is possible to say that ecofeminism has a lot to contribute in this twenty-first century and that in this context humanity will have to face a profound socioeconomic and cultural transformation to achieve equality and eco-justice and survive.

In this sense, it is worth noting that discussions about ecofeminism, despite the diversity of approaches and positions, have been reflected in international programs and agendas that increasingly seek to include the issue of gender in its principles, guidelines and programs of action.

4. The Issue of Gender and the Environment on the International Agendas

Siliprandi (2000) mentions that the height of social and political visibility of ecofeminist positions took place in the early 1990s with the holding of the United Nations Conference on Environment and Development in Rio de Janeiro - Rio 1992. According to Iraci (2005), this event put the relationship between the population and the environment at the center of the debates; in this context, the interdependence between these two poles of the equation and how specific segments of the population affects and are affected in this reality take different sectors of organized social groups to reflect on this issue.

In June 1992, the conference brought together legislators, diplomats, scientists, the media and representatives of non-governmental organizations (NGOs) from 179 countries in a concerted effort to reconcile the interactions between human development and the environment. Adopted by heads of state from 179 countries, the result of this mobilization can be translated into the 173 recommendations contained in the Agenda 21 document – the Platform for Action on Sustainable Development (Iraci 2005). More specifically, Chapter 24 brings together a set of recommendations, mechanisms and goals for integrating women and the gender issue at all levels of government and in the related activities of all UN agencies (Iraci 2005, UN Women2012, Mouro 2017).

Principle 20 of the Rio Declaration stated that "women have a vital role in environmental management and development. Their full participation is therefore essential to achieve sustainable development" (UN Women 2012, p. 6). Also, Chapter 21 of Agenda 21 recommended that national governments develop strategies to "eliminate constitutional, legal, administrative, cultural, behavioural, social and economic obstacles to women's full participation in sustainable development and in public life" (UN Women2012).

According to Iraci (2005) at this event, organizations such as the Network for the Defense of Human Species (REDEH) and the Woman Education Network (RME) were part of the coordination of the Female Planet at the Global Forum. The Female Planet occupied a significant space in the debates on environment and development at the NGO Social Forum, an event parallel to Rio-92. Women from all over the world for 12 days, discussed the problems experienced on the planet and formulated and adopted their platform – the Women's Action Agenda 21 which addressed issues, such as governance, militarism, globalization, poverty, land rights, food security, women's rights, reproductive rights, science and technology and education. New forms of education, preservation of natural resources and participation in planning a sustainable economy were some of the recommendations included in that document.

The Female Planet defended "a feminine look on the world", the participating organizations criticized the predatory style of consumption coming from the North, which aggravated the poverty of the South and environment and the problems of women, who were excluded from these great discussions, and suffered the consequences of these processes (Siliprandi 2000, Tornquist et al. 2010).

It is important to highlight that the mobilization of women at the Rio-92 NGO Forum has opened space for their participation in all UN conferences. Thus, gender relations have been incorporated into international agendas and have been instrumental in building the vision of sustainability between society and the environment (Iraci 2005). According to UN Women (2012), Rio 92, among other fundamental issues such as poverty, highlighted the areas in which specific work was needed to advance gender equality, including mainstreaming the gender perspective in all policies and strategies, the elimination of all forms of violence and discrimination against women, ensuring full and equal access to economic opportunities, credit, education, health services, land and agricultural resources and promoting equal access and full participation of women in decision-making at all levels.

According to Iraci (2005), the document Agenda 21 of Women for a Healthy Planet "marked the intervention of the feminist movement in the UN conferences that took place in the 1990s: Human Rights (Vienna 1993); International Conference on Population and Development (Cairo 1994); Fourth World Conference on Women (Beijing 1995); World Conference for Social Development (Copenhagen 1995) and Habitat II (Istanbul 1996); UN Conference on Food Security; International Conference on Youth and Adult Education (Hamburg 1997); World Conference against Racism" (Durban 2001, Iraci 2005).

The mobilization of women from Rio-92 consolidated a view that feminism and ecology were intrinsically linked since both proposed profound changes in the natural order based on social justice (Iraci 2005). The conference also contributed to the creation of three significant environmental agreements, which became known as the Rio Conventions: the United Nations Convention on the Combat and Control of Desertification, the Convention on Biological Diversity and the United Nations Framework Convention on Climate Change (UNFCCC). The first two promoted the gender perspective from the beginning; the UNFCCC did not originally incorporate it. The awareness of the links between gender equality and climate change has increased since then (UN Women 2012), culminating its inclusion in the Paris Agreement, approved in 2015 by 195 countries as part of the United Nations Framework Convention on Climate Change. According to the document:

> "Acknowledging that climate change is a common concern of humankind, Parties should, when taking action to address climate change, respect, promote and consider their respective obligations on human rights, the right to health, the rights of indigenous peoples, local communities, migrants, children, persons with disabilities and people in vulnerable situations and the right to development, as well as gender equality, empowerment of women and intergenerational equity [...]" (United Nations 2015a).

Ten years after Rio-92, governments and civil society mobilized to participate in the Sustainable Development Conference, Rio + 10, which took place in Johannesburg, South Africa in September 2002. This conference aimed at reviewing policies and assessing successes and failures in the implementation of sustainability plans defined in Rio-92. It also sought to work on the new issues that have emerged in the last decade, especially the current economic system and model of neoliberal globalization, responsible for the increase in poverty and environmental breakdown (Iraci 2005).

Rio + 10 had the challenge of revising the Women's Agenda 21, to recover the values and proposals contained in this document and monitor its results, as qualitative statistics demonstrate that in general women's living conditions have not improved or remained the same. In this analysis, a negative balance was observed and evidencing that the countries did not fulfill the commitments assumed in the UN Conferences, in particular those concerning women (Iraci 2005). However, after ten years of the extraordinary exercise of daring that was the Female Planet, the Women, Environment and Development Organization (WEDO) and the Human Development Network (REDEH), they sought to rescue the strategy that generated this movement, relaunching the proposal for a new version of Agenda 21 for Women 2002 (Iraci 2005).

This new proposal was discussed by international leaders and networks of women during the preparatory processes for Rio + 10 at regional consultation meetings in all parts of the world: Rio de Janeiro (Brazil), Pittsburgh (United States), Prague, (Czech Republic), Bangkok (Thailand) and Venice (Italy). As a result, the consultation and articulation process resulted in the Women's Action Agenda for Peace and a Healthy Planet 2015 (Figure 1). The document, which gathers a series of recommended actions for governments, national and international institutions and civil society organizations, presents as a basic five points considered strategic for women: peace and human rights, globalization and sustainability, access to control and resources, environmental security and health and governance for sustainable development (Iraci 2005).

Figure 1: Elaborated from Women's Action Agenda for a Healthy and Peaceful Planet 2015
Source: WEDO (2015)

Considering the ecofeminism relationship and the search for sustainability, it is essential to mention Agenda 30 for Sustainable Development, which came into force in January 2016. The document presents 17 objectives (Figure 2), successors to the eight-millennium development goals and 169 goals. These objectives are integrated and indivisible and among other factors seek to realize the human rights of all and achieve gender equality and the empowerment of women and girls, to balance the three dimensions of sustainable development: economic, social and the environmental (United Nations 2015b).

Agenda 30 considers that reaching human potential and sustainable development is not possible if half of humanity continues to be denied their full human rights and opportunities. Thus, women and girls must enjoy equal access to quality education, economic resources and political participation, as well as equal opportunities with men and boys in terms of employment, leadership and decision-making at all levels.

Figure 2: Sustainable Development Goals
Fonte: UN (2020b)

Objective 5 of the document specifically addresses gender equality and women's empowerment as shown in Table 2.

Table 2: Goal 5 of the Agenda 30 for Sustainable Development

5.1	End all forms of discrimination against all women and girls everywhere.
5.2	Eliminate all forms of violence against all women and girls in public and private spheres, including trafficking and sexual and other exploitation.
5.3	Eliminate all harmful practices, such as premature, forced and child marriages and female genital mutilation.
5.4	Recognize and value unpaid assistance and domestic work through the provision of public services, infrastructure and social protection policies, as well as promoting shared responsibility within the home and family, according to national contexts.
5.5	To guarantee the full and effective participation of women and equal opportunities for leadership at all levels of decision-making in political, economic and public life.

Source: The authors based on United Nations (2015b)

According to Seixas and Hoefel (2020), the 17 objectives of the 2030 Agenda must be considered as a platform for inspiration and action to improve the living conditions of humanity. In relation, specifically to the SDG 5, for the authors, this objective is fundamental to achieve sustainability and the fullness of human rights "and invites society to think globally, to eliminate all forms of violence, and to work hard in a project of quality education for women and girls. Without removing these tremendous obstacles of life from countless women and girls on the planet, this goal cannot be achieved " (Seixas and Hoefel 2020).

Because of the Agenda 30 proposal, it is essential to mention that the achievement of more equal personal interrelations between men and women provides the outline of human development in line with the environment in which it operates. For Carmo et al. (2016), the guarantee of this ideal is based on the substantial recognition of people, regardless of their representation as part of the male or female sign, that is distanced from any differences in their rights.

For Bohler-Muller and Olorunju (2020) the women's movements played the role of critics and defenders of the States and the UN, with the majority recognizing that woman's human rights and gender equality can only be achieved in one state functional and human rights-based democratic system. However, in this context, there are still many feminist movements that wish to reformulate traditional policies and programs that do not consider feminist views on development agendas and the importance of women's voices.

In addition to the events mentioned, there are still many others at a global and local level that seek to include gender in their approaches given the various environmental problems experienced by today's society and which have directly interfered in the quality of life, safety, health and guarantee of a healthy and socially just environment for women worldwide. Such reflections and discussions are possible and feasible thanks to movements such as ecofeminism which, in its different nuances, have allowed this theme to be considered and addressed at different scales.

5. Conclusions

Considering all the advances achieved since the emergence of ecofeminism that brought up the issue of gender in face of the oppressive and predatory relationship present in the current development model, it is clear that there are still many challenges to be overcome in the ecofeminist agenda. In this context, Bohler-Muller and Olorunju (2020) cite a list produced in 2919 by the UN Women of disturbing areas of inequality that women still face. This list, among other factors, states that the fact that more than 12 million girls are forced to be child brides worldwide, one every two seconds; it is said that it will take about 108 years to close the gender gap that exists today and that only six countries worldwide offer women the same labor rights as men.

These and other challenges present in the relationship of women with environmental issues and sustainability tend to be overcome based on the understanding of the complexity of the concept of the environment itself that has been obtained through the discussions and reflections promoted by ecofeminism, whose trajectory has demonstrated the interconnection of women with the environment. The dimension of the social, economic, political, cultural and philosophical configurations inherent to the concept of the environment has been shown amid the structural concept of ecofeminism, which is dynamic and constantly changing.

From this perspective, it is possible to conclude that sustainability will only be viable when the social factors that are part of its scope are taken into account together with economic and ecological factors. Thus, there is no way to promote a more sustainable society without including gender issues and based on this process, review the current development model that has placed women in a situation of vulnerability, domination and exploitation.

Finally, ecofeminism has contributed to the awareness of women themselves regarding the situation of oppression in which they find themselves and how this reality is reflected in environmental issues as well as how women have been most affected in different ways by environmental impacts or how they have actively participated in modifying this reality based on their participation in the design and implementation of public policies that make it possible to transform gender relations and promote sustainability.

Acknowledgments

The author would like to thank São Paulo Research Foundation (FAPESP, Brazil) for financial support to the research that leads to this entry (FAPESP n. 2019/08044-3).

References

Agarwal, Bina. 2004. El debate sobre género y medio ambiente: lecciones de la India. pp. 239-280. *In*: V. Vázquez and M. Velásquez [eds.]. Miradas al Futuro, Hacia la Construcción de Sociedades Sustentables con Equidad de Género. Universidad Autónoma de México, México.

Ballew, M., J. Marlon, A. Leiserowitz and E. Maibach. 2018. Gender Differences in Public Understanding of Climate Change. Yale Programo n Climate Change Communication. Available at: https://climatecommunication.yale.edu/publications/gender-differences-in-public-understanding-of-climate-change/

Barcellos, G.H. 2013. Mulheres e lutas socioambientais: as intersecções entre o global e o local. R. Katál. 16: 214-222.

Bohler-Muller, N. and N. Olorunju. 2020. Feminist movements for a sustainable future. pp. 1-11. *In*: W. Leal Filho, A. Azul, L. Brandli, P. Özuyar and T. Wall [eds.]. Gender Equality. Encyclopedia of the UN Sustainable Development Goals. Springer Nature, Switzerland, AG.

Camargo, W. and R. Fontoura. 2014. Mulheres e Sustentabilidade. Gazeta do Povo. Available at: [http://www.gazetadopovo.com.br/blogs/giro-sustentavel/mulheres-e sustentabilidade/ Accessed 12 oct. 2020.

Carmo, J.C., M.M. Pires, J.G. Jesus, A.L. Cavalcente and S.D.P. Trevizan. 2016. Voz da natureza e da mulher na Resex de Canavieiras-Bahia-Brasil: sustentabilidade ambiental e de gênero na perspectiva do ecofeminismo. Revista Estudos Feministas. 24: 155-180.

Castro, M.G. and M. Abramovay. 2005. Questões introdutórias e metodológicas. pp. 35-42. *In*: M.G. Castro and M. Abramovay [eds.]. Gênero e meio ambiente. Cortez, São Paulo, Brasil.

Couto, S.F.M. and J.N.S. Wivaldo. 2017. Aspectos da relação entre empoderamento feminino e desenvolvimento sustentável. Anais Congresso Internacional de Políticas Públicas para a América Latina. Brasil. 32: 21-23. Available at:https://siaiap32.univali.br/seer/index.php/aemv/article/view/11279/6365.

Foladori, G. 2002. Avanços e limites da sustentabilidade social. Revista paranaense de desenvolvimento. 102: 103-111.

Gaard, G. and L. Gruen. 1993. Ecofeminism: toward global justice and planetary health. Society Nature. 2: 1-35.

Garcia, S.M. 1992. Desfazendo os vínculos naturais entre gênero e meio ambiente. Estudos Feministas. 0: 163-167.

Gimenes, T., M.K. Machado and E.B.R. Vernalha. 2019. Empowerment in sustainability. pp. 1-7. *In*: W. Leal Filho [ed.]. Encyclopedia of Sustainability in Higher Education. Springer Nature, Switzerland, AG.

Iraci, N. 2005.Gênero e meio ambiente: qual a sustentabilidade possível? pp. 11-24. *In*: M.G. Castro and M. Abramovay [eds.]. Gênero e meio ambiente. Cortez, São Paulo, Brasil.

Jacobi, P., V. Empinotti and R.F. Toledo. 2015. Gênero e meio ambiente. Ambient. Soc. 18: 1-4. Available at: http://www.scielo.br/scielo.php?script=sci_arttext&pid=S1414 753X20 15000100001&lng=en&nrm=iso.

Kuhnen, T.A. 2017. Conservação da natureza e manutenção do patriarcado: apontamentos ecofeministas. pp. 73-92. *In*: C. Ferri, A.M.P. Camardelo, M. Oliveira [orgs.] Mulheres, desigualdade e meio ambiente. Educs, Caxias do Sul, Brasil.

Lamim-Guedes, Valdir and A.G. Inocêncio. 2018. Mulheres e sustentabilidade: uma aproximação entre movimento feminista e a educação ambiental. Educação Ambiental em Ação. 12: 1-10.

Leff, E. 2002. Epistemologia Ambiental. 2nd. Cortez, São Paulo, Brasil.

Martínez-Alier, J. 2007. O Ecologismo dos Pobres. Contexto, São Paulo, Brasil.

Mouro, H.H. 2017. Gênero e Ambiente: Reflexões sobre o papel da mulher na questão socioambiental. Doutorado. Tese, Universidade Nova Lisboa, Lisboa, Portugal.

Pelicioni, M.C.F. and A. Philippi Jr. 2005. Bases bolíticas, conceituais, filosóficas e ideológicas da educação ambiental. pp. 3-12. *In*: A. Philippi Jr. and M.C.F. Pelicioni [eds.]. Educação Ambiental e Sustentabilidade. Manole, Barueri, SP, Brasil.

Pilon, F. 2005. Ocupação existencial do mundo: uma proposta ecossistêmica. pp. 305-352. *In*: A. Philippi Jr. and M.C.F. Pelicioni [eds.]. Educação Ambiental e Sustentabilidade. Manole, Barueri, SP, Brasil.

Puleo, A. 2012. Feminismo y ecologia. Mujeres en Red – El periódico Feminista. Available at: http://www.mujeresenred.net/spip.php?article2060

Raimi, M.O., R.M. Suleiman, O.E. Odipe, J.T. Salami and M. Oshatunberu. 2019. Women role in environmental conservation and development in Nigeria. Ecology & Conservation and Science 1: 1-15.

Ruether, R.R. 1996. Ecofeminism: symbolic and social connections of the oppression of women and the domination of nature. pp. 135-154. *In*: R.S. Gottlieb [ed.]. This Sacrad Earth: Religion, Nature and Environment. Routledge, London.

Sachs, I. 2000. Sociedade, cultura e meio ambiente. Mundo & Vida. 2: 7-13.

Seixas, S.R.C. and J.L.M. Hoefel. 2020. Human rights and gender equity: building sustainable development. pp. 12-13. *In*: W. Leal Filho, A. Azul, L.Brandli, P. Özuyar, T. Wall [eds.]. Gender Equality. Encyclopedia of the UN Sustainable Development Goals. Springer Nature, Switzerland, AG.

Shiva, V. 1995. Abrazar la vida: mujer, ecología y supervivencia. Instituto del Tercer Mundo, Montevideo, Uruguay.

Shiva, V. 2006. Manifiesto para una Democracia de la Tierra. Justicia, sostenibilidad. Paidós, Barcelona.

Siliprandi, E. 2000. Ecofeminismo: contribuições e limites para a abordagem de políticas ambientais. Revista Agroecologia e Desenvolvimento Rural Sustentável 1: 61-71.

Tornquist, C.S., T.K. Lisboa and M.F. Montysuma. 2010. Mulheres e Meio Ambiente. Estudos Feministas. 16: 865-869. Available at: https://www.scielo.br/pdf/ref/v18n3/v18n3a12.pdf

UN (United Nations). 2015a. Paris Agreement. Available at: https://unfccc.int/files/meetings/paris_nov_2015/application/pdf/paris_agreement_english_.pdf

UN (United Nations). 2015b. The 2030 Agenda for Sustainable Development. Available at:https://sustainabledevelopment.un.org/content/documents/21252030%20Agenda%20for%20Sustainable%20Development%20web.pdf

UN (United Nations). 2020a. Introduction to Gender and Climate Change. Available at: https://unfccc.int/gender

UN (United Nations). 2020b. Social Development for Sustainable Development. Available at: https://www.un.org/development/desa/dspd/2030agenda-sdgs.html

UN Women (United Nations Entity for Gender Equality and the Empowerment of Women). 2012. The future women want: a vision of sustainable development for all. Available at: https://www.unwomen.org/ /media/headquarters/media/publications/en/thefuturewomenwant.pdf?la=en&vs=947

Valencia, E.C. 2008. Ecofeminismo y ambientalismo feminista: Una reflexión crítica. Argumentos (Méx.). 21: 183-188.

WEDO (Women's Environment & Development Organization). 2015. Women's Action Agenda for a Healthy and Peaceful Planet 2015. Available at: https://www.wedo.org/wp-content/uploads/agenda2015_eng.pdf

Gender Equity, Sustainability and Human Rights: Considerations About Gender and Race in the COVID-19 Pandemic Scenario

Sônia Regina da Cal Seixas[1]*, João Luiz de Moraes Hoefel[1,2], Amasa Ferreira Carvalho[1], Luana Aparecida Ribeiro Javoni[3], Gianlucca Consoli[1,2] and Waldo Emerson de Souza Nascimento[1,2]

[1] Center for Environmental Studies and Research, NEPAM, State University of Campinas, UNICAMP, Rua dos Flamboyants, 155, Cidade Universitária, Campinas, SP 13083-867, Brazil
[2] Center for Sustainability and Cultural Studies, UNIFAAT University Center (NESC/CEPE/UNIFAAT), Estrada Municipal Juca Sanches 1050, CEP 12954-070, Atibaia, São Paulo, Brazil
[3] Postgraduate Program in Energy Systems Planning, School of Mechanical Engineering, State University of Campinas (FEM – UNICAMP), Rua Mendeleyev, 200 Cidade Universitária, Campinas, SP, CEP 13.083-860, Brazil

"Many people know that capitalist societies are, by definition, class societies that allow a small minority to accumulate private profits through the exploitation of a much larger group, which must work for wages. What is less widely understood is that capitalist societies are also, by definition, the source of gender oppression. Far from being accidental, sexism is embedded in its structure" (Arruzza, Bhattacharya and Frazer 2019).

1. Introduction

A closer look at the society's structure, let us recognize at its base and the origin of gender oppression. The authors mentioned above in their work analyze 11 theses about a new proposal of feminism for the whole society. Through criticism of liberal feminism, which in the authors' view seeks to meet exclusively the demands of capitalism and only 1% of the community, they propose real feminism that they call anti-capitalist feminism and where all people can be included, implying 99% of

*Corresponding author: srcal@unicamp.br

society such as women, blacks, indigenous, transgenders, lesbians, environmentalists and human rights defenders.

These considerations and approaches are what will allow us to question and elaborate our perspective on gender equity and human rights.

One of the biggest challenges in current social and planetary structure is to offer gender equity to contemporary society. Historically, there is already confirmation that when gender equality is achieved, socially, it means that individuals, family and objective living conditions reach another level, more appropriate and that provides social well-being.

This chapter analyzes the relationships between gender equity, sustainability and human rights, and some aspects of the crisis caused by the COVID-19 pandemic and its impacts on black women's life situations in positions of increased socio-environmental vulnerability. It seeks to reflect on the current conditions of gender equality from a Brazilian and global perspective. It also analyzes different proposals that suggest new ways of implementing concrete actions to achieve the objectives of Agenda 2030, in particular, the Sustainable Development Goal (SDG) 05.

An Analysis of the Global Dimensions of Gender Equity and Human Rights

Do we start from the fundamental question of trying to answer what gender equity is? How can we understand this concept within a holistic and integrated approach and one that is directly related to the Human Rights perspective? And yet, is gender equity present for all women on all continents?

Based on these initial questions, our perspective is to present an international panorama from the possibilities inherent to the concept on a global level, and how the relationship between gender equity and human rights on different continents is presented.

Tedeschi and Colling (2016) warned that gender and human rights have proven to be a difficult problem to solve in recent years, and gender inequality is an insult to equalization proposed by human rights. For the authors, the issue of gender or how they highlight the claim of human rights for women is still a long process and under construction. Violence against women, lesbians, gays and transgenders presents itself as a challenge to be overcome to achieve a more just and equal society for all.

Alves (2016) seeks to discuss the limitations and advances in gender relations that have occurred in Brazil and the world in the last 70 years since the creation of the United Nations (UN). The author points out that we have achieved many advances, especially after the 6th World Conference on Women, which took place in 1995, but we still have a long way to go to achieve gender equity.

From the Progress of the World's Women 2015-2016 Transforming Economies, Realizing Rights (UN Women 2015) report the author explores, in general, the data that reveals substantive achievements for women in the past seven decades. But they also show the existence of an incomplete revolution, maintaining the sexual division between productive and reproductive work, which limits the autonomy and empowerment of women in the family and society.

Despite advances and countless setbacks, Sen (2019) recognizes that feminist mobilization has played a crucial role in formulating gender equality guidelines. And it has contributed enormously to being present directly or indirectly in all 17 objectives and in the 169 goals of Agenda 2030 (UN 2015). SDG 05 explicitly states that "achieving gender equality and empowering all women and girls" (UN 2015) will enable sustainable development to be built.

The author points out that, although feminist mobilization has led to significant advances in the construction of the Agenda 2030 SDGs, the current historical context cannot be ignored. There are still present today in the persistence of gender inequality and human rights violations against women and girls.

Some places with economic, social and political environments that are more open to progressive social changes may have advanced. But the current global situation is not allowing even mobilization, let alone overcome unresolved barriers to the financing of programs and projects and the persistence of political opposition to women's rights and gender equality, which remain and will require continued feminist mobilization (Sen 2019).

The remarkable observation of this moment is that, in terms of gender equality and the achievement of women's emancipation, we still have a long way to go. As Odera and Mulusa (2020) warn sustainable development and the political objective of reducing gender inequalities remain unmet. However, Agenda 2030 presents numerous qualities and advances, as mentioned previously, but significant positive indicators from a global point of view, are still incipient.

The authors warn that many of the first highlights on the reduction of inequalities were inaugurated from a debate on the exclusion and discrimination of women in the labor market that are still valid today. Based on the discourse on poverty, they warn that this is a fundamental issue for SDGs, but that the feminization of poverty puts more women at risk.

Indeed, an institutional perspective on the theme also suggests that the SDGs need to fulfill their transversal objectives, as all of them have questions, information and actions to achieve gender equity for women, in addition to SDG 05. It cannot be overlooked that although progress has been made at the normative level, overall progress has been unacceptably slow. With stagnation and even regression in some contexts, especially if we consider the advance of conservative governments in several countries in the world, especially in Latin America. The shift towards profound gender equity is, despite all the opposition movements, irreversible (UN 2015).

The great paradox that can be overcome to achieve gender equity at the global level is to create the objective social and economic conditions to overcome the feminization of poverty. As several studies show reducing the gender gap creates objective conditions for more significant economic growth (Odera and Mulusa 2020).

We need to create the objective conditions for the education of women and girls to address the SDG 05. Brabo (2019), points out that, when we mention education, we talk about all the processes in which the individual builds himself as a person as an agent of history, of its past, which involves in addition to school, work, the arts, the family, professional association, leisure, etc. Educating women fully and for full citizenship will only be achieved by recognizing the history of their struggles, their actions and their rights.

Two issues are fundamental to this achievement: global social policies and the mobilization of women. It cannot be forgotten that the role of women's action in the world has promoted the difference in the achievements that have taken place and especially in the part of implementing the desired social policies. The movement of Brazilian women was oriented toward the construction of a new public space in which they should be included and taking part. Blay (2002) recalls that, at the same time, they also questioned the omissions of unions, class associations, discrimination spread by the press and schools, seeking profound changes within the union structure, the political-party organization and the laws themselves governing civil rights.

2. The Different Dimensions of the Lack of Gender Equity in the Face of the Covid-19 Pandemic

The World Health Organization (WHO) declared the new coronavirus, COVID-19, as a pandemic on March 11, 2020 (Buss 2020, Gonzaga 2020), which was already spreading on the planet since December 2019 after originating from China. As respiratory droplets are the primary mode of transmission from one human to another (Allaerts 2020, Bandyopadhyay 2020, Srivastava et al. 2020, Zhang et al. 2020), the essential recommendations for preventing the spread of serious illness are the maintenance of personal hygiene, physical distance and respiratory hygiene (Bandyopadhyay 2020). Various government agencies across the world are taking different steps to reduce the spread of the virus, mainly through increased physical distancing, close down factories, offices, educational institutions and in general avoid crowding people. These actions and restrictions are drastically changing social behaviors, economic activities and environmental and health issues, having profound effects on private and professional lives (Allaerts 2020, Buss 2020, Bradbury-Jones and Isham 2020, Marques et al. 2020, Srivastava et al. 2020, Ventura et al. 2020).

The current pandemic also evokes in many people existential anxiety experienced as a threat to their normal sense of identity and their sense of place in the world. Many suffer from uncertainty, fear of infection, moral distress and grief, often experienced alone. There is increasing concern about coping with the resulting anxiety as well as with its long-term individual and collective impacts (Bradbury-Jones and Isham 2020, Buss 2020, Peteet 2020).

According to Marques et al. (2020), Nery and Mattos (2020), Pires (2020) and Vieira et al. (2020) another relevant issue to consider is that the social isolation imposed by the pandemic of COVID-19 brings up some worrying indicators about domestic and family violence against women. Thus, organizations focused on combating domestic violence have observed a significant increase due to forced coexistence, economic stress and fears about the coronavirus.

For Vieira et al. (2020), these problems as well as other inequalities are not news brought only by the pandemic of COVID-19, but old issues that are accentuated at this moment. In a tense way, society is experiencing an exacerbation of several problems, reinforced by backward-thinking models, which are reflected on public policies that would be fundamental to face the pandemic context more fairly.

Bradbury-Jones and Isham (2020), Marques et al. (2020), Pires (2020) and Viveiros and Bonomi (2020) emphasize that for many women, the emergency measures needed to fight COVID-19 increase domestic work and beware of children, the elderly and the sick family members. Thus, movement restrictions, financial limitations and widespread insecurity also encourage abusers, giving them additional power and control, which is reflected in the increase in violence against women and girls.

It is essential to highlight as Tiburi (2019) reminded us the issue of domestic violence is still one of the main flags of feminist movements. Violence against women is mainly domestic violence but not only. The inequality of domestic work, the role of motherhood and the whole logic of marriage itself as the submission of women to men, has much of a type of violence, which is also symbolic. It also represents extra domestic violence. It will be present in countless situations of exclusion of women, such as in parliaments, in positions of power and of management, to remain in some. With this, the author warns: "on the one hand are women and domestic violence, on the other hand, there are men and public power (...) while women suffer violence, men exercise power" (Tiburi 2019).

In Brazil, according to Okabayashi et al. (2020), violence against women has increased annually, both concerning the number of cases of intentional bodily injury and domestic violence (194,273 cases in 2016; 252,895 cases in 2017; 263,067 cases in 2018) and the number of cases of femicide (929 cases in 2016; 1,151 cases in 2017; 1,206 cases in 2018). According to the authors (Okabayashi et al. 2020), the most prevalent types of violence in women treated by the Unified Health System (SUS) are physical violence in 48.7% of the visits, followed by psychological violence in 23% of the cases. Sexual violence responsible for 11% of visits and reports of violence against women, increased by approximately 9% after the establishment of social isolation to contain the COVID-19 pandemic.

The factors related to social isolation that contributed to an increase in this crime are isolation of the victim, which makes him more vulnerable, consumption of alcohol or illicit drugs by the aggressor, which increases the violence, greater ease of the aggressor in controlling the victim and unemployment (Okabayashi et al. 2020). Thus, the increase in cases of femicide is directly linked to the worldwide pandemic and with this, the imposition of social isolation which was established to prevent the spread of the disease since women victims of violence remain for long periods inside their home in contact with the aggressor. Financial limitations and insecurities are aggravated by less social contact, which makes aggressions worse.

Women and men are both affected by COVID-19, but women carry a different kind of charge from COVID-19. Inequities disproportionately affect their wellbeing and economic resilience during lockdowns. Households are under strain, but childcare, elderly care and housework typically fall on women and concerns over increased domestic violence are growing (Editorial 2020).

Therefore, during pandemics, women may worry about their physical safety or experience additional mental or emotional distress, making it even more challenging to mitigate immediate risk for violence. Women working in humanitarian actions may be at even higher risk, but their protection depends on specific organization policies and procedures (Sharma et al. 2020). The COVID-19 outbreak may make existing protective strategies identified by women and girls, such as moving in groups or ensuring aid workers are accompanied when visiting refugee households more challenging to implement. At the same time, shortages in goods mean women and girls face more pressure to access these items for themselves and their homes (Peterman et al. 2020). Therefore, family violence during pandemics, according to Usher et al. (2020), is associated with a range of factors including economic stress, disaster-related instability, increased exposure to exploitative relationships and reduced options for support.

The world has undergone significant economic, social, environmental and cultural transformations, and the emancipation of women and the reduction of gender inequalities represent an essential step in civilizing progress at this decisive stage in the history of humanity (Alves 2016). For Alves (2016), there was an improvement in women's living conditions, but this occurred differently in the different areas of human activity and was not uniform in national and regional terms. The emancipation of women involves mobilization and social struggle, the achievement of diverse and essential rights and the equality of opportunity between the sexes in the family and society.

In the last decades, there have been advances in several social sectors, such as in education, in the labor market, in the spaces of power and in the leadership functions in public and private spheres. Still, it is observed that these gains were partial and that there are several barriers to be overcome, obstacles that need to be recognized and placed at the center of public policies (Alves 2016, Tedeschi and Colling 2014).

Thus, according to Tedeschi and Colling (2016), gender and human rights issues have proven to be a problem that is difficult to solve. For Brabo (2015), the current society lives in a moment of capitalist reorganization marked by the neoliberal ideology with feminist demands not yet reached, despite the achievements already achieved. Among the various problems experienced, gender inequality and violence against women stand out, which are serious social problems and still unsolved. Another challenge mentioned is, according to Brabo (2015), the realization of education for gender equality at all levels of education because although schooling does not have the strength to change society alone, without it there is no transformation. Even though gender education, there is the prospect of changes concerning violence against women.

For Tedeschi and Colling (2016), the constitutions establish equality as a fundamental principle, vetoing all distinctions. Still, it turns out that constitutional equality does not end the discrimination between men and women that has accompanied the history of civilization. Inequality between the sexes is historically constructed, and its most cruel face is the violence practiced against women.

Zamora et al. (2018) emphasize the importance of international human rights obligations to prevent discrimination of any kind based on ethnicity, color, sex, language, religion, political or another opinion, national or social origin, property, birth or another status, such as disability, age, marital and family status, sexual orientation and gender identity, health status, place of residence, economic and social situation.

To Rudolf (2020), discrimination against women is embedded in gendered societal power relations. They influence individual conduct, and more importantly, they permeate structures, procedures, and institutions of the state as well as within society and the family. These power relations are upheld by gender stereotypes, which express societal expectations of women's (and men's) proper conduct and whose violations are sanctioned. They ensure a hierarchy between men and women and the domination of women by men.

3. Gender and Race Issues in the Brazilian Scenario

In Brazil, the last three years proved to be even more complicated for women, especially for black women, poor and living on the outskirts of large Brazilian cities. Thus, it is worth analyzing the current situation in more detail.

Besides having an SDG entirely dedicated to the cause of promoting the human rights of women, the exceptionally high rates of domestic violence, femicide, aggression and sexual harassment in Brazil indicate that violence against women is still present in our society and presents itself in a multifaceted (Pinho 2020). According to the author mentioned above, some indicators point out that these forms of aggression still place Brazilian women in a situation of the constant threat to their physical and psychological integrity and establish barriers so that they are on an equal footing with men.

The Atlas of Violence (2019) shows there was an increase in female homicides in Brazil in 2017 with around 13 murders per day wherein all 4,936 women were killed, the highest number recorded since 2007 with 66% being the proportion of black women among victims of lethal violence in 2017. While the homicide rate of non-black women grew by 1.6% between 2007 and 2017, the homicide rate of black women grew by 29.9%.

Considering the year 2017, the homicide rate of non-black women was 3.2 per 100 thousand non-black women, whereas among black women the rate was 5.6 for every 100 thousand women in this group. In 2018, according to the Atlas of Violence (2020), a woman is murdered in Brazil every two hours, totaling 4,519 victims with 68% of the women murdered being black. While among non-black women, the homicide mortality rate in the last year was 2.8 per 100 thousand, while among black women the rate reached 5.2 per 100 thousand, which is practically double.

As for the outsourced workforce in Brazil, 70% are women and in precarious working conditions 39.9% are black women, 31.6% are black men, 26.9% white women and 20.6% white men (IBGE 2018).

Black women form the largest group of the population, totaling almost 60 million people and making up 28% of the Brazilian people. And of the total informal jobs created between 2014 and 2017, they occupied 82%. As for the average monthly income of the Brazilian population, black women receive R $ 1,476, while black men receive R $ 1,849, white women get R $ 2,529 and white men make R $ 3,364 (Machado et al. 2020).

In Brazil, in 2011, almost 15% of the black population was among the lowest 10% poorest; only 7% of whites were in the same segment and 20% of black women were among the lowest 10%, which makes black women overrepresented among the poorest and underrepresented among the richest (OXFAM BRASIL 2017).

About political representativeness, the World Economic Forum (2020) points out that Brazilian women continue to face a marked discrepancy concerning men, elements that were decisive for the country to be downgraded in the ranking by the

Global Gender Gap report. The item 'Political Empowerment' points out how much the political participation of women remains the most vulnerable aspect of promoting gender equality in Brazil. At this point, the fall was even more pronounced and the country went from 86th to 104th (Figure 1), given the shallow representation of women in the National Congress and the Federal Executive.

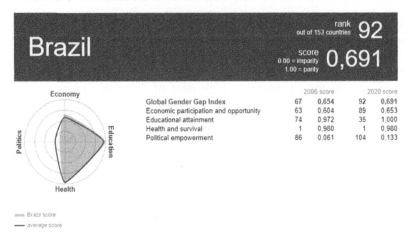

Figure 1: Global Gender Gap Index
Source: World Economic Forum 2020

According to the Global Gender Gap report (World Economic Forum 2020), the ranking of the World Economic Forum that analyzes equality between men and women, Brazil, which in 2016 ranked 79th, dropped to 90th in 2017 and now to the 92nd in 2020 (Figure 1). In the first edition of the research, in 2006, Brazil was in the 67th position, which means that there was a setback, even in face of advances such as the promulgation of Maria da Penha Law and the typification of femicide in the Penal Code.

Reflecting on gender equality in the world of work, women's economic autonomy has been one of the most significant challenges of the beginning of the twenty-first century. In the Global Gender Gap report (World Economic Forum 2020), the average income of women corresponds to 58% of that received by men. Thus, it is not enough to insert women into the job market or guarantee their economic and financial autonomy, it is essential to reflect the time management of women because of the accumulation of functions exercised daily.

According to the IBGE (2019), women spend 18.5 hours a week with work related to the care of children, parents, grandparents and housework, while men only 10.4 hours (Figure 2) since within families men do not take on these tasks equally, which leads to yet another burden for women.

4. Impacts with the COVID-19 Pandemic for Gender and Race Equity

Since March 2020, when the Covid-19 pandemic was decreed worldwide, several countries have adopted measures of social isolation to minimize contamination of the

Sex and occupation situation	Week Hours					
	Brazil	North	Northeast	South	Southeast	Midwest
Man						
Busy	10,4	10,8	10,0	10,7	10,7	9,3
Not busy	12,1	12,1	11,1	12,6	13,2	10,9
Woman						
Busy	18,5	18,4	19,1	18,8	17,7	16,8
Not busy	24,0	22,2	23,6	25,4	22,9	21,4

Figure 2: Average hours devoted to household chores and/or caring for people
(weekly hours)
Source: IBGE (2019)

population by the new virus. Although these measures are crucial and necessary, the situation of home isolation has the possible side effect of consequences for thousands of women (Bohoslavsky and Rulli 2020).

On March 23, 2020, approximately 154 million children and adolescents (over 95% of those enrolled) in Latin America were out of school due to the government's decision to temporarily close these establishments due to the pandemic (Bohoslavsky and Rulli 2020). These measures implied an overload of care work with the house, food, monitoring of distance education, new standards of domestic cleaning and teleworking among others for women who were already dedicated to working hours more than men.

The women who managed to keep their jobs with the adaptation to the teleworking model also had the overload of daily schedules with video conferences, and not all employers were able to adapt with the flexibility of the plan so that it was possible to reconcile the tasks in addition to those feelings of discomfort with the presence of the family at work or with work within the family.

In the context of jobs, the economic and health crisis resulting from the pandemic affects women more than men since women are usually represented in the sector of commerce, cleaning, health services, restaurants and hotels, all with low-income jobs and low remuneration and more likely to lose their jobs because the sectors are positively affected by the crisis and with high levels of informality (Bohoslavsky and Rulli 2020).

In the paid domestic work sector, the vulnerability of female workers, in the context of the economic crisis and the current pandemic is even more significant due to the lack of supervision and fewer opportunities for association and the exercise of labor rights and collective bargaining (thus leaving more exposed to workers), abuses by their employers that in many cases compel them to work even during quarantine and the devaluation and invisibility of this type of work (CEPAL 2020).

The pressure of the COVID-19 pandemic on health systems exposes women unequally as they represent 72% of the total employed in this sector in Latin America, which means the highest percentage in the world. Health system workers work under extreme conditions of pressure, and exhaustion that include long working hours, risk of being infected by the virus, stress in emergencies and distress. And, in turn, the workers return to their homes, where the burden of unpaid work related to dependents or who need care in their homes awaits them (CEPAL 2020).

Many women lost their jobs, thus submerging their financial independence and started to stay longer in their homes, thus being subject to the pressures of abusive relationships and making it difficult to access help channels and identify cases of aggression. According to the Brazilian Public Security Forum (FBSP 2020), it appears that month after month, there is a reduction in a series of crimes against women in several Brazilian states. But it is an indication that they are finding it more challenging to report the violence suffered during this period, the only exception being the lethal violence.

In contrast, in several states, periodic surveys have shown that in all months, there have been increases in the rates of femicides. Similarly, the data also indicate a reduction in the distribution and granting of emergency protective measures, a fundamental instrument for the protection of women in situations of domestic violence (FBSP 2020).

All Federation Units present in the survey showed a reduction in the record of intentional bodily injury between March and May 2020 compared to the same period in the previous year (FBSP 2020). There was a fall of 27.2% in the accumulated period with the largest decreases in the states of Maranhão (84.6%), Rio de Janeiro (40.2%) and Ceará (26%) (Table 1).

Table 1: Average Hours Devoted to Household Chores and/or Caring for People (Weekly hours)

Sex and occupation situation	Week hours					
	Brazil	*North*	*Northeast*	*South*	*Southeast*	*Midwest*
Man						
Busy	10,4	10,8	10,0	10,7	10,7	9,3
Not busy	12,1	12,1	11,1	12,6	13,2	10,9
Woman						
Busy	18,5	18,4	19,1	18,8	17,7	16,8
Not busy	24,0	22,2	23,6	25,4	22,9	21,4

Source: IBGE (2019)

There was a reduction in the rape and vulnerable rape records for the survey states, and in May 2020 the records fell 31.6% from 2,116 in 2019 to 1,447 in 2020. In the accumulated period between March and May 2020, there was a 50.5% reduction in records of rape and rape of the vulnerable with female victims compared to the same period in 2019. The registration of threats against women has also been falling since the beginning of the isolation period in the states analyzed. There was a 26.4% reduction in threat records in May 2020 compared to the same last year period.

Intentional homicides with female victims, on the other hand, increased 7.1% in May from 127 in 2019 to 136 in 2020. The most significant increases were in Ceará (208.3%), in Acre (100%) and the Rio Grande do Norte (75%). Looking exclusively at femicides, in the period between March and May 2020, there was an increase of

2.2% in registered cases compared to the same period in 2019. There were 189 cases in 2020 against 185 in the previous year. In the accumulated period, the state of Acre showed an increase of 400% in registrations, which went from 1 in 2019 to 5 in 2020. In Mato Grosso, the 157.1% rise in registrations, went from 7 to 18. Maranhão showed an increase from 11 cases to 20, a rise of 81.8% in the records (Table 2).

Broadening the look to countries neighboring Brazil, the situation is very similar, for example in Argentina alone, from March 20 to April 16, 2020, 21 femicides of women and girls were committed. In the same period of the previous year, there were 14 femicides. In other words, in a context of isolation, 50% more femicides occurred compared to last year. One in 5 of these women had made previous complaints. 65% were murdered in their homes in a context of isolation (Bohoslavsky and Rulli 2020).

In an interview with V (formerly Eve Ensler), indigenous leader Célia Xakriabá reports that saying that Covid-19 does not choose race, class or gender, is a lie, as the only choice is the state and he is the one who decides who will die of this disease (V, 2020). In this sense, it is very significant that the first death in Brazil by Covid-19 was of a black woman, a domestic worker, whose employer had recently returned from a trip from Italy (Magacho Filho and Santos 2020).

It is worth mentioning that the form of infection is the same for all people; however, the difference is in the way of preventive measures and possibilities of treatment and worsening of signs and symptoms. As for those who work informally, they have difficulties accessing the health system, which lacks financial resources to maintain hygiene and isolation among others (Estrela et al. 2020).

According to Viñas, Duran and Carvalho (2020), based on data released in the epidemiological bulletins of the Ministry of Health, 57% of those killed by the disease in Brazil are black and brown. According to Célia Xakriabá, indigenous peoples are 1% of the Brazilian population, but they are almost 9% of the victims of Covid-19 (V 2020).

Therefore, although necessary data is still lacking and the surveys carried out by the control bodies in governments and municipalities still have many inconsistencies, there are notes where race, class and gender cross as vital generalizing elements for Covid-19 (Estrela et al. 2020).

5. Ways to Implement Gender Equity: A Global Platform

The analysis that we have sought in this chapter, part of the need to highlight the importance of the global agendas and multilateral agreements, not only that proposed by Agenda 2030 but recognizing their significance and their strong link with the Universal Declaration of Human rights.

Where they intermingle, in the importance and recognition of human dignity as a universal right, while recognizing that human dignity involves gender equality and the exclusion of violence against women and girls, of all of them, regardless of race, skin color and vulnerability conditions.

In this sense, all 17 SDGs are fundamental, but especially recognizing the role that SDG 05 plays, because it is in its content that the existence of the problem is founded, at the same time that it presents objective proposals to overcome it.

Table 2: Records of Domestic Violence (Intentional Bodily Injury) Selected States, March to May 2019-March to May 2020

Federation unity	Intentional Bodily Injury											
	Mar/19	Mar/20	Variation (%)	Apr/19	Apr/20	Variation (%)	May/19	May/20	Variation (%)	2019 Mar–May	2020 Mar–May	Variation (%)
Acre	14	10	-28.6	...	...	...	...	...	...	...	...	...
Amapá	74	36	-51.4	26	29	11.5	25	27	8	125	92	-26.4
Ceará	462	365	-21	483	329	-31.9	467	351	-24.8	1.412	1.045	-26
Espírito Santo	...	...	...	613	431	-29.7	556	420	-24.5	...	...	...
Maranhão	223	6	-97.3	10	3	-97.2	84	55	-34.5	415	64	-84.6
Mato Grosso	953	744	-21.9	818	731	-10.6	896	729	-18.6	2.667	2.204	-17.4
Minas Gerais	2.108	1.807	-14.3	1.900	1.653	-13	...	...	...	...	...	...
Pará	607	527	-13.2	643	126	-80.4	357	704	97.2	1.607	1.357	-15.6
Rio de Janeiro	3.796	2.750	-27.6	3.641	1.875	-48.5	3.117	1.686	-45.9	10.554	6.311	-40.2
Rio Grande do Norte	287	385	34.1	286	121	-57.7	62	78	25.8	635	584	-8
Rio Grande do Sul	1.949	1.799	-7.7	1.719	1.259	-26.8	1.499	1.216	-18.9	5.167	4.274	-17.3
São Paulo	4.753	4.329	-8.9	4.937	3.244	-34.3	4.439	3.237	-27.1	14.129	10.810	-23.5
Total	15.226	12.758	-16.2	15.174	9.801	-35.4	11.502	8.503	-26.1	36.711	26.741	-27.2

Source: Fórum Brasileiro de Segurança Pública (FBSP 2020)

Table 3: Femicides Selected States, March to May 2019-March to May 2020

Federation unity	Mar/19	Mar/20	Variation (%)	Apr/19	Apr/20	Variation (%)	May/19	May/20	Variation (%)	2019 Mar–May	2020 Mar–May	Variation (%)
							Intentional Bodily Injury					
Acre	1	2	100	0	2	-	0	1	-	1	5	400.0
Amapá	0	0	-	0	0	-	1	0	-100	1	0	-100
Ceará	2	3	50.0	1	1	0	4	2	-50	7	6	-14.3
Espírito Santo	2	3	50	4	0	-100	1	1	0	7	4	-42.9
Maranhão	1	8	700	5	8	60	5	4	-20	11	20	81.8
Mato Grosso	2	7	250	4	5	25	1	6	500	7	18	157.1
Minas Gerais	8	8	0	14	9	-35.7	14	10	-28.6	36	27	-25
Pará	4	4	0	1	6	500	3	4	33.3	8	14	75.0
Rio de Janeiro	9	5	-44.4	9	3	-66.7	7	6	-14.3	25	14	-44.0
Rio Grande do Norte	1	4	300	3	0	-100	2	1	-50	6	5	-16.7
Rio Grande do Sul	11	11	0	1.719	10	66.7	11	6	-45.5	28	27	-3.6
São Paulo	13	20	53.8	16	21	31.3	19	8	-57.9	48	49	2.1
Total	54	75	38.9	63	65	3.2	68	49	-27.9	185	189	2.2

Source: Fórum Brasileiro de Segurança Pública (FBSP 2020)

Thus, we must resume in a more concrete way the theses that we started at the beginning of our chapter, where Arruzza et al.(2019) highlight that the criticism of liberal feminism and the construction of a new feminist wave is highlighted, mainly considering the importance of anti-capitalist feminism as has been pointed out previously. For this, it must be recognized that we are experiencing a crisis in society as a whole and that its cause is capitalism as it is implanted in contemporary society. Gender oppression in capitalist societies is rooted in the subordination of social reproduction to produce a profit and that this also needs to be overcome and better addressed.

Gender-based violence takes many forms, but it is always subordinated and linked to capitalist relations. The authors also point out that capitalism was born out of racist and colonial violence and the need to regulate sexuality as a form of power and ultimately the destruction of natural resources. For them, capitalism is incompatible with democracy and peace but that as they emphasize, the feminism proposed by them calls on all radical movements to unite in an anti-capitalist action (Arruzza et al. 2019).

The COVID 19 pandemic only highlighted our inequalities, injustices and the absence of possible alternatives. Mainly, in countries like Brazil and the USA, with their conservative and denialist governments, they do not offer real conditions for the confrontation. Thus, the theses presented by the authors are fundamental to rethink this moment.

Sen (2019) points to the difficulties besetting the path toward greater equality. At the same time, the author (Sen 2019) mentions that the depth of the gender inequality problem suggests that treating the achievement of gender equality as a mostly technical exercise will be entirely inadequate.

According to Sen (2019) to achieve momentum and to overcome the existing tendencies toward retrogression and backlash will need political mobilization on a significant scale. Sen (2019) suggests targets and indicators achieve gender equality and empower all women and girls as proposed by SDG5.

The main targets and indicators suggested are:

- End all forms of discrimination against all women and girls everywhere, and as an indicator, it is about whether or not legal frameworks are in place to promote, enforce and monitor equality and non-discrimination on the basis of sex.
- Eliminate all forms of violence against all women and girls in the public and private spheres, including trafficking and sexual and other types of exploitation.
- Indicators: Proportion of ever-partnered women and girls aged 15 years and older subjected to physical, sexual or psychological violence by a current or former intimate partner in the previous 12 months by a form of violence and by age, and proportion of women and girls aged 15 years and older subjected to sexual violence by persons other than an intimate partner in the previous 12 months by age and place of occurrence.
- Eliminate all harmful practices, such as child, early and forced marriage and female genital mutilation.

- Indicators: Proportion of women aged 20-24 years who were married or in a union before age 15 and before age 18, and proportion of girls and women aged 15-49 years who have undergone female genital mutilation/cutting by age.
- Recognize and value unpaid care and domestic work through the provision of public services, infrastructure and social protection policies and the promotion of shared responsibility within the household and the family as nationally appropriate.
- Indicators: Proportion of time spent on unpaid domestic and care work by sex, age and location
- Ensure women's full and effective participation and equal opportunities for leadership at all levels of decision-making in political, economic and public life.
- Indicators: Proportion of seats held by women in (a) national parliaments and (b) local governments, and proportion of women in managerial positions.
- Ensure universal access to sexual and reproductive health and reproductive rights as agreed in accordance with the Programme of Action of the International Conference on Population and Development and the Beijing Platform for Action and the outcome documents of their review conferences.
- Indicators: Proportion of women aged 15-49 years who make their own informed decisions regarding sexual relations, contraceptive use and reproductive health care, and number of countries with laws and regulations that guarantee full and equal access to women and men aged 15 years and older to sexual and reproductive health care, information and education.
- Undertake reforms to give women equal rights to economic resources, as well as access to ownership and control over land and other forms of property, financial services, inheritance and natural resources, in accordance with national laws.
- Indicators: Proportion of total agricultural population with ownership or secure rights over agricultural land by sex; share of women among owners or rights bearers of agricultural land by type of tenure, and proportion of countries where the legal framework (including customary law) guarantees women's equal rights to land ownership and/or control.
- Enhance the use of enabling technology in particular information and communications technology to promote the empowerment of women.
- Indicators: Proportion of individuals who own a mobile telephone, based on sex.
- Adopt and strengthen sound policies and enforceable legislation for the promotion of gender equality and the empowerment of all women and girls at all levels
- Indicators: Proportion of countries with systems to track and make public allocations for gender equality and women's empowerment.

The ability of feminist organizations to hold their own, to defend human rights, and to advance economic, ecological and gender justice in Sen (2019) perspective, will require not only clarity of vision and a track record of analysis and advocacy but also stronger communications skills, greater organizational resilience and effectiveness, and the ability to build and nurture effective alliances in which younger people play strong roles.

6. Conclusions

Seixas and Hoefel (2020) emphasize that efforts to achieve a fair and sustainable future must recognize the rights, dignity and capacities of the population worldwide, considering the role of gender equity to accomplish these goals as fundamental. To be effective, political actions for sustainability must correct the disproportionate impacts on women and girls in situations that lack economic, social and environmental changes. Knowledge, agency and collective action are the potential to improve resource productivity, improve conservation of ecosystems and use of natural resources and create more sustainable, low-carbon food, energy, water and efficient health systems.

In this sense, in addition to the mobilization of all, not only women and the one pointed out by all the authors referred to above, a synthesis which we consider essential, is worth changing this panorama of gender inequality and violence, especially in the Brazilian case. This short synthesis includes the following challenges: (1) incorporating the perspective of gender, race, ethnicity and sexual orientation in the formal and informal educational process; (2) guarantee a non-discriminatory educational system that does not reproduce stereotypes of gender, race and ethnicity; (3) promote access to primary education for young and adult women; (4) boost the visibility of women's contribution to the construction of human history; and (5) combat gender, race and ethnicity stereotypes in culture and communication (Brasil 2005). Or better yet, a phrase from Tiburi (2019) that summarizes our proposal:

> "The transformation of society needs to be thought of towards a better life for all people. It implies thinking about another project. Another policy, another power, another ethics, another economy."

Acknowledgments

We would like to thank Sao Paulo State Foundation for Research Support (FAPESP Brazil) for financial support for the research that led to this article (FAPESP n. 2019/08044-3) and for the scholarships to the fourth and fifth authors (FAPESP n. 2019/26490-0; 2020/03603-1). The Coordination for the Improvement of Higher Education Personnel - CAPES, Brazil for the scholarship to the third author and The National Council for Scientific and Technological Development, CNPq, Brazil for the Fellowship granted to the first author.

References

Allaerts, W. 2020. How could this happen? *Acta Biotheoretica*. https://doi.org/10.1007/s10441-020-09382-z

Alves, J.E.D. 2016. Desafios da equidade de gênero no século XXI. Dossiê Economia, Direitos Humanose Igualdadede Genêro: Uma Nova Agenda? Rev. Estud. Feministas. 24(2). Florianópolis mai./ago.

Arruzza, C., T. Bhattacharya and N. Fraser. 2019. Feminismo para os 99%: um manifesto. São Paulo: Boitempo.

Atlas of Violence. 2019. Instituto de Pesquisa Econômica Aplicada; Fórum Brasileiro de Segurança Pública (orgs). Brasília: Rio de Janeiro: São Paulo: Instituto de Pesquisa Econômica Aplicada; Fórum Brasileiro de Segurança Pública. Disponível em: https:// forumseguranca.org.br/wp-content/uploads/2019/06/Atlas-da-Violencia-2019_05jun_ vers%C3%A3o-coletiva.pdf. Acesso em: 01 Set. 2020.

Bandyopadhyay, S. 2020. Coronavirus Disease 2019 (COVID-19): we shall overcome. Clean Technologies and Environmental Policy. 22: 545-546. https://doi.org/10.1007/s10098-020-01843-w

Blay, E.A. 2002. Igualdade de oportunidades para as mulheres: um caminho em construção. São Paulo: Humanistas.

Bohoslavsky, Juan Pablo and Mariana Rulli. 2020. Covid-19, instituciones financieras internacionales y continuidad de las políticas androcéntricas en América Latina. Revista de Estudos Feministas, Florianópolis. 28(2): e73510. Disponível em: https://www.scielo. br/scielo.php?script=sci_arttextandpid=S0104-026X2020000200201andlng=ptandnrm=i soandtlng=es

Bradbury-Jones, Caroline and Louise Isham. 2020. Editorial – The pandemic paradox: The consequences of COVID-19 on domestic violence. Journal of Clinical Nursing 29: 2047-2049. DOI: 10.1111/jocn.15296

Buss, Paulo. 2020. De pandemias, desenvolvimento e multilateralismo. Available at: https:// agencia.fiocruz.br. Accessed 25 July 2020.

Brabo, T.S.A.M. 2015. Movimentos sociais e educação: feminismo e equidade de gênero. pp. 109-128. *In*: Neusa Maria Dal Ri, N.M. and Brabo, T.S.A.M. (organizadoras). Políticas Educacionais, Gestão Democrática E Movimentos Sociais: Argentina, Brasil, Espanha e Portugal /– Marília: Oficina Universitária; São Paulo: Cultura Acadêmica. ISBN 978-85-7983-682-4

Brasil. 2005. Plano Nacional de Políticas para as Mulheres. Brasília: Secretaria de Políticas para as Mulheres.

CEPAL. 2020. La pandemia del COVID-19 profundiza la crisis de los cuidados en América Latina y el Caribe. 2 de abril de 2020. Disponible en https://repositorio.cepal.org/ handle/11362/45335

Editorial – The gendered dimensions of COVID-19. 2020. The Lancet. 395(10231): 1168, April 11. DOI: https://doi.org/10.1016/S0140-6736(20)30823-0

Estrela, F.M., C.F. Soares e Soares, M.A. Cruz, A.F. Silva, J.R.L. Santos, T.M.O. Moreira, A.B. Lima and M. Silva. 2020. Pandemia da Covid-19: Refletindo as vulnerabilidades a luz do gênero, raça e classe. Disponível em: http://www.cienciaesaudecoletiva.com.br/ artigos/pandemia-da-covid-19-refletindo-as-vulnerabilidades-a-luz-do-genero-raca-e-cla sse/17581?id=17581andid=17581.

FBSP. 2020. Fórum Brasileiro de Segurança Pública. Violência doméstica durante a pandemia de Covid-19 – ed. 3. Disponível em: https://forumseguranca.org.br/publicacoes_posts/ violencia-domestica-durante-pandemia-de-covid-19-edicao-03/

Gonzaga, Eunir Augusto Reis, Isabella do Carmo Lacerda, Tuila Tachikawa de Jesus and Samuel do Carmo Lima. 2020. Equidade, justiça social e cultura de paz em tempos de pandemia: um olhar sobre a vulnerabilidade municipal e a COVID-19. Hygeia-Revista Brasileira de Geografia Médica e da Saúde, Edição Especial: Covid-19, Jun./2020, p.111 – 121. DOI: http://dx.doi.org/10.14393/Hygeia0054569

IBGE – Instituto Brasileiro de Geografia e Estatística. 2018. Desigualdade Sociais por Cor ou Raça no Brasil. Disponível em:https://biblioteca.ibge.gov.br/visualizacao/livros/ liv101681_informativo.pdf

IBGE – Pesquisa Nacional por Amostra de Domicílios Contínua. 2019. Outras formas de Trabalho. Disponível em: https://biblioteca.ibge.gov.br/visualizacao/livros/liv101722_ informativo.pdf

Machado, Machado, Fabiane, Roberta da Silva Gomes and Caroline Bertolino. 2020. Saúde Mental das Mulheres e a Covid-19: Um Recorte de Gênero, Raça e Classe. Disponível em:https://www.youtube.com/watch?v=SBB_5qh2mGE

Magacho Filho, M.R. and L.C.N. Santos. 2020. Covid-19: gênero e raça no trabalho de cuidado. Disponível em: https://www.conjur.com.br/2020-jul-07/opiniao-covid-19-genero-raca-trabalho-cuidado

Marques, Emanuele Souza, Claudia Leite deMoraes, Maria Helena Hasselmann, Suely Ferreira Deslandes and Michael Eduardo Reichenheim. 2020. A violência contra mulheres, crianças e adolescentes em tempos de pandemia pela COVID-19: panorama, motivações e formas de enfrentamento. Cad. Saúde Pública. 36(4): 1-6 e00074420. doi:10.1590/0102-311X00074420

Nery, Déa Carla Pereira and José Renato Oliva de Filho Mattos. 2020. A violência contra a mulher em tempos de pandemia. pp. 190-203. *In*: Hirsch, Fábio P. de A. COVID-19 e o Direito na Bahia. Salvador: Editora Direito Levado a Sério. Odera, J.A. and J. Mulusa. 2020. SDGs, gender equality and women's empowerment: what prospects for delivery? pp. 95-118. *In*: Kaltenborn, M., Krajewski, M. and Kuhn, H. [eds.]. Interdisciplinary Studies in Human Rights 5: Sustainable Development Goals and Human Rights. https://doi.org/10.1007/978-3-030-30469-0

Okabayashi, Nathalia Yuri Tanaka, Izabela Gonzales Tassara, Maria Carolina Guimarães Casaca, Adriana de Araújo Falcão and Márcia Zilioli Bellini. 2020. Violência contra a mulher e feminicídio no Brasil – impacto do isolamento social pela COVID-19. Brazilian Journal of Health Review. 3(3): 4511-4531. may./jun. 2020. ISSN 2595-6825 DOI:10.34119/bjhrv3n3-049

Oxfam Brasil/INESC/Center for Economic and Social Rights (2017). Brasil. Direitos humanos em tempos de austeridade. Available at: https://www.oxfam.org.br/publicacao/direitos-humanos-em-tempos-deausteridade/

Peteet, J.R. 2020. COVID-19 Anxiety. Journal of Religion and Health. https://doi.org/10.1007/s10943-020-01041-4

Peterman, A., A. Potts, M. O'Donnell, K. Thompson, N. Shah, S. Oertelt-Prigione and N. van Gelder. 2020. Pandemics and violence against women and children. CGD Working Paper 528. Washington, DC: Center for Global Development. https://www.cgdev.org/publication/pandemics-and-violence-against-women-and-children

Pinho, Tássia Rabelo de. 2020. Debaixo do tapete: a violência política de gênero e o silêncio do conselho de ética da câmara dos deputados. Revista Estudos Feministas, Florianópolis. 28(2): e67271. Disponível em: https://www.scielo.br/scielo.php?script=sci_arttextandpid=S0104-026X2020000200202andlng=ptandnrm=iso

Pires, Roberto Rocha C. 2020. Os efeitos sobre grupos sociais e territórios vulnerabilizados das medidas de enfrentamento à crise sanitária da covid-19: propostas para o aperfeiçoamento da ação pública. Nota Técnica 33. Brasília: Ipea – Instituto de Pesquisa Econômica Aplicada

Rudolf, Beate. 2020. Freedom from violence, full access to resources, equal participation, and empowerment: the relevance of CEDAW for the implementation of the SDGs. pp. 73-94. *In*: Kaltenborn, Markus, Krajewski Markus and Kuhn, Heike [eds.]. Sustainable Development Goals and Human Rights. Interdisciplinary Studies in Human Rights 5, https://doi.org/10.1007/978-3-030-30469-0_5

Seixas, S.R.C. and J.L.M. Hoefel. 2020. Human rights and gender equity: building sustainable development. *In*: Leal Filho W. et al. [eds.]. Gender Equality, Encyclopedia of the UN Sustainable Development Goals. Springer Nature, Switzerland AG, https://doi.org/10.1007/978-3-319-70060-1_60-1

Sen, G. 2019. Gender equality and women's empowerment: feminist mobilization for the SDGs. Global Policy. 10(Suppl. 1), https://doi.org/10.1111/1758-5899.12593

Sharma, V., J. Scott, J. Kelly and M.J. VanRooyen. 2020. Prioritizing vulnerable populations and women on the frontlines: COVID-19 in humanitarian contexts. International Journal of Equity Health. 19(66): 1-3. https://doi.org/10.1186/s12939-020-01186-4

Srivastava, N., P. Baxi, R.K. Ratho and S.K. Saxena. 2020. Global trends in epidemiology of Coronavirus Disease 2019 (COVID-19). *In*: Saxena, S. [ed.]. Coronavirus Disease 2019 (COVID-19). Medical Virology: From Pathogenesis to Disease Control. Springer, Singapore.

Tedeschi, L.A. and A.M. Colling. 2016. Os Direitos Humanos e as questões de Gênero. História Revista. 19(3): 33-58. https://doi.org/10.5216/hr.v19i3.32992

Tiburi, M. 2019. Feminismo em Comum. Para Todas, Todes e Todos. 12 ed. Rio de Janeiro: Rosa dos Tempos.

UN (United Nations). 2015. Understanding Human Rights and Climate Change. Submission of the Office of the High Commissioner for Human Rights to the 21st Conference of the Parties to the United Nations Framework Convention Climate Change. Available at: https://www.ohchr.org/Documents/Issues/ClimateChange/COP21.pdf.

UN Women (United Nations Entity for Gender Equality and the Empowerment of Women). 2015. Progress of the World's Women 2015-2016 Transforming Economies, Realizing Rights. ISBN: 978-1-63214-015-9, View the Report at: http://progress.unwomen.org

Usher, Kim, Navjot Bhullar, Joanne Durkin and Naomi Gyamf. EDITORIAL – Family violence and COVID-19: increased vulnerability and reduced options for support (2020). International Journal of Mental Health Nursing (2020). 29: 549-552. doi: 10.1111/inm.12735

Ventura, Deisy de Freitas Lima, Ribeiro Helena, Giulio Gabriela Marques di, Jaime Patrícia Constante, Nunes João, Bógus Cláudia Maria, Antunes José Leopoldo Ferreira, and Eliseu Alves Waldman. 2020. Desafios da pandemia de COVID-19: por uma agenda brasileira de pesquisa em saúde global e sustentabilidade. Cad. Saúde Pública. 36(4): e00040620. Epub Apr 22, 2020. doi: 10.1590/0102-311X00040620

Vieira, P.R., L.P. Garcia and E.L.N. Maciel. 2020. Isolamento social e o aumento da violência doméstica: o que isso nos revela? Revista Brasileira de Epidemiologia. 23: e200033. Epub April 22, 2020 https://www.scielo.br/pdf/rbepid/v23/1980-5497-rbepid-23-e200033.pdf

Viñas, D., P. Duran and J. Carvalho. 2020. Morrem 40% mais negros que brancos por coronavírus no Brasil. Disponível em: https://www.cnnbrasil.com.br/saude/2020/06/05/negros-morrem-40-mais-que-brancos-por-coronavirus-no-brasil

Viveiros Nelia and E. Bonomi Amy. 2020. Novel Coronavirus (COVID-19): violence, reproductive rights and related health risks for women, opportunities for practice innovation. Journal of Family Violence, published on line 06 June 2020. https://doi.org/10.1007/s10896-020-00169-x

World Economic Forum. 2020. The global gender gap report. Geneva: World Economic Forum. Available at: http://reports.weforum.org/global-gender-gap-report-2020/the-global-gender-gap-index-2020/

Zamora, Gerardo, Koller Theadora Swift, Rebekah Thomas, Mary Manandhar, Eva Lustigova, Adama Diop and Veronica Magar. 2018. Tools and approaches to operationalize the commitment to equity, gender and human rights: towards leaving no one behind in the sustainable development goals. Global Health Action. 11(1): 75-81, DOI: 10.1080/16549716.2018.1463657

Zhang, Y. 2020. The epidemiological characteristics of an outbreak of 2019 novel Coronavirus Diseases (COVID-19) – China, 2020. Chinese Centre for Disease Control and Prevention – CCDC Weekly. 2(8), 113-122.

Transformation to Sustainable, Healthy and Just Food Systems

Angelina Sanderson Bellamy

Sustainable Places Research Institute, Cardiff University, 33 Park Place, Cardiff, CF10 3BA

1. Introduction: Food System in Crisis

The food system is a vital component for current and future generations to better safeguard their health and livelihoods, yet it also represents a threat as production, distribution and consumption practices can erode ecosystem resilience, contribute to the climate emergency and endanger health. Ecosystem resilience and a strong, healthy environment underpin the entire food system. Without this, our capacity, both globally and regionally, to produce food is greatly reduced. Our current food environment erodes public health, as cheap and highly processed foods, high in sugar, fat and salt, are widely promoted, cheaply accessible and strategically placed within supermarkets and within cities to maximise sales. The increasing health cost of diet-related chronic non-communicable diseases (NCDs) also pose a public health threat.

The food landscape has changed dramatically over the last 60 years with supply chains from farm-to-fork growing longer and more complex and the distribution of power becoming more concentrated through vertical integration of agri-food business activities. Seed companies and procurement organisations for supermarkets exert a major influence on what happens in other parts of the food system. A disproportionate amount of economic value is captured by corporate food processors and retailers, which squeezes the margins of local producers, independent retailers and consumers alike.

Political, economic and sociological critiques of current agri-food dynamics identify intensification, specialisation, distancing, homogenisation and concentration of power as the key processes resulting in food insecurity, health concerns and environmental degradation. These five trends have configured a 'place-less' foodscape disconnected from diverse social demands and the ecological basis of distinct territories. This has led to a food system, from global to regional scales,

*Email: BellamyA1@cardiff.ac.uk

with interlinked crises of obesity (The Lancet Commission on Obesity 2019), undernutrition, poverty (FAO et al. 2019), climate change (IPCC 2019 Climate Change and Land) and ecosystem degradation and biodiversity loss (Hayhow et al. 2019, IPBES 2019).

There is a growing understanding of the links between industrial food systems, ill-health, environmental damage and injustice (Frison and Clément 2020). As indicated in the EAT Lancet Commission Report: "food systems have the potential to nurture human health and support environmental sustainability; however, they are currently threatening both" (Willett et al. 2019). Agriculture is a major contributor of greenhouse gas emissions and the largest consumer of global freshwater resources, whilst food production and supply account for one-quarter of global energy use (Ritchie and Roser 2020). At present, climate change and malnutrition play out in ways that affect most of those who are vulnerable.

We currently have a food system where there is no place at the table for civil society and agricultural workers to participate in policy-making and develop democratic food governance. Intensive systems of farming are continuing to destroy biodiversity; both farmers and consumers face increasing costs and food insecurities, whilst those working in the food industry receive low pay and low esteem (Food Manufacture 2019, Taylor and Loopstra 2016) and 64% of farmers earn less than £10,000 a year (Defra 2015). Working in the food and farming sector is characterised by insecure, precarious and unpredictable labour conditions. Farmers are highly dependent upon current EU subsidies whilst still responsible for unsustainable soil, water and carbon management. Neither producers nor consumers are being protected in the current food system, whilst most of the profits from the processes of food production, processing and retailing are contained in the privatised and highly concentrated 'black-box' which delinks producers and consumers. Eight supermarkets control almost 95% of the food retail market (Kantar Worldpanel 2017), and farmers receive less than 10% of the value of their produce sold in supermarkets (People Need Nature 2017).

The above-mentioned long-term stressors within the food system have slowly eroded the resilience of the food system and made it more vulnerable to short-term shocks, as we are witnessing currently with the COVID-19 pandemic. The COVID-19 pandemic illustrates the food system's inability to meet key sustainable and just objectives. In terms of the UK population, the groups at risk of food insecurity have been identified as those economically, socially or medically vulnerable (Food Foundation 2020a). The Food Foundation YouGov survey conducted in early April 2020 found that households with children were 50% more likely to be experiencing food insecurity than those without (Food Foundation 2020b). Another survey conducted by both The Food Foundation and YouGov and published on May 4 shows 5 million people in the UK living in households with children under 18 have experienced food insecurity since the lockdown started (Food Foundation 2020c). Approximately 1.8 million of these experienced food insecurities solely due to the lack of supply of food in shops, leaving 3.2 million people (11% of households) suffering from food insecurity due to other issues such as loss of income or isolation.

The COVID-19 pandemic has highlighted the various weaknesses in the UK food system, such as lack of diversity in supply chain structure, lack of diversity in retail outlets, just-in-time supply chains and heavy reliance on food imports, which have resulted in food shortages in some supply chains and food waste in others. This, in combination with widespread job losses and reduced household income, has led to increased severity of household food insecurity in the UK.

However, COVID-19 only worsens and makes visible existing societal inequalities and levels of food poverty which have been growing steadily after more than a decade of austerity (O'Connell et al. 2019, Power et al. 2020). And while the nature and extent of the COVID-19 crisis may be rare, climate change research indicates that extreme climatic events will increase in severity and frequency and may impact societies in a similar manner (IPCC 2019). Droughts and floods, for example, are likely to bring much greater economic and supply chain disruption in the future, and loss of natural capital is a major issue. On this point, the United Nation's Special Report on Global Warming of 1.5°C (IPCC 2018) and Global Assessment Report on Biodiversity and Ecosystem Services (IPBES 2019) identify unsustainable agriculture and land use as one of the main drivers of environmental degradation.

The status quo is unsustainable and a radical redesign of the food system is urgently required. There is little doubt that markets that are currently regulated are failing to deliver healthy diets or sustainable food systems. Within the current system, public health initiatives 'compete' with market forces and cultural norms. This chapter argues that a human rights approach that foregrounds food equality, affordability, and accessibility is fundamental to achieving a food system that is healthy and sustainable. Such an approach would mobilise state, private sector and community engagement with our food system in order to meet targets set by UN Sustainable Development Goals, specifically:

- SDG 2: End hunger, achieve food security and improved nutrition and promote sustainable agriculture.
- SDG 3: Ensure healthy lives and promote well-being for all at all ages, including reducing premature mortality from non-communicable diseases (NCDs) through prevention and treatment.
- SDG 6: Ensure availability and sustainable management of water and sanitation for all, including protecting and restoring water-related ecosystems such as mountains, forests, wetlands, rivers, aquifers and lakes.
- SDG 8: Promote sustained, inclusive and sustainable economic growth, full and productive employment and decent work for all.
- SDG 12: Ensure sustainable consumption and production patterns.
- SDG 13: Take urgent action to combat climate change and its impacts.
- SDG 14: Conserve and sustainably use the oceans, seas and marine resources for sustainable development.
- SDG 15: Protect, restore and promote sustainable use of terrestrial ecosystems, sustainably manage forests, combat desertification and halt and reverse land degradation and halt biodiversity loss.

The current food system is broken and the time for engaging a transformation is now urgent.

To find solutions, we need food system stewardship, which involves the active shaping of pathways of social-ecological change to enhance ecosystem resilience, healthy diets and accessibility of food. In so doing, society, and more specifically, food citizens can actively shape place-based food systems that address shortcomings and build a foundation for a food system that delivers societal objectives of sustainability, resilience and justice.

This chapter focuses on the UK context and briefly discusses current actions taken to transform the UK food system in England, Scotland and Wales. Using this as a departure point, it then raises a number of policy approaches for transforming food systems to deliver SDGs, focusing on delivering a national universal food framework. Finally, it considers the role that agroecology can play across the food system to transform relationships across the food system, to move food system relations from transactional interactions to relational interactions and reflects on how these actions have the power to transform the way our food system functions so that it delivers environmental sustainability, healthy diets and accessibility.

2. The UK Context: Current Actions to Transform the UK Food System

The context and current actions for transforming the UK food system vary across the devolved territories with a widespread recognition that we cannot continue operating under a scenario of business as usual. In England, 2019 saw the initiation of a campaign to build a National Food Strategy, led by Henry Dimbleby, and in part informed by citizen food assemblies held in cities across England. The Strategy consists of a review of the current food system and will make a series of recommendations for delivering a food system transformation in England (National Food Strategy 2019). In addition, the 2020 Agriculture Bill moved away from European Union Common Agricultural Policy (CAP) procedures of paying farmers based on the size of the landholders and instead proposed that payments to farmers be based on the provision of public goods, such as clean water, biodiversity and soil conservation. Additionally, for the first time, the Agriculture Bill mentions the term 'agroecology'; it states "The Secretary of State may give financial assistance for...better understanding of the environment...'better understanding of the environment' includes better understanding of agroecology" (UK Parliament 2020: 2-3). Such a statement can be interpreted as an indication of the direction of travel for the development of future agricultural practices and shifts in farming systems. While these changes mark a significant and important shift from current policy practices, the bill arguably does not go far enough to encourage changes in farming practices that will be necessary if the UK is to reach zero carbon emission targets set for 2050.

Arguably, however, these shifts represent some of the most conservative actions taken in the UK to promote a transformation in the current food system. In Scotland, for example, due in part to the highly-organised work of civil society, there is pressure to develop policies and practices that shift the food system from a market-based approach to a rights-based approach to food. The UK government provides tax funding for the provision of universal health care because healthcare is viewed as a universal right that everyone should be able to access. The International Covenant on Economic, Social and Cultural Rights (ICESCR) is an international treaty that aims to ensure the protection of economic, social and cultural rights, such as the rights to work, social security, health and education (UN 1976). Article 11(1) of the Covenant recognises the right of everyone to an adequate standard of living, including adequate food, clothing and housing and the continuous improvement of living conditions. Article 11(2) guarantees the fundamental right of everyone to be free from hunger and obliges State Parties (i.e., those countries that have ratified the Covenant) to take steps in this regard, including the improvement of methods of distribution of food and dissemination of knowledge concerning the principles of nutrition. The UK ratified ICESCR in 1976 and is therefore legally bound by its articles. Similarly, the UN Special Rapporteur on the Right to Food defines the right to food as "the right to have regular, permanent and free access, either directly or by means of financial purchases, to quantitatively and qualitatively adequate and sufficient food corresponding to the cultural traditions of the people to which the consumer belongs, and which ensures a physical and mental, individual and collective, fulfilling and dignified life free of fear". Therefore, the Government is legally required under international human rights law to secure the human right to adequate food for everyone in the UK.

The Scottish charity Nourish Scotland collaborated with other food and farming organisations to form the Scottish Food Coalition and together they published the report Plenty (2016), which called for the right to food to be enshrined in Scots law and a Good Food Nation Act. This grew out of the Scottish Government's publication of their national food and drink policy Becoming a Good Food Nation in 2014 (Scottish Government 2014). The policy set a new vision for Scotland's food system, that by 2025 Scotland would be "a Good Food Nation, where people from every walk of life take pride and pleasure in, and benefit from, the food they produce, buy, cook, serve, and eat each day" (Scottish Government 2014).

The Programme for Government 2019 to 2020 contained a commitment to publish a Good Food Nation Bill but was delayed due to the COVID-19 crisis (Scottish Government 2020). However, in June 2020 the Right to Food Bill was introduced for public consultation. This proposal argues that establishing a statutory right to food for all is an urgent matter and should be part of a separate, expedited process, as the food sector impacts on so many policy areas. The COVID-19 pandemic has highlighted the urgent necessity for such an approach.

In other parts of the devolved territory, Wales, in 2015, led with another innovative policy approach by establishing the Well-Being of Future Generations Act, which forms overriding core legislation that should inform and guide all public service bodies. The Act creates a legal obligation to improve Wales' social, cultural, environmental and economic well-being. It requires public bodies in Wales to think about the long-term impact of their decisions, to work better with people, communities and each other and to prevent persistent problems, such as poverty, health inequalities and climate change. The Act is unique to Wales, attracting interest from countries across the world as it offers a huge opportunity to make a long-lasting, positive change to current and future generations and embodies the spirit of the UN SDGs. The Act defines seven well-being goals, namely, to create:

a prosperous Wales, a resilient Wales, a more equal Wales, a healthier Wales, a Wales of cohesive communities, a Wales of vibrant culture and thriving Welsh language and a globally responsible Wales (Welsh Government 2015). Under this legislation, there is an opportunity to build policy approaches for transforming the Welsh food system to build a food system fit for future generations (Sanderson Bellamy and Marsden 2020).

3. Introducing a Rights-Based Approach to Food in National Policy

From a governance perspective, the growing food crisis has made clear that food system outcomes are affected by a complex range of determinants and that traditional governmental efforts to steer these determinants through monocentric command and control strategies get stranded in 'siloed' administrative systems, intractable controversies between opposing value systems and power struggles between constellations of interests. In order to give a strong steer and coordinate action, the UK needs to build a National Vision for the Food system (Sanderson Bellamy and Marsden 2020). Sanderson Bellamy and Marsden (2020) outline the arguments and develop a framework for taking a national-scale food systems approach in food governance. A national vision for the food system and a national framework for the food system are born from the realisation that it is no longer sufficient to talk about a more sustainable food system; but rather there is a need for a radical transformation of the system so as to deliver sustainable and healthy food for all in the face of shrinking and increasingly degraded natural resources and a growing global population. The lessons learned from the impacts of COVID-19 on the food system only serve to more clearly demonstrate the need for an integrated governance and policy approach across all actors in the food system.

Taking a food system approach would thus mean understanding food as more than a marketable commodity, which creates problems for certain departments. Government has a responsibility to the public to ensure basic rights to food and food security. In specific circumstances, especially where private markets are clearly failing to deliver sustainable forms of food security, the food needs to be delivered through both public and private means. These structural changes can be created in the form of a National Universal Food Framework.

We know that food and diet are tightly coupled to health and well-being. A National Universal Food Framework would ensure universal access to a healthy diet, as defined by the Eatwell Guide and WWF's Livewell plate. A National Universal Food Framework would engage with food producers, food hubs, schools, local authorities and the NHS to re-connect people to the sources of food, to nature and each other. With the increasing trend toward globalised long food supply chains, interactions between actors in the food system have moved from being relational to being purely transactional. Transactional relationships are by nature optimized around getting the most you possibly can in exchange for as little as possible on your part. They are all about you and what you can get and not about what you can give. Transactional relationships: (1) focus on self-benefits, (2) are results-oriented, (3) use both positive and negative reinforcement, (4) are based on expectations and judgement and (5) consist of actors that compete against each other. When based in a market economy, the outcomes of such an interaction are optimized for the short-term gains rather than considering the long-term benefits and thus can be contradictory to sustainability and well-being goals.

However, relational interactions are based on building long-term relationships between actors, which are based upon mutual interests (or values), what each actor gives rather than what they get and resolving conflict instead of winning the conflict. Relational interactions have the power to be transformative. It can be argued that actors are stronger if acting collectively and in an optimized way to advance each other's goal, which would, in turn, lead to an outcome that can deliver greater well-being (this will depend on the objectives of actors in the first place). Building relationships and relational interactions back into the food system can foster environmental sustainability and positive mental and physical health outcomes. The ongoing UKRI-funded research project TGRAINS (Transforming and Growing Relationships within regional food systems for Improved Nutrition and Sustainability) is investigating the impact of interventions that build relationships in regional food systems, and initial results indicate the above listed positive outcomes of building relational interactions back into food systems. Relational interactions are not only limited to regional food systems where distances between actors are short and enable face-to-face interactions. Online digital platforms, operating to support community food hubs, for example, the Open Food Network (https://www.openfoodnetwork.org.uk/) or individual farmers operating in other countries, such as Crowd farming (https://www.openfoodnetwork.org.uk/), also work to build relational interactions back into interactions between food system actors.

A National Universal Food Framework can create the mechanisms to empower communities to engage with their food system so as to design and build through bottom-up participation and with facilitation support, community-based universal food programmes that utilise a capabilities approach and meet the unique needs of each community.

Where food is produced locally, cross-sectoral partnerships and policy approaches can lead to programmes where school children and community members are connected with local farms and community gardens, where they would spend time with other individuals, on activities on the farm/garden to grow food. It could also incorporate community kitchens, which would be stocked with basic store cupboard goods, where people could go to cook food in a social and supportive environment. This would grow the scheme supported by the Soil Association in the Food for Life programme (https://www.foodforlife.org.uk/) to award funding to groups for multigenerational social gatherings based around food. This is just to illustrate the ways in which Government could build a National Universal Food Framework and benefit from existing programmes at community, regional and national scales.

A National Universal Food Framework could coordinate and use social and nature prescribing to have individuals spend time on farms/gardens growing food. The National Universal Food Framework would incorporate education and school curriculum through partnering farmers and community gardens with schools to incorporate regular farm visits into health and well-being learning objectives under the new school curriculum. Adult learning programs for nutrition and cooking skills would incorporate learning about how food grows (which encourages a greater appreciation of the food). It would establish lifelong education that reconnects people to the source of their food. Agri-environment payments for the provision of public goods would reward farmers for their participation in education and social prescribing programmes.

A National Universal Food Framework would grow a connection between individuals and the food they eat, it would create health benefits in terms of increasing physical activities, and it would improve mental health due to both outdoor physical activity and socialising with other people also on the farm/garden. It would do so by integrating education, agriculture, food, procurement and health strategies and action plans and would drive the National Food System Strategy.

A National Universal Food Framework would draw upon already existing strategies across the above-mentioned sectors but create a structure for unifying policy strategies. A food system policy approach presents opportunities for cost savings by creating synergies across the food system to achieve the transformation needed for a sustainable and just food system. A National Universal Food Framework would also improve costs through improved health outcomes that reduce the burden of non-communicable and dietary-related diseases and improve education outcomes (Weitkamp et al. 2013). While such an approach seems radical, a recent report released by the RSA Food, Farming and Countryside Commission advocates similar approaches, for example through their proposed Beetroot Bonds (RSA 2019). Implement such an approach successfully would require a broader systemic infrastructure of support such that a National Universal Food Framework could offer.

A National Universal Food Framework disrupts the current system and proposes taking a food system approach to policy by integrating and aligning a National Universal Food Framework to other national service systems, such as the NHS, the Education System and all national and local government procurement processes. The National Universal Food Framework, embedded within these institutions and mechanisms would provide: a) important opportunities for norm-setting, b) advocacy and accountability and c) behavioural change across local, regional and national levels. Rather than precluding existing community interventions that operate in a piecemeal fashion, it would provide a national framework under which interventions may operate in a systematic manner to drive behaviour change at all system levels to improve health outcomes for all, help address the climate and ecological crises, increase resilience and ensure sustainability. Such a system would also align with international guidelines, such as the UN Decade on Nutrition and UN Right to Food. Moreover, a National Universal Food Framework would utilise government buying power to drive systemic changes in business, the third sector and policy-making to ensure that all sectors and members of society benefit.

4. Transforming Food Systems to Deliver SDGs: Policy Approaches Moving Forward

Empty shelves in supermarkets during the COVID-19 pandemic drove many people to seek out alternative markets, such as local box schemes, farm shops and food hubs to secure food and thus illustrated the importance of diversity of routes to markets. Consumers responded to stock-outs and the food insecurity it brought by diversifying and balancing market share across retail models (Sanderson Bellamy et al.; forthcoming). There are indications that this diversification can help build food security resilience to supply chain shocks. These alternative retail models, independently owned (e.g., market traders such as farmers markets and farm shops) or cooperatively owned (e.g., food cooperatives, food hubs and buyer cooperatives), made food more accessible to consumers who needed it. This expanding retail helped the urban poor, who are most exposed to insecurity, particularly when they were already a part of a community.

Supply chains thus need to evolve to include a greater diversity of retail outlets; government policy can be used to support the development of alternative routes to market through business grants that support establishing, such as community food hubs and cooperative food enterprises. Greater diversity of markets creates resilience to crises that may impact one type of retail stream more than others. For example, if the hospitality sector is impacted, farms and other food provisioning businesses can turn to other developed routes to markets to continue to bring food from the field to the fork.

Data collected on Covid-19 impact and responses indicate that those businesses who were able to respond flexibly and reroute food through alternative market channels have thrived during the Covid-19 pandemic (Pitt et al. 2020). Additionally, supporting a greater number of suppliers within the varied routes to markets also creates greater resilience of food supply chains. This means that if anyone (or several) business fails to adapt during times of crisis, there are several other businesses that can continue to carry on and manage to meet market demand overall. Because businesses are not operated identically, they have the differing capacity to respond to different crises. Diversity and redundancy are key principles of resilience and in an above-illustrated way can confer robustness to the food system. A resilient food system requires a diversity of business models and modes of production. Welsh food producers' response to the COVID-19 crisis demonstrates their vital contribution to supply chains and the advantages of relatively small producers which are flexible and adaptable. In order to convert crisis response into long-term economic, environmental and social resilience there is a need for:

- Financial support for small food producers and those which focus on supplying local demand to ensure they remain economically active.
- Facilitation of cooperation between growers and coordination across the supply chain to enhance efficiency and enable expansion.
- Monitoring and investigation of consumption trends in order to share intelligence with producers to inform their business planning.
- Ongoing dialogue with diverse types of food producers to remain alert to their needs.
- National consumer campaigns to promote local produce and producers.

Local food also requires local food processing and manufacturing facilities. Hidden costs along the entire supply chains, e.g. high food miles, can mislead consumers. Investment in this area and the development of local infrastructures (e.g. co-ops, local food packing and processing centres, local abattoirs and on-farm butchering and manufacturing) is important. Given this, there is a clear role for government to be proactive in providing primary producers with a business environment that enables them to secure a fair return from the market and a fair share of the profit that exists in the supply chain. Transparency is key to fair supply chains and better regulation and enforcement throughout should be put in place to prevent abuses of power.

Government policies that address these concepts and support activity to build a more structurally complex and diverse supply chain can support a more resilient food system.

Supply chain complexity would include:

• variable length pathways, i.e. shorter supply chains based on regional and national production to balance the dominant role of the global supply chains;
• varying numbers of actors involved at different steps of the supply chain;
• multiple and diverse routes to market, such as retail chains, community food hubs, box schemes and buyer collectives; food processing and manufacturing facilities, such as mills, abattoirs, dairies and washing centres.

A more structurally complex supply chain with the above elements also has a greater potential to distribute power and profit more equitably across food system actors.

With the right policies in place, re-localising food production and consumption, and promoting shorter supply chains can generate multiple benefits for the local economy, environment, health and well-being. Public sector procurement is one policy area with the power to exert change by ensuring the procurement of locally-sourced and sustainably-produced food that supports public health objectives. Practical steps to enable small-scale producers to supply large public contracts via an intermediary or through a processing centre should also be explored to increase the feasibility of such an approach. Small-scale vegetable producers may not have the facilities to wash and prepare vegetables to the standard required by a local school; however, a local processing centre that multiple growers can access would create more market opportunities for growers.

Creating more brand awareness around local, nature-friendly, sustainable produce and encouraging wildlife/conservation/environmentally friendly food certification could help food producers to become more profitable and sustainable, improving efficiency, reducing input costs and enhancing income streams through diverse activities. Labels such as Fair to Nature, Leaf Marque, Organic and Pasture Fed and Marine Stewardship Council are a good way to determine if the food purchased has been produced in a more nature/environmentally friendly way. Government, industry and corporate backing for these emerging market systems can help promote awareness. Furthermore, future UK trade policy and global supply chains must be based on high environmental and welfare standards with imports of food produced to UK standards. Unfortunately, an amendment to the UK Agriculture Bill calling for lower standard imports to be kept out of future trade deals has been voted down, risking exporting our food-related emissions and reducing the competitiveness of Welsh producers producing to higher standards. It will be important for Welsh

Government to influence the (UK) National Food Strategy in those areas that will impact Wales, for example the trade and labelling.

The Covid-19 pandemic has also illustrated (and will likely continue to illustrate as horticulture production in countries like Spain is disrupted) that the UK cannot over-rely on imported food to meet UK consumption needs and should consider means for increasing the production of healthy foods in Wales (see the case for investment in horticulture under lessons learnt), coupled with a policy to increase accessibility. Both approaches are needed to improve UK food security and many examples are given through the Sustainable Food Places Network (https://www.sustainablefoodplaces.org/).

5. Building an Ecologically Resilient Food System

Resilience also requires policies that do not favour intensive agricultural practices, which negatively impact ecosystems or greenhouse gas emissions and instead promote mixed and agroecological horticulture and practices that capture carbon and enhance ecosystems (Francis et al. 2003). As the UK prepares to leave the European Union and the Common Agricultural Policy, new legislation should direct public investment to support the expansion of agroecological production (Welsh Government 2019, UK Parliament 2020).

Agroecology is, for some in the agricultural community, a frightening and complicated term; this is understandably so as there is much history wrapped up in the meaning of agroecology and it can also be used in a number of different ways. Agroecology can be thought of as a science, as a practice and as a political movement, and there are elements of all three present in the UK (Sanderson Bellamy and Ioris 2017). Hence, it is important to be clear about what we mean when we use the word agroecology. When advocating for a Welsh and UK transition to agroecology, I am referring to a set of practices that make up a holistic approach to the farm's ecosystem. Core techniques and practices include: building the soil organic matter by recycling and retaining nutrients on the farm; using a diverse range of crops and varieties; focusing on diversity, rotations and polyculture to enhance beneficial interactions; using native seeds, plants and livestock; and using holistic techniques for fertilisers and pest control such as introducing natural predators of pests. However, an agroecological approach can also be considered more comprehensively from a food system perspective, such as The UN Food and Agriculture Organisation does through their Ten Elements of Agroecology (FAO 2018):

- **Diversity:** Diversification is key to agroecological transitions to ensure food security and nutrition while conserving, protecting and enhancing natural resources.
- **Co-creation and Sharing of Knowledge:** Agricultural innovations respond better to local challenges when they are co-created through participatory processes.
- **Synergies:** Building synergies enhances key functions across food systems, supporting production and multiple ecosystem services.
- **Efficiency:** Innovative agroecological practices produce more using less external resources.

- **Recycling:** More recycling means agricultural production with lower economic and environmental costs.
- **Resilience:** Enhanced resilience of people, communities and ecosystems is key to sustainable food and agricultural systems.
- **Human and Social Values:** Protecting and improving rural livelihoods, equity and social well-being is essential for sustainable food and agricultural systems.
- **Culture and Food Traditions:** By supporting healthy, diversified and culturally appropriate diets, agroecology contributes to food security and nutrition while maintaining the health of ecosystems.
- **Responsible Governance:** Sustainable food and agriculture require responsible and effective governance mechanisms at different scales – from local to national to global.
- **Circular and Solidarity Economy:** It reconnects producers and consumers and provides innovative solutions for living within our planetary boundaries while ensuring the social foundation for inclusive and sustainable development.

Agroecology is about working together with nature and with each other to produce food and bring it in our homes. On every farm, ecological conditions can vary, for example soil type, condition and quality (including for example pH, salinity, presence of biotic and abiotic elements, compactness, rates of water infiltration), depth of topsoil, amount of rainfall, temperatures and micro-climates and insect community dynamics. Within the farm, these conditions can vary from one field to the next. Agroecology emphasises the importance of farmer knowledge in determining the best practices for their farm. There are no one-size-fits-all solutions. It emphasises independent experimentation rather than dependence on high-tech equipment from external suppliers with a high degree of dependency on support services.

In this context, agroecology is distinct from mainstream and industrial agriculture techniques, which rely on mono-crops, formulaic application of chemical fertilisers and pesticides and commodified inputs such as patented seeds. Agroecology can be similar in nature to organic production, but this is not always the case; in other words, agroecological production will always be organic, but organic production does not imply agroecological production. Agroecology is distinguished from organic in that it emphasises a whole-system approach with minimal external inputs. For example, organic farming still relies on external inputs such as organic fertiliser, may still produce single or few varieties of crops or livestock and may not necessarily prioritise other holistic principles like water conservation or use of renewable energy.

Agroecology relies on the use of local resources and therefore focuses on low energy inputs (for example recycled wastes and organic wastes instead of synthetic fertilizers), family and community work and short supply chains. A well-functioning agroecological system will operate as a closed system that is able to recycle energy and resources within the production system.

Practice-based agroecology can also be seen in more overtly social terms. Several of agroecology's significant proponents have consistently emphasised the transformative potential of agroecology as a practice-based social movement, primarily in Latin America (Altieriand Toledo 2011, Gliessman 2007). De Molina (2013) defines agroecology as "a disciplinary field responsible for designing and producing actions, institutions and regulations aimed at achieving agrarian sustainability". Altieri and Toledo (2011) view it as a "paradigm based on the revitalisation of small farms and social processes that value community involvement and empowerment" (Altieri and Toledo 2011: 589).

These definitions share a commitment to using alternative modes of food production to bring about broader changes for the better in social and ecological outcomes. Empowerment and participation, based on techniques and ideas pioneered by Robert Chambers (1983, 1997), are viewed as central to bringing about positive change (Uphoff 2013, Warner 2007). These approaches explicitly position practical agroecology as a social movement for change and theorise it as a counter-movement to mainstream agri-food systems (Wezel et al. 2011, Petersen et al. 2013, Altieri and Toledo 2011, Sevilla and Woodgate 2013), which leads to a more radical branch of agroecology and political agroecology.

Political agroecology considers agriculture and food production as inherently political and calls for concepts of agroecology which foreground power and politics. This approach is related to the sub-discipline of political ecology and draws attention to power relations, such as class and gender which produce uneven access to natural resources and which produce ecological degradation (Peet and Watts 2004). Political agroecology is concerned with broader food systems, especially the conventional agri-food system dominated by large corporations, market ideologies and governments.

Agroecology is about transforming food systems towards sustainability. As such, an agroecological approach, as outlined herein, also addresses all of the UN SDGs outlined in the first section of this chapter (UN SDG 2, 3, 6, 8, 12-15). This is confirmed by the results of a review of agroecological farms conducted by the Ecological Land Cooperative (ELC 2018).The main findings of benefits achieved by the farms are as follows: profitability; efficiency; generating 'multiplier effects'; creating fresh, local and healthy food; high employment figures per land area; promote, incorporate and inspire biodiversity; pesticide and chemical-free; low carbon emissions; positive social impact – small farms focus on local economies and communities; provide open spaces to learn; provide employment and other opportunities.

Case Study: Slade Farm

For much of the last 50 years we've been moving towards easily consumable food with permanent availability. For farming that has meant predictability of supply leading to specialisation, intensification and consolidation. This is at odds with the natural system, which depends on diversity, and it also creates an imbalance of power between supply chain producers and buyers.

A resilient Wales Farming practices can be used to ensure increased biodiversity as well as the production of nutritious food. Slade Farm uses cyclical nutrition and fertility building within the farm to provide a definitively sustainable system within no external inputs. The system is supported via management practices, such as areas of permanent pasture, spring sown cereals and hay meadow maintenance. Taken together these practices ensure the maintenance and growth of biodiversity above and below ground.

(Contd.)

(Contd.)

A healthier Wales An overwhelming driver of poor outcomes in health and nutrition is the level of processed foods consumed. Slade Farm connects people to primary products through local supply of meat and vegetables. This creates the micro opportunities within households to make different choices. The shape of our food system influences our behavioural patterns; changing its shape can have positive consequences for health and nutrition. Wales has an opportunity to enshrine the benefits of regenerative agriculture within its food system. It is clear that when such practices are in place they provide enormous public and common goods that contribute directly to a better Wales.

A more cohesive Wales Using farms as a 'place' provides the opportunity to engage communities in food production and nature. Slade Farm has regular public events to showcase food production and runs a small Community Supported Agriculture scheme. This brings the local community together and creates both bonding and bridging social capital as well as reconnecting people to the land.

Agroecology is smart farming and can enhance more independence for farm families in promoting diversification of their land and resources. In many parts of Europe, it is exploding as a diversified way of farming and agro-forestry. In 2012, France based its agricultural policy explicitly on agroecology through its implementation of the 'Agroecological Project for France' (French Ministry of Agriculture, Agri-food and Forestry 2016). Instead of using regulation to promote agroecology, in the spirit of agroecology, the Project incentivises voluntary bottom-up approaches, for example through reduced taxes to encourage organic approaches, demonstration farms and on-farm experiments. New legislation for farmer training and introducing agroecology in agricultural education as well as facilitating the emergence of collective agroecological projects involving farmers and other local actors and help for them to apply for funding has been credited with increasing farmer awareness of agroecology from 50% in 2014 to 79% in 2016 and 92% of farmers are either engaged in or planning to engage in, one or more agroecological activities (Ajates Gonzalez et al. 2018).

Agroecology is not dogmatic, and it is not necessarily linked directly to particular certification schemes. Rather it celebrates place-specific, diversified and nature-enhancing practices. Using a diversity of practices can create a more robust and resilient farm system that is less vulnerable ecologically and to market fluctuations. As such now is the time to mainstream these practices with growing consumer demands for more local and nature-based foods and production systems, such as pasture and grazing-based livestock systems, less intensive cereals and planting more horticultural crops for local markets. A new agri-environment scheme based on policies put forward in the Sustainable Farming and Our Land (Wales) consultation report and the UK Agriculture Bill, namely to pay farmers for public goods can create positive financial incentives for farmers to experiment with agroecological practices.

Agroecology represents a more people-centred approach as it empowers farmers and values their knowledge. Agroecology empowers farmers by explicitly recognising the value of their knowledge. Agroecology also focuses on collaborative and communal social practices, such as knowledge sharing. Knowledge sharing is seen as one of the most fundamental components of agroecology: developing human capital to build healthier, well-functioning food production systems. Agroecology moves us away from transactional interactions by emphasising the relational interactions between farmers and their land, other farmers and their wider communities. It represents a transformation in the way food system actors interact with each other and explicitly integrates values back into the management, distribution and consumption practices. The people-centred approach relies on dynamics capable of stimulating more just outcomes across the food system.

In light of the social disconnect between consumers and the land and food production, there is an urgent need to understand the feasibility and potential of alternative 'place-based' approaches to the food system, where consumers are more connected to where their food comes from and understand the implications of their choices. Research suggests that by engendering a shared responsibility for environmental and social burdens (Seyfang 2006, Mills 2016, Pike 2013, Parker and Sinclair 2001), stakeholders in the food chain are more likely to take positive action. In the case of the consumer, a shared commitment to sustainability has been argued to affect purchasing decisions and is exemplified by an increasing number of consumers whose awareness of the impacts of their choices on the environment and animal welfare is leading them to buy local foods (many through supermarkets) and engage in alternative food networks (Weatherell 2003). In the case of producers, advice and support networks play a critical role in the up-take of participatory activities such as agri-environmental schemes (Seyfang 2006, Mills 2016, Pike 2013). Social networks provide a valuable source of information for producers and also help to create a sense of shared responsibility and personal and social norms that lead to positive outcomes (Woolcock and Narayan 2000, Meador et al. 2016).

6. Conclusion

Our food systems are in a state of crisis. They fail to deliver environmental sustainability and healthy outcomes in all places for all people. The UN Human Rights International Covenant on Economic, Social and Cultural Rights (ICESCR), signed by 175 nations, includes the human right to adequate food. Countries that are ratified are subject to investigations on their current situation by the Special Rapporteur on the Right to Food. In 2018, Professor Philip Alston, the UN Special Rapporteur on extreme poverty and human rights visited the UK to investigate and wrote a harsh critique of the state of poverty in the UK, the world's fifth wealthiest nation. He said, "For almost one in every two children to be poor in the twenty-first century Britain is not just a disgrace, but a social calamity and an economic disaster, all rolled into one... The Government has remained determinedly in a state of denial" (Alston 2018).

Globally, the number of people affected by hunger has been slowly on the rise since 2014 (FAO 2020). The report concludes that the food security and nutritional status of the most vulnerable population groups are likely to deteriorate further due to the health and socio-economic impacts of the COVID-19 pandemic. The time to act is now. We need a radical transformation of our food systems in order to ensure every individual's right to live a healthy life. There are seeds of change planted, but these must be nurtured and encouraged with governments leading through changes in policies and legislation and people re-engaging with their food and with each other to build a vision for our food system that can deliver these critical objectives: healthy, sustainable and affordable food.

References

Ajates, Gonzalez R., J. Thomas and M. Chang. 2018. Translating agroecology into policy: the case of France and the United Kingdom. Sustainability. 10(8): 2930.

Alston, P. 2018. Statement on Visit to the United Kingdom, by Professor Philip Alston, United Nations Special Rapporteur on extreme poverty and human rights. UNHR, Geneva. Available online at: https://www.ohchr.org/en/NewsEvents/Pages/DisplayNews. aspx?NewsID=23881

Altieri, M. and V.M. Toledo. 2011. The agroecological revolution in Latin America: rescuing nature, ensuring food sovereignty and empowering peasants. The Journal of Peasant Studies. 38(3): 587-612. DOI: 10.1080/03066150.2011.582947

Chambers, R. 1997. Whose Reality Counts. London: Intermediate Technology Publications.

Chambers, R. 1983. Rural Development-Putting the Last First: Longman Scientific and Technical. Harlow, UK.

De Molina, M.G. 2013. Agroecology and politics. How to get sustainability? About the necessity for a political agroecology. Agroecology and Sustainable Food Systems. 37(1): 45-59.

Defra (Department for Environment, Food and Rural Affairs) 2016. Annual Report and Accounts 2015–16. Available online at: https://www.gov.uk/goverment/publications

Ecological Land Cooperative. 2018. Small Farm Profits. ELC, UK. Available online at: https:// www.agricology.co.uk/resources/small-farm-profits

FAO. 2018. The 10 elements of agroecology: guiding the transition to sustainable food and agricultural systems. UN FAO. Available online at: http://www.fao.org/agroecology/ knowledge/10-elements/en/

FAO, IFAD, UNICEF, WFP and WHO. 2019. The state of food security and nutrition in the world 2019. Safeguarding against economic slowdowns and downturns. Rome, FAO. Licence: CC BY-NC-SA 3.0 IGO.

Food Foundation. 2020a. COVID-19: Latest impact on food. March. 2020. Food Foundation, UK. Available at: https://foodfoundation.org.uk/covid-19-latest-impact-on-food/

Food Foundation. 2020b. New food foundation survey: three million Britons are going hungry just three weeks into lockdown. April 2020. Food Foundation, UK. Available at: https:// foodfoundation.org.uk/new-food-foundation-survey-three-million-britons-are-going-hungry-just-three-weeks-into-lockdown/

Food Foundation. 2020c. New poll data: more than five million people in households with children have experienced food insecurity since lockdown began. May 2020. The Food Foundation, UK. Available at: https://foodfoundation.org.uk/vulnerable_groups/new-poll-data-more-than-five-million-people-in-households-with-children-have-experienced-food-insecurity-since-lockdown-began/

Food Manufacture. 2019. Food processing employers underpay 6,700 workers. 29-Apr-2019 by James Ridler. Available at: https://www.foodmanufacture.co.uk/Article/2019/04/29/Food-processing-employers-underpay-6-700-workers

Francis, C., G. Lieblein, S. Gliessman, T.A. Breland, N. Creamer, R. Harwood, L. Salomonsson, J. Helenius, D. Rickerl, R. Salvador and M. Wiedenhoeft. 2003. Agroecology: the ecology of food systems. Journal of Sustainable Agriculture. 22(3): 99-118.

French Ministry of Agriculture, Agri-Food and Foresty. 2016. The agroecology project in France. French Government. Available online at: https://www.google.com/url?sa=t&rct=j&q=&esrc=s&source=web&cd=&ved=2ahUKEwjuy424tNvsAhW1RBUIHTBXDbQQFjAOegQIARAC&url=https%3A%2F%2Fagriculture.gouv.fr%2Fsites%2Fminagri%2Ffiles%2F1604-aec-aeenfrance-dep-gb-bd1.pdf&usg=AOvVaw2y0Y2MCLwqWbO1aZ5mhUmO

Frison, E. and C. Clément. 2020. The potential of diversified agroecological systems to deliver healthy outcomes: making the link between agriculture, food systems & health. Food Policy. 101851.

Gliessman, S.R. 2014. Agroecology: The Ecology of Sustainable Food Systems. CRC Press.

Hayhow, D.B., M.A. Eaton, A.J. Stanbury, F. Burns, W.B. Kirby, N. Bailey, B. Beckmann, J. Bedford, P.H. Boersch-Supan, F. Coomber, E.B. Dennis, S.J. Dolman, E. Dunn, J. Hall, C. Harrower, J.H. Hatfield, J. Hawley, K. Haysom, J. Hughes, D.G. Johns, F. Mathews, A. McQuatters-Gollop, D.G. Noble, C.L. Outhwaite, J.W. Pearce-Higgins, O.L. Pescott, G.D. Powney and N. Symes. 2019. The State of Nature 2019. The State of Nature Partnership. 107 pp.

IPBES. 2019. Global assessment report on biodiversity and ecosystem services of the Intergovernmental Science-Policy Platform on Biodiversity and Ecosystem Services. E.S. Brondizio, J. Settele, S. Díaz, and H.T. Ngo (editors). IPBES secretariat, Bonn, Germany. Available at: https://ipbes.net/global-assessment

IPCC. 2018. Global Warming of 1.5°C. An IPCC Special Report on the impacts of global warming of 1.5°C above pre-industrial levels and related global greenhouse gas emission pathways, in the context of strengthening the global response to the threat of climate change, sustainable development, and efforts to eradicate poverty [Masson-Delmotte, V., P. Zhai, H.-O. Pörtner, D. Roberts, J. Skea, P.R. Shukla, A. Pirani, W. Moufouma-Okia, C. Péan, R. Pidcock, S. Connors, J.B.R. Matthews, Y. Chen, X. Zhou, M.I. Gomis, E. Lonnoy, T. Maycock, M. Tignor and T. Waterfield (eds.)]. Available at: https://www.ipcc.ch/sr15/

IPCC, 2019. Special Report on climate change, desertification, land degradation, sustainable land management, food security, and greenhouse gas fluxes in terrestrial ecosystems. 2019. *In*: The approved Summary for Policymakers (SPM) was presented at a press conference (Vol. 8).

Kantar Worldpanel, 2017. Grocery Market Share - Kantar Worldpanel. Available at https://www.kantarworldpanel.com/en/grocery-market-share/great-britain

Meador, J.E., D.J. O'Brien, M.L. Cook, G. Grothe, L. Werner, D. Diang'a and R.M. Savoie. 2016. Building sustainable smallholder cooperatives in emerging market economies: findings from a five-year project in Kenya. Sustainability. 8(7): 656.

Mills, J., P. Gaskell, J. Ingram, J. Dwyer, M. Reed and C. Short. 2017. Engaging farmers in environmental management through a better understanding of behaviour. Agriculture and Human Values. 34(2): 283-299.

National Food Strategy. 2019. Our approach and principle. National Food Strategy, UK. Available at: https://www.nationalfoodstrategy.org/approach-2020/

O'Connell, R., A. Knight and J. Brannen. 2019. Below the Breadline: Families and Food in Austerity Britain. pp. 167-185. *In*: [eds.]. Families in Economically Hard Times. Emerald Publishing Limited. doi:10.1108/978-1-83909-071-420191012

Parker, C. and M. Sinclair. 2001. User-centred design does make a difference. The case of decision support systems in crop production. Behaviour & Information Technology. 20(6): 449-460.

Peet, R. and M. Watts. 2004. Liberation Ecologies: Environment, Development, Social Movements. Psychology Press.

People need Nature. 2017. Pebble in the Pond: Opportunities for Farming, Food and Nature after Brexit. p. 13. Available at http://peopleneednature.org.uk/wp-content/uploads/2016/12/A-Pebble-in-the-Pond-final.pdf

Petersen, P., E.M. Mussoi and F. Dal Soglio. 2013. Institutionalization of the agroecological approach in Brazil: advances and challenges. Agroecology and Sustainable Food Systems. 37(1): 103-114.

Pike, T. 2013. Farmer engagement: an essential policy tool for delivering environmental management on farmland. Aspects of Applied Biology. 118: 187-191.

Pitt, H., A. Wheeler, K. Palmer and S. Gould. 2020. C19 Horticulture Summit Results from edible producers in Wales April 2020. Cardiff University. Available online at: https://www.cardiff.ac.uk/news/view/2369940-growers-of-fresh-produce-in-wales-feeling-the-pressure-of-covid-19

Power, M., B. Doherty, K. Pybus and K. Pickett. 2020. How COVID-19 has exposed inequalities in the UK food system: The case of UK food and poverty. Emerald Open Research, 2.

Ritchie, H. and M. Roser. 2020. Environmental impacts of food production. Published online at OurWorldInData.org. Available at: https://ourworldindata.org/environmental-impacts-of-food

Royal Society of the Arts. 2019. Our Future in the Land. London: Food, Farming and Countryside Commission.

Sanderson Bellamy, A. and T. Marsden. 2020. A Welsh food system fit for future generations. World wide fund for Nature and Cardiff University. Cardiff, UK. Available at: https://www.wwf.org.uk/updates/welsh-food-system-fit-future-generations

Sanderson Bellamy, A. and A. Ioris. (2017) Addressing the knowledge gaps in Agroecology and identifying guiding principles for transforming conventional agri-food systems. Sustainability. 9(3): 330.

Sanderson Bellamy, A., E. Furness, P. Nicol, H. Pitt and A. Taherzadeh. In press. Shaping more resilient and just food systems: lessons from the COVID-19 Pandemic. Ambio.

Scottish Food Coalition. 2016. Plenty: food, farming and health in a New Scotland (Scottish Food Coalition). Available at http://www.foodcoalition.scot/uploads/6/2/6/8/62689573/plenty_complete.pdf

Scottish Government. 2020. Protecting Scotland's Future: the Government's programme for Scotland 2019-2020. 3 Sept 2019. Scottish Government, Scotland, UK. Available at: https://www.gov.scot/publications/protecting-scotlands-future-governments-programme-scotland-2019-20/

Sevilla Guzmán, E. and G. Woodgate. 2013. Agroecology: foundations in agrarian social thought and sociological theory. Agroecology and Sustainable Food Systems. 37(1): 32-44.

Seyfang, G. 2006. Ecological citizenship and sustainable consumption: examining local organic food networks. Journal of Rural Studies. 22(4): 383-395.

Taylor, A. and R. Loopstra. 2016. Too poor to eat: food insecurity in the UK (The Food Foundation). Available at http://foodfoundation.org.uk/wp-content/uploads/2016/07/FoodInsecurityBriefing-May-2016-FINAL.pdf

UK Parliament. 2020. The UK Agriculture Bill. UK Parliament. Available at: https://services.parliament.uk/Bills/2019-21/agriculture/documents.html

United Nations. 1976. International Covenant on Economic, Social and Cultural Rights. United Nations Human Rights Office of the High Commissioner. Available at: https://www.ohchr.org/en/professionalinterest/pages/cescr.aspx

Uphoff, N. (ed.). 2013. Agroecological Innovations: Increasing Food Production with Participatory Development. Routledge.

Warner, K. 2007. Agroecology in Action: Extending Alternative Agriculture through Social Networks. MIT Press.

Weatherell, C., A. Tregear and J. Allinson. 2003. In search of the concerned consumer: UK public perceptions of food, farming and buying local. Journal of Rural Studies. 19(2): 233-244.

Weitkamp, E., M. Jones, D. Salmon, R. Kimberlee and J. Orme. 2013. Creating a learning environment to promote food sustainability issues in primary schools? Staff perceptions of implementing the food for life partnership programme. Sustainability. 5(3): 1128-1140.

Welsh Government. 2015. Well-being of Future Generations Act. Welsh Government, Wales, UK. Available at: https://www.legislation.gov.uk/anaw/2015/2/contents/enacted

Welsh Government. 2019. Sustainable Farming and our land. Consultation report. Available online at: https://gov.wales/sites/default/files/consultations/2019-07/brexit-consultation-document.pdf

Wezel, A., S. Bellon, T. Doré, C. Francis, D. Vallod and C. David. 2009. Agroecology as a science, a movement and a practice. A review. Agronomy for Sustainable Development. 29(4): 503-515.

Willett, W., J. Rockström, B. Loken, M. Springmann, T. Lang, S. Vermeulen, T. Garnett, D. Tilman, F. DeClerck, A. Wood and M. Jonell. 2019. Food in the Anthropocene: the EAT–Lancet Commission on healthy diets from sustainable food systems. The Lancet. 393(10170): 447-492.

Woolcock, M. and D. Narayan. 2000. Social capital: implications for development theory, research, and policy. The World Bank Research Observer. 15(2): 225-249.

Part II
Natural Resources and Sustainability

Challenges and Opportunities for Water Security in São Paulo Metropolitan Region, Brazil

Luana Dandara Barreto Torres[1]* and Gabriela Farias Asmus[2]

[1] Universidade Federal do ABC - Centro de Engenharia, Modelagem e Ciências Sociais Aplicadas. Avenida dos Estados, 5001 - Bairro Santa Terezinha - Santo André. CEP 09210-580

[2] Universidade Federal do ABC - Centro de Engenharia, Modelagem e Ciências Sociais Aplicadas. Avenida dos Estados, 5001 - Bairro Santa Terezinha - Santo André. Torre A, Sala 614-b. CEP 09210-580

1. Introduction

Water security has been a topic of increased interest in the past years, especially due to its relevance to the major global problems currently faced – hunger, food insecurity, climate crisis, the spread of waterborne diseases, unsustainable production and consumption; all of which have been addressed by the United Nations 17 Sustainable Development Goals (SDGs).

The quest for water security is directly addressed by Goal 6 – Clean Water and Sanitation with most of the other SDGs being closely related to this one, such as Goal 1 (No Poverty), Goal 2 (Zero Hunger), Goal 3 (Good Health and Well-Being) and Goal 10 (Reduced Inequalities). In an attempt to reinforce this purpose, the United Nations has declared the current decade (2018-2028) as the "International Decade for Action on Water for Sustainable Development" (Haie 2020).

The importance of water as means for achieving ideal living standards and environmental health is unquestionable. However, as discussed by Naim Haie (2020), water management so far has been addressed mostly as an intermediate for achieving other ends, such as food security, sustainable land use, economic flow, health and sustainable ecosystem functioning. Hardly, water availability has been put in the center of the problem probably due to its historical abundance. Hence, the author highlights the importance of looking into the water itself and the concepts of management and governance bound to it.

*Corresponding author: luana.torres@aluno.ufabc.edu.br

The United Nations has defined the term 'water security' as follows, aiming at providing a common framework for collaboration:

"Water security is defined as the capacity of a population to safeguard sustainable access to adequate quantities of acceptable quality water for sustaining livelihoods, human well-being, and socio-economic development, for ensuring protection against water-borne pollution and water-related disasters, and for preserving ecosystems in a climate of peace and political stability" (UN Water 2013).

According to this definition, water security interconnects diverse challenges and is the common ground for security, sustain ability, development and human well-being. All of these relate to factors ranging from biophysical to infrastructural, institutional, political, social and financial conditions, enhancing the need for an interdisciplinary approach toward sustainable global water management (UN Water 2013).

Under this standpoint, we attempt to report the latest discussions and political deeds on water security regarding the Metropolitan Region of São Paulo (MRSP), considering that it is the largest metropolitan area in Latin America, the most densely occupied and economically developed area in Brazil (SEADE Foundation 2020), and it is located within one of the largest watersheds of the continent – the Paraná water basin (ANA 2020). Looking upon the protagonal of this region in the nexus food-water-energy within the continent, it seems daunting that only a few studies on water management/security/scarcity in the MRSP have taken place in the past two decades (1-3 studies per year in average), though an ascending trend was observed after the 2014-2015 water crisis – the harshest faced in the region so far[1].

On this count, we hope that this review will help further develop and integrate scientific knowledge on the social, environmental and economic fronts of water management in the MRSP, as well as serve as a tool for sustainable political decision-making in the region.

2. Water Security in the Metropolitan Region of São Paulo (MRSP)

São Paulo (Brazil) is the most populated city in Latin America, reaching 21 million inhabitants when considering its metropolitan region (SEADE Foundation 2020). The Latin America and Caribbean region is considered one of the most urbanized in the world, with 81% of its population living in urban areas and an expected increase of 18% of its population up to 2030 (UN 2019). Latin America is home to six megacities, urban areas holding more than 10 million inhabitants (UN 2019),

---

[1] In order to assess the extent to which water security and water management have been addressed within the Metropolitan Region of São Paulo throughout the past two decades, a Boolean search was executed in three of the most worldly recognized scientific databases (Scopus, Web of Science and SciELO). The set of combined keywords chosen for the search consisted of *(São OR Sao) AND Paulo AND metropolitan AND "water security" OR "water management" OR "water scarcity"* in the title, abstract and keywords; only articles or reviews.

indicating that the pressure over water resources and their management poses a difficult challenge.

São Paulo is the biggest economic polo in Brazil, resulting in large outdraws of water for industry, agriculture and domestic consumption (SEADE Foundation 2020). While the biggest numbers on water consumption in Brazilian municipalities usually account for irrigation, São Paulo has the urban human supply as its predominant source of consumption with an estimated outdraw of 40.2 m^3/s for the city in 2019 (ANA 2019a). At the same time, its metropolitan region is located at the headwaters of the Tietê river basin, a region of low water availability. Most of the water used by São Paulo's inhabitants (around 32 m^3/s) has to be imported from adjacent river basins to support the increasing water demand (ANA 2010), forming a complex water supply system that requires an integrated and meticulous approach for adequate management.

The MRSP urbanization process is quite similar to other megacities in Latin America. The opportunity to live in an environment served with social equipment and essential services, such as health access, transportation lines and sanitation facilities, is limited to few individuals, mostly to those who sustain some wealth. The central neighborhoods usually hold the highest living standards combined with diverse job opportunities. The housing market strongly explores these advantages, generating a real estate speculation that frequently pushes the poorest to the peripheries. Devoid of any chance of owning or even renting a house, many families occupy irregular places, such as hills (slope risk areas), areas of riparian vegetation within watersheds and other environmental protection areas in the peripheries. It is estimated that 2 million people live in watershed protection areas in the south wing of the MRSP (Martins 2006, Maricato 2015).

Considering that environmentally protected areas are not prone to or allowed for urban settlements, they do not qualify for receiving sanitation, energy, health equipment or other urban facilities, according to Brazilian municipal legislation. Thus, many families lack canalized water in their homes, contributing to the incidence of diarrhea and other waterborne or hygiene-related diseases. Also, the absence of sanitation will most likely occasion soil and water contamination by litter and sewage, which increases water treatment costs before it can be distributed to the population (Zhang 2019).

Another problem that should be largely addressed, but is often neglected by the municipalities, is the occurrence of Pollutants of Emerging Concern (PECs) in supply water, especially in urbanized areas. PECs are relatively new pollutants that result from modern human life and have been reaching environmental compartments in alarming quantities over the past decades, such as personal care products, pharmaceutical products, hormones, pesticides and products containing microplastics among others (Zhang 2019). Those products are usually not removed from water destined for human consumption by conventional treatment; flocculation and coagulation, which compose the system mostly used in Brazil, aim at removing sediment and pollutants associated with sediment and microorganisms (Jain et al. 2014).

Provisioning safe water to the growing MRSP population may be especially challenging when considering that the reservoirs have been continuously reaching their operational capacity since 2013. During the 2014 winter-spring season, they reached exceptionally low volumes, accounting for less than 10% of their total capacity (Coutinho et al. 2015, Lima et al. 2018). The episode was considered a serious crisis in the region, as many people suffered from water shortages and cuts (Cavalcante 2015). Although many researchers tried to explain the crisis in terms of climatic anomalies (Marengo et al. 2015), it is now clear that the unusual drought from 2013-2015 just worsened the situation. There is no simple answer to explain the water crisis in the MRSP and the multiple factors involved may include the increasing demand in domestic, industrial and agricultural water consumption, the waste of hydric resources in its various unsustainable uses, the lack of basic sanitation and water losses in the supply system due to lack of maintenance and adequate planning, the growing deforestation rates that jeopardize aquifer recharge in the urban water, encumbering the supply system, and the water company management decision-making process (Buckeridge and Ribeiro 2018).

3. Water Availability in the Residences

Most of the MRSP residents are connected to the general water supply network. São Paulo city is usually celebrated for its drinking water coverage index, which reached 99.3% of the municipality's population in 2018 (Instituto Trata Brasil 2020). Household water tariffs depend on the consumption range established by the main service provider company in the region (SABESP). A residence using between 21 and 30 m^3 of water per month, for instance, would pay 2 US\$/m^3 (SABESP 2019). The company offers a different tariff for low-income residents or unemployed consumers, although the reduction in value is far from substantial.

The social water tariff applies to one single family houses who live in buildings smaller than 60 m^2 with family income of up to three minimum wages and reduced use of electric energy (up to 130 kWh per month). Families living in collective buildings as urbanized favelas may also ask for the benefit. The social tariff also depends on the consumption range. A beneficiary using between 21 and 30 m^3 of water per month would pay 1.5 US\$/m^3 (SABESP 2019), 25% less than the full tariff for the same consumption range.

It is important to consider that the possibility of being connected to the water network does not mean that residents afford the fees charged for services, especially in times of economic recession. Then, it is plausible to assume that many residents are led to explore irregular forms of supply, such as clandestine connections or clandestine wells, increasing the risk of consuming contaminated water and disfavoring the proper functioning of the system (Britto 2020).

Water from artesian wells or delivered by an alternative water supply system (for example water trucks) may be contaminated by viruses, bacteria, helminths or other parasites associated with fecal coliforms. Waterborne and food borne diseases, considered preventable, still bring many people into hospitalization in the MRSP (Figure 1). Diarrhea and gastroenteritis of presumed infectious origin (ICD-10[2]: A-09) is the principal waterborne/food borne disease in terms of hospitalization.

---

[2] ICD-10: International Classification of Diseases, 10th Revision.

Acute diarrhea also motivates many visits to public health care facilities, recording thousands of cases every year in São Paulo city. Given that the Brazilian program responsible for monitoring acute diarrhea diseases only notifies data from selected health facilities, the records are considered under reported despite the diseases' high incidence. Moreover, incidence rates have increased in recent years (Figure 2).

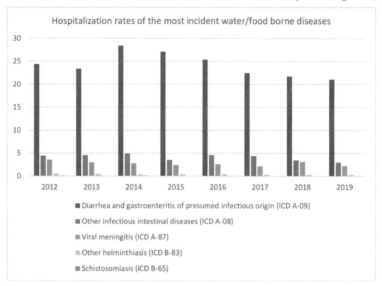

Figure 1: Hospitalization Rates of the most Incident Water-borne/Food-borne Diseases Recorded in the São Paulo Metropolitan Region from 2012-2019. Rates are Calculated for Every Thousand Inhabitants
Source: Data from DATASUS/SP (2020) and SEADE FOUNDATION (2020). Elaborated by the authors

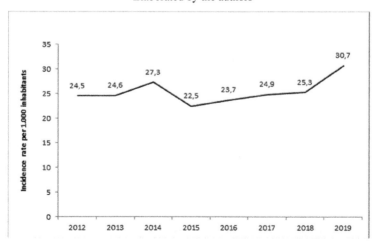

Figure 2: Acute Diarrhea Diseases Recorded in São Paulo City from 2012-2019. Rates Calculated for Every Thousand Inhabitants
Source: Data from DATASUS/SP (2020) and SEADE FOUNDATION (2020). Elaborated by the authors.

The year 2014 recorded the highest incidence of acute diarrhea and hospitalizations due to diarrhea, gastroenteritis, and other intestinal infections. It is possible that the water crisis that took place in this same year contributed to this rise in the records. The low water pressure in the supply system and an eventual interruption of the service may have led more people to fetch water from the alternative systems mentioned above. Simultaneously, the low rainfall levels of the period increased the concentration of pollutants in the reservoirs, requiring the application of more aggressive chemicals to ensure potability that, in turn, may have increased the concentration of water disinfection by-products (DBP) in supply water (Fajersztajn and Saldiva 2018). DBP is a chemical compound formed by a reaction between a water disinfectant (e.g. chlorine) with a precursor (e.g. natural organic matter) present in the supply water. There are some human epidemiological pieces of evidence connecting high levels of DBP exposure and increased risk of cancer and reproductive effects (Spivey 2009). Therefore, the water crisis may also contribute to the increase of other chronic diseases, which will be noticed as a long-term health effect.

The lack of access to safe water due to a momentaneous crisis in the supply system or due to the impossibility of paying for the water service also favors the dissemination of infectious diseases, such as conjunctivitis, scabies, measles among other diseases preventable through household cleaning and personal hygiene. Considering that washing hands and reinforcing hygiene habits are the main resources in combating the expansion of the new covid-19, SABESP exempted residences with social/favela tariffs benefits from paying for water and sewage bills during the pandemic's critic period. This can be taken as an example of good practice adopted, especially in a country where there is no legal prohibition against cutting water supply for users who cannot afford to pay for the service (Britto 2020). SABESP has also installed public lavatories in socially vulnerable communities and donated several domestic water tanks to ensure water supply during the pandemic crisis in the MRSP (SABESP 2020). Regardless, these emergency measures should be replaced by permanent public policies that ensure the human right to safe water, prioritizing water uses related to health and the maintenance of life (Casazza 2020).

It is estimated that 11% of the population in the RMSP (2,162,368 people) live in irregular settlements where public basic sanitation services are practically nonexistent or precarious (Trata Brasil n/d). Yet, there is no clear estimative of the MRSP population who cannot pay for water service despite living in regular houses. As the housing and economic crisis that is taking place in the MRSP has not been a priority for a political resolution, it is expected that the number of families excluded from access to safe water will increase in the coming years, as well as the diseases associated with water, food or hygiene transmission. Thus, one challenge to ensure water security in SPMR is to prioritize this human right in the political agenda, which implies tackling land regularization as well.

4. Water Quality in the Reservoirs

While the water supply is considered 'universalized' in households regularly registered in the municipality of São Paulo, the rates related to the collection and treatment of sewage can still be improved. In São Paulo, 96% of the sewage generated is collected. However, 64% of the water that is consumed and collected by the sewage system is effectively treated (Trata Brasil 2020). According to data collected in 2006, 15 of the 39 municipalities in the MRSP usually transport and release wastewater—without any previous treatment—in to bodies of water farther away (Whately and Diniz 2009). This implies that a large part of domestic, industrial or agricultural sewage reaches water bodies without proper treatment, increasing the amount of organic matter, nutrients, heavy metals and other pollutants in the reservoirs destined to urban supply.

Concurrently, it is recognized that even effectively treated wastewater in one of the sewage treatment station (STS) in the MRSP does not guarantee the removal of some relevant particles, such as divalent and monovalent ions (e.g., potassium, sodium, magnesium and calcium). According to the 'direct potable reuse system', adopted within the MRSP, the wastewater treated at STS is to be released downstream of the water bodies serving the municipalities. Therefore, pollutants that are not removed by the STS travel towards the reservoirs, contaminating the water destined for supply (Cruz and Mierzwa 2019).

The inadequate release of industrial and domestic effluents associated with the growing use of industrialized chemical compounds has favored the occurrence of emerging pollutants in many classes of water bodies. Those pollutants have yet limited regulatory status and their effects on the environment and human health are yet poorly understood (Alda and Perez 2017). Added to the chemicals previously mentioned, the following are also considered emerging pollutants: phthalates, Polycyclic Aromatic Hydrocarbons (PAH), Polychlorinated Biphenyls (PCBs) and Bisphenol (Deblond et al. 2011). Emerging compounds such as these have been widely detected in waters and have been appointed as a potential cause for infertility, sperm quality reduction, diabetes, cancer and endocrine disruptions (Zhang 2019), thus requiring a further and more thorough investigation.

Water security is closely related to water quality, which permeates the preservation of environmentally sensitive areas, sanitation, adequate housing, well-coordinated governance and environmental justice. Thus, just, equitable and sustainable water management depends not only on adequate infrastructure and preservation of water resources but also on social and political factors that should be managed harmoniously and democratically.

References

Alda, Miren Lopez De and Sandra Perez. 2017. Environmental analysis: Emerging pollutants. Liquid Chromatography, 2nd ed. 15: 451-477. https://doi.org/10.1016/B978-0-12-805392-8.00015-3.

ANA. 2010. Atlas Brasil. Abastecimento Urbano de Água. Região Metropolitana de São Paulo. 2010. Available at: http://atlas.ana.gov.br/atlas/forms/analise/RegiaoMetropolitana.aspx?rme=24. Access: 15 June 2020.

ANA. 2019 Conjuntura dos recursos hídricos no Brasil 2019. Informe Anual. Agência nacional de Águas. Brasília: ANA:, 2019. Available at: http://conjuntura.ana.gov.br/static/media/conjuntura-completo.bb39ac07.pdf. Access: 15 June 2020.

ANA. 2019b. Manual de uso consuntivo da água no Brasil. Agência Nacional de Águas (Brasil). Brasília: ANA, 2019. Available at: http://www.snirh.gov.br/portal/snirh/centrais-de-conteudos/central-de-publicacoes/ana_manual_de_usos_consuntivos_da_agua_no_brasil.pdf/@@download/file/ANA_Manual_de_Usos_Consuntivos_da_Agua_no_Brasil.pdf. Access: 15 June 2020.

Britto, A.L. 2020. As tarifas sociais de abastecimento de água e esgotamento sanitário no Brasil: seus impactos nas metas de universalização na garantia dos direitos humanos à água e ao esgotamento sanitário. Available at: https://ondasbrasil.org/wp-content/uploads/2020/05/As-tarifas-sociais-de-abastecimento-de-%C3%A1gua-e-esgotamento-sanit%C3%A1rio-no-Brasil.pdf Access: 23 July 2020.

Buckeridge, M. and W.C. Ribeiro. 2018. Uma visão sistêmica das origens, consequências e perspectivas das crises hídricas na Região Metropolitana de São Paulo. pp. 14-21. *In*: Buckeridge, M. and W.C. Ribeiro [eds.]. Livro branco da água. A crise hídrica na Região Metropolitana de São Paulo em 2013-2015: Origens, impactos e soluções. Instituto de Estudos Avançados, São Paulo, SP, BRA.

Casazza, I.F. 2020. O acesso à água e os excluídos da prevenção à Covid-19. 2020. Fundação Oswaldo Cruz. Notícias. Available at: http://www.coc.fiocruz.br/index.php/pt/todas-as-noticias/1789-o-acesso-a-agua-e-os-excluidos-da-prevencao-a-covid-19.html#.XyLN7bB7mM9. Access: 28 July 2020.

Coelho, C.A.S., D.H.F. Cardoso and M.A.F. Firpo. 2015. Precipitation diagnostics of an exceptionally dry event in São Paulo, Brazil. Theoretical and Applied Climatology. 1: 769-784.

CVE-SP - Centro de vigilância Epidemiológica "Prof. Alexandre Vranjac". Available at: http://www.saude.sp.gov.br/cve-centro-de-vigilancia-epidemiologica-prof.-alexandre-vranjac/ Access: 22 July 2020.

Cruz, N. and J.P. Mierzwa. 2020. Public health and technological innovations for public supply. Saúde e Sociedade. 29(1): 1-8.

Deblond, T., C. Cossu-Leguille and P. Hartemann. 2011. Emerging pollutants in wastewater: A review of the literature. International Journal of Hygiene and Environmental Health. 214(6): 442-448.

Fajersztajn, L. and P. Saldiva. 2018. Impactos da crise hídrica em São Paulo na saúde. *In*: Buckeridge, M. and Ribeiro, W.C. [eds.]. Livro branco da água. A crise hídrica na Região Metropolitana de São Paulo em 2013-2015: Origens, impactos e soluções. São Paulo: Instituto de Estudos Avançados, 2018.

Haie, Naim. 2020. Transparent Water Management Theory. Edited by Asit K. Biswas and Cecilia Tortajada. Singapore: Springer Nature.

Jain, Ravi, Mary Kay Camarillo and William T. Stringfellow. 2014. Threats. Drinking Water Security for Engineers, Planners, and Managers. 45–67. https://doi.org/10.1016/b978-0-12-411466-1.00003-3.

Lima, G.N., M.A. Lombardo and V. Magaña. 2018. Urban water supply and the changes in the precipitation patterns in the metropolitan area of São Paulo – Brazil. Applied Geography. 94: 223-229.

Marengo, J.A., C.A Nobre, M.E. Seluchi, A. Cuartas, L.M. Alves, E.M. Mendiondo, G. Obregon and G. Sampaio. 2015. A seca e a crise hídrica de 2014-2015 em São Paulo. Dossiê Crise Hídrica. Revista USP. 106: 31-44.

Maricato, Ermínia. 2015. Para Entender a Crise Urbana. São Paulo: Expressão Popular.

Martins, Maria Lúcia. 2006. Moradia e Mananciais Tensão e Diálogo Na Metrópole. São Paulo: FAUUSP.

SABESP. 2019a. Comunicado 5/19. Available at: http://site.sabesp.com.br/site/uploads/file/asabesp_doctos/Comunicado%205-19.pdf Access: 22 Jul 2020.

SABESP. 2020b. Notícias. Coronavírus. Available at: http://site.sabesp.com.br/site/imprensa/noticias-detalhe.aspx?secaoId=65&id=8299. Access: 22 Jul 2020.

SEADE Foundation. 2020. Fundação Sistema Nacional de Análise de Dados. Sistema Seade de Projeções populacionais. Available at: https://produtos.seade.gov.br/produtos/projpop/. Access: 19 June 2020.

Spivey, A. 2009. Better biomarker of DBP exposure. Environmental Health Perspectives. 117(11): A487. Available at: https://link-gale.ez42.periodicos.capes.gov.br/apps/doc/A212769434/AONE?u=capes&sid=AONE&xid=b31d8275. Accessed 29 July 2020.

Trata Brasil. Ranking do saneamento Instituto Trata Brasil 2020 (SNIS 2018). Disponível em http://www.tratabrasil.org.br/images/estudos/itb/ranking_2020/Relatorio_Ranking_Trata_Brasil_2020_Julho_.pdf. Acesso em 30 July 2020.

Trata Brasil. Saneamento básico em áreas irregulares: pesquisa aponta a situação em 13 grandes municípios do Estado de São Paulo. Available at: http://www.tratabrasil.org.br/datafiles/estudos/areas-irregulares/Release-SANEAMENTO-NAS-AREAS-IRREGULARES-DO-ESTADO-DE-SP.pdf. Access: 29 July 2020.

UN Water. 2013. Water Security & the Global Water Agenda - Analytical Brief. United Nations (UN). 2019. World Population prospects. The 2018 Revision. 2019. New York, 126 p. Available at https://population.un.org/wup/Publications/Files/WUP2018-Report.pdf Access: 15 June 2020.

Whately, M. and L.T. Diniz. 2009. Água e esgoto na grande São Paulo: situação atual, nova lei de saneamento e programas governamentais propostos. Instituto Socioambiental: São Paulo, SP, BRA.

Zhang, Yunhui. 2019. Emerging Chemicals and Human Health. Edited by Yunhui Zhang. Singapore: Springer Nature. https://doi.org/10.1007/978-981-32-9535-3.

For a New Relationship with Nature: The Paradigm Shifts between Society and the Environment in the Light of Human Rights

Ana Paula Leal Pinheiro Cruz* and Sônia Regina da Cal Seixas

NEPAM/University of Campinas – UNICAMP, Rua dos Flamboyants, 155, Cidade Universitária, 13083-867 Campinas, SP, Brazil

1. Introduction

Warnings about the severity of nature degradation are becoming more frequent, configuring part of the environmental collapse[1] that we currently experience. The rise in Amazon burning and deforestation, the tailings dams rupture, or even the oil spill on northeastern Brazil beaches are part of the current environmental agendas with ecological, social, and political repercussions worldwide. Although it is not a recent concern, questioning our relation to nature seems increasingly pertinent and necessary. The epistemological division between society and nature, as something separate and external to it, to support distinct research fields has repercussions to the present day. Its unfoldings are in the way we inhabit, appropriate, and modify the natural environment, with different consequences.

The globalizing character of modern society also has an impact on the risks[2] produced. For example, climate change, directly related to anthropic action, will

---

[1] This term is adopted based on Lourenço (2019). "...perhaps the vocabulary "crisis" is inappropriate to designate the current environmental scenario, as it traditionally carries the sense of an unpredictable and temporary event, characteristics that definitely are not part of this state of affairs, which is why I believe it is more pertinent to talk about environmental collapse" (Lourenço 2019). Freely translated by the author.

[2] Reference to Risk Society: Towards a New Modernity (1986) by Ulrick Beck. In advanced modernity the social production of wealth is systematically accompanied by the social production of risks. Accordingly, the problems· and conflicts relating to distribution in a society of scarcity overlap with the problems and conflicts that arise from the production, definition and distribution of techno-scientifically produced risks.

(Contd.)

---

*Corresponding author: a264455@dac.unicamp.br

increase global temperature and tide levels. Among the results are increased food insecurity, irregular periods of rain and intensified drought in certain regions. Consequently, it causes losses in the fauna and flora biodiversities, resulting in diseases and other problems. Currently, the spread of the novel coronavirus is another demonstration of the global aspect of the risks. In December 2019 and January 2020, the virus seemed distant, restricted to eastern countries. However, it took over Europe in a few months and it spread among the Americas and Africa, overcoming continental barriers. Since the first days of March, we have watched and experienced the pandemic's fear and consequences[3]. On August 8, Brazil surpassed 100,000 COVID-19 victims (the second largest, behind only the United States[4]). Therefore, it is indispensable to emphasize that, among all the problems and peculiarities of each country, these scenarios happen under denialists' administration—not only of the pandemic but also of climate change—that is, Trump in the US and Bolsonaro in Brazil. The mention of the COVID-19 pandemic is present not only because of its contemporaneity and global character but also because of its origin. As in the recent outbreaks[5] of SARS (2002), Ebola (2014), swine flu (2009), and avian flu (2004), the current pandemic origin possibly relates to the interaction between man and the environment in an excessive way.

Thus, this chapter seeks to relate the terms sustainability, human rights, and climate change, treated by a social bias, with the problem surrounding the value of nature, which is repeatedly threatened and fated to meet human demands. Besides, this section discusses the mining topic, which is this author's research field. The paper is structured in two parts, in addition to this introduction. First, some aspects of Human Rights character and its applicability are presented, showing the expansion of the UN's role concerning environmental issues and the relationship between environmental preservation and the dignity guarantee. Furthermore, to support this section, analyzes that contemplate the neo-liberalization movement of environmental governance and the pertinent aspects of vulnerability measurement are highlighted, showing the impact of structural imperatives in promoting human and environmental safeguard.

(Contd.)

This change from the logic of wealth distribution in a society of scarcity to the logic of risk distribution in late modernity is connected historically to (at least) two conditions. First, it occurs—as is recognizable today—where and to the extent that genuine material need can be objectively reduced and socially isolated through the development of human and technological productivity, as well as through legal and welfare-state protections and regulations. Second, this categorical change is likewise dependent upon the fact that in the course of the exponentially growing productive forces in the modernization process, hazards and potential threats have been unleashed to an extent previously unknown" (BECK, 1986).

[3] The World Health Organization declared on March 11, 2020, that the COVID-19 outbreak became a pandemic. Source: https://www.paho.org/pt/covid19

[4] Source: https://noticias.uol.com.br/ultimas-noticias/rfi/2020/06/19/com-brasil-liderando-novas-vitimas-pandemia-contabiliza-meio-milhao-de-mortos-no-mundo.htm

[5] A report warns about the possibility of new outbreaks of zoonotic diseases if countries do not prevent their spread. Source: https://nacoesunidas.org/relatorio-da-onu-defende-abordagem-que-une-saude-humana-animal-e-ambiental-para-evitar-futuras-pandemias/

In the second part, some considerations by the authors Gudynas (2019) and Lourenço (2019) are gathered about the rights and values of nature, pointing briefly to the paths made possible by Andean neo-constitutionalism, represented by the Ecuadorian Charter. It attempts to build a dialogue between the issues raised with part of the mining dynamics on a large scale. During the last decades in Latin America, this activity has increased the appropriation of nature, as well as the life and rights of those who inhabit its territories.

Three questions move this paper: (1) Has the path of environmental preservation associated with human welfare been sufficient to ensure nature's protection? (2) Is it possible to recognize nature's importance beyond its utilitarian character and man's service? (3) Would the understanding of nature as a subject of rights redirect actions that directly interfere with it?

2. Of the Formulation of Human Rights to the Care for the Environment

Recent news reminded that 75 years ago, a bomb was dropped on Hiroshima, which left more than 140,000 people dead and many others injured. The atomic bombing, which was later released on Nagasaki, was carried out by the USA at the end of World War II, and the victims were mainly civilians[6]. Also, 75 years ago, the United Nations was being established, after a period of elaboration through the Charter of the United Nations (1945), to react to the terror and destruction experienced during World War II. Atrocities such as those committed by the Nazi dictatorial regime and atomic bombing have left wounds in society. These injuries indicated the need to build instruments capable of guaranteeing peace between nations through conflict mediation, avoiding future confrontation. The Universal Declaration of Human Rights (1948) defines a new code of conduct and collective action among nations. By consensus, the 58 delegates of the represented states showed not only repudiating the barbaric acts committed during World War II through the conception of instruments capable of preventing the repetition of these scenarios but also to promote more harmonious relations between nations.

The Human Rights promulgation established a universal ethic for the protection and valorization of human life. Therefore, the international community is committed to ensuring the integrity of any individual, establishing one of the UN's primary purposes: promoting Human Rights and citizens' fundamental freedom. Its non-restriction to national competence contributed to launching a reformulation on the traditional notion of sovereign Nation-States. Equally, it added to the individual's construction as an international subject of rights, guaranteeing and protecting human dignity. In the middle of the twentieth century, therefore the international order started to be conducted by a new paradigm based on the elaboration of an effective

---

[6] Source: https://www.nexojornal.com.br/expresso/2016/05/27/Por-que-as-bombas-sobre-o-Jap%C3%A3o-foram-criminosas-segundo-especialistas-em -Right of War

international protection system[7], committed to guaranteeing a future free from brutal acts such as those previously experienced (Piovesan 1998, Tozo 2018).

It was also after the end of WWII, and due to the fear of new nuclear attacks (in the cold war period) that the population feared possible pollution by radiation[8] and its destructive power. Since the 1960s, environmental concerns among scientists, politicians, and civil society have intensified. The warning about the agricultural use of chemical pesticides, promoted by Rachel Carson's work *Silent Spring*, denounced the intensification of nature's degradation and, consequently, of human life, impelling contesting environmental movements. The counterculture's social mobilization promoted libertarian ideals and aimed for new social and political behavioral parameters (LOURENÇO 2019).

Still, in the late 1960s, the first photograph of the Earth taken by an astronaut would impact human perception aesthetically and politically. As its name suggests, the Earthrise marked an awakening of the relationship with the Earth and human thinking. The angle provided from space instigated several questions about human unity, life on a single planet, and the sharing of an interdependent ecosystem threatened in diverse ways. The foundation of the Club of Rome, which assembled first in 1968, expressed concerns about natural resource exploitation. The release of the iconic report *Limits to Growth* (1972) warned of humanity's devastating future. According to it, economic and populational growth would be the central factors in depleting planetary resources and the collapsing economy. These episodes were responsible for part of a global collective conscience development on the environment and the emergence of the sense of responsibility for protecting the Earth as well as promoting human welfare (UN).

The environmental concern, which is understood as a global phenomenon, became part of the UN agenda in 1972. The United Nations Conference on the Human Environment, held in Stockholm (Sweden) was the first government leaders meeting to discuss environmental degradation. It is considered a milestone in the relationship between man and the environment. From this conference, the United Nations Environment Program, UNEP, commenced acting for environmental issues, responding to catastrophes and conflicts, managing ecosystems, promoting

---

[7] Part of the contemporary conception of Human Rights comes from its universality and indivisibility, covering the civil, political, economic, cultural, and social fields to safeguard the values of freedom, equality, and solidarity among individuals. The collaboration between the global and regional normative protection systems made it possible to create an institutional universe for protecting human rights at the international level, which considers regional realities, expanding and strengthening its action. The appropriation of the International Treaty guidelines can be observed, for example, in the Brazilian Constitution (1988) formulation. After years of a military dictatorship in Brazil, the drafting of the Constitution during the re-democratization process, besides marking a significant government transition period, also meant an adaptation movement towards the international agenda contributing to the improvement of the country's image to the eyes of other nations. Therefore, the two documents carry remarkable similarities, coincidences, and concept extensions (Piovesan 1998).

[8] The establishment of the International Atomic Energy Agency (IAEA) in 1957.

environmental governance, using natural resources efficiently, and paying attention to the climate change impacts.

> "A point has been reached in history when we must shape our actions throughout the world with a more prudent care for their environmental consequences. Through ignorance or indifference we can do massive and irreversible harm to the earthly environment on which our life and well being depend. Conversely, through fuller knowledge and wiser action, we can achieve for ourselves and our posterity a better life in an environment more in keeping with human needs and hopes. [...]To defend and improve the human environment for present and future generations has become an imperative goal for mankind [...]" (excerpts from the Declaration of the United Nations Conference on the Human Environment, 1972, paragraph 6).

Thus, the World Commission on Environment and Development started to convene conferences to discuss the planet's natural resources use. One of these conferences, the Brundtl and Commission, set in 1987, led to the publication of *Our Common Future*. The report defined 'sustainable development' as the ability to meet current demands without compromising future generations and the environment, proposing to balance economic development and environmental conservation objectives. The breadth of the term and the lack of a precise definition of the necessary factors to guide sustainable development's practical achievement led to a certain trivialization. For example, countless companies worldwide adopted this term multiplying their profits, without genuinely taking actions to generate less social and environmental impacts.

> "Sustainable development is development that meets the needs of the present without compromising the ability of future generations to meet their own needs. [...] A world in which poverty is endemic will always be prone to ecological and other catastrophes. Sustainable development requires meeting the basic needs of all and extending to all the opportunity to satisfy their aspirations for a better life. [...] Yet many of us live beyond the world's ecological means, for instance in our patterns of energy use [...] At a minimum, sustainable development must not endanger the natural systems that support life on Earth: the atmosphere, the waters, the soils, and the living beings. [...] In essence, sustainable development is a process of change in which the exploitation of resources, the direction of investments, the orientation of technological development; and institutional change are all in harmony and enhance both current and future potential to meet human needs and aspirations" (Excerpts from the Brundtl and Report, *Our Common Future*).

From *Our Common Future* emerged the conference that became known as the 'Earth Summit' (1992), which took place in Rio de Janeiro, Brazil. This event established Agenda 21. Besides questioning unsustainable development models, the conference agenda also brought attention to its considerations about poverty and the necessary equity in income distribution, indicating the urgency of formulating environmental policies that could guarantee common goods access to those who directly depend on them. To this end, it was recommended to empower diverse groups such as labor union organizations, women, farmers, indigenous people, the scientific community, local authorities, NGOs, and other groups, as important actors in seeking sustainable development. As Gonçalves (2019) points out, after

500 years of domination, the Europeans' modern-colonial system, which ignored the geographical and cultural distinctions regarding life and nature values, showed its limits and offered room to a different knowledge that so far had been disregarded. Subsequently, Rio+10 and Rio+20 were held in Johannesburg, South Africa, in 2002, and Rio de Janeiro in 2012.

> "Therefore, a strategy specifically aimed at combating poverty is an essential requirement for the existence of sustainable development. For a strategy to be able to face poverty, development, and environmental problems simultaneously, it is necessary to begin by considering resources, production, and people, as well as, simultaneously, demographic issues, the improvement of health care and education, women's rights, the role of youth, indigenous people and local communities, and, at the same time, a participative democratic process, associated with an improvement in its management" (Part of section 3.2, taken from Agenda 21, 3 Combat poverty. Available at: http://www.direitoshumanos.usp.br/index.php/Agenda-21-RIO-92-ou-ECO-92/capitulo-03-combate-a-pobreza.html) [UNCED - Conferência das Nações Unidassobre o Meio Ambiente e Desenvolvimento (1992), Agenda 21 (global)].

In 2015, the Sustainable Development Goals (SDGs) were adopted at the Sustainable Development Summit held at the UN headquarters in New York. The meeting sought to join forces to elaborate a global agenda[9] for sustainable development, establishing 169 goals to be achieved by 2030. The objectives and goals served as nations' action guide to all social life dimensions and their relation to the environment. The discussions concerning poverty and its relationship with development became central agendas, showing an interdependence hitherto neglected by international policies. If at the end of the 1980s, the idea that a population low economic condition was accepted as the cause of environmental degradation, especially in developing countries, historical analyzes of ecological and social contexts revealed that most of the environmental degradation originated from developed countries actions due to their life and consumption habits. The major developed powers are responsible for the consequences arising from 'development' and 'progress' imposed in an excluding and predatory manner over several nations, leaving communities subject to different vulnerabilities (Giller et al. 2018).

As noted, international attention has turned to environmental issues, associating the protection of human dignity with nature preservation. The Climate Changes and Human Rights report (2015) makes this connection even more explicit by showing that climate change constantly threatens individual rights. Moreover, it highlights that human actions themselves have the primary responsibility for such phenomena. The impacts which increase primarily on communities that are already vulnerable and devoid of their rights, impair or prevent access to land and water, imposing changes in their daily lives that interfere in food security, in the practice

---

[9] The Agenda 2030 includes the need for poverty eradication; the achievement of zero hunger through sustainable agriculture; guarantee of quality education to children and young people; guarantee of health care and welfare; the seek for gender equality; access to drinkable water and sanitation as well as to affordable clean energy; decent work combined with economic growth; innovation and infrastructure; inequality reduction; transformation of sustainable cities and communities; promotion of actions against global climate change; and other topics.

of work and the guarantee of education. Hence, the report points out the need for coordinated actions, thought globally, but possible to be implemented according to each region's peculiarities. To this end, the need for international financial assistance from countries with better conditions in the so-called cooperation between nations (Knox 2015, Schapper and Lederer 2014) is highlighted. In this scenario, a migration increase and, consequently, climate refugees[10] are also expected. It worsens the challenges and efforts to safeguard human dignity globally, especially in the so-called underdeveloped countries. In addition to physically affecting individuals, the consequences also affect them psychologically having women and children as their primary victims (Hogget 2019, Mapp and Gabel 2019, Swin et al. 2018).

The Paris Agreement (2015) and the treaty signed by 195 nations, intended to express a global commitment to reduce greenhouse gas emissions (GHG) from 2020. It aimed to replace fossil fuels with more sustainable energy matrix alternatives to maintain the temperature increasing goal below 2°C. The signing of a commitment between countries to reduce gas emissions, which earlier imposed stricter rules on industrialists, ended up being modified by voluntary actions, arousing strong criticism from the scientific community. In 2019, the United States also formally announced the agreement's abandonment, concerning the international community greatly as it is the second-largest emitter of CO_2 as China is the first. Combating climate change is crucial to ensure human life and biodiversity on the planet continues. As mentioned at the beginning of this chapter, its global consequences demand urgent economic and political actions (Clipet and Roberts 2017).

According to Clipet and Roberts (2017), there is a robust neo-liberalization movement in international environmental governance, especially in the last decades, which has guided the programs, directing them to prioritize specific economic interests. The parameters to gas emission reduction, which are now voluntarily accepted by the countries, are no longer equitable, emphasizing even more how hard it is to establish effective transparency systems regarding carbon counting and measuring impacts, as well as guaranteeing communities' rights. Dynamics of the past, such as the polluter pays principle to compensate for carbon emission, have not produced real benefits in reducing greenhouse gas (GHG) emissions. The actions did not have concrete reports regarding the results. However, they allowed the companies to present a 'green' aspect before society's eyes through marketing strategies that boosted consumption even more. The 'carbon credits buying', for example, is a practice that tends to affect the credit market mainly to highly polluting industries such as airlines that find in these programs more advantageous alternatives but less efficient ones than reducing gas polluters effectively.

Some scientists defend the unprecedented impact of human activities on the environment as responsible for the transition from the Holocene to the Anthropocene era. In this era, humankind is mainly responsible for changing the environment. Although the consequences and risks of the environmental collapse that we experience in this era are global, it is also necessary to warn that they are felt differently by the populations. The economic condition, the available infrastructure,

---

[10] Available at: https://www.theguardian.com/commentisfree/2020/mar/24 covid-19 -climate
-crisis-governments-coronavirus

and the possibility of obtaining aid and treatment are some aspects that influence the individual's capacity and the possibility of recovery and adaptability in adverse situations. That is, there are differences in the degree of vulnerability and population resilience capacity (Schiwartz 2019).

In addition to that Schiwartz (2019) also points out how in the Anthropocene era neoliberal societies have methods to measure the vulnerability indexes to which specific populations are subject. In most cases, they are not satisfactorily representative, warning that several aspects are deliberately neglected. According to his analysis, contemporary ways of measuring the degree of vulnerability, whether ecological, humanitarian, or fiscal, are focused on privileging the capital perspective. It ends in paralysis in the face of global responsibility, which directs actions related to the climate, for example, to a temporal leap to a hypothetical and distant future. Still, according to his analysis, this attitude, which would be related entirely to the capitalist system, brings a sense of risk normalization in favor of asymmetric economic growth, which guarantees privileges to the most developed countries through the maintenance of inequality. Thus, the risk becomes a commodity, capable of being quantified and priced (Schiwartz 2019, Barnett 2020).

> "The concept of vulnerability has a wide spectrum of connotations and has been extensively theorized. While advocates for overcoming and embracing vulnerability sit comfortably together on bookshelves, my concern is the neoliberal entanglement of vulnerability with subsequent possibilities, that is, vulnerability as exposure to adverse futures and for whom these futures are considered adverse. Drawing from work in ecology, literary criticism, and queer scholarship, I argue that the capitalization process defers vulnerability and responsibility (and thus justice, rights, and equity) from an experienceable present to a hypothetical future. In conjunction with this, experiences of suffering are transformed into deterministic outputs of governance calculations. This deterministic framing leads to the idea that if better or more accurate (quantified) data can be produced, more desirable futures could be output. As such, within neoliberal formulations, being vulnerable is simply to be probabilistically deviant; to be outside of a trajectory or trend" (Schiwartz 2019).

3. Nature as a Subject of Rights: The Possibilities of Another Valuation

As explained before, there is a narrow relationship between guaranteeing human rights to populations and preserving the environment. The expansion of this international interrelationship and effort on promoting sustainable development alternatives, which are a way of mitigating the impacts arising from climate change, encounter economic and above all political barriers to its effectiveness. Over time, several trends aimed at environmental conservation have sought support on promoting the quality of human life and the moral obligations of its surroundings (Lourenço 2019).

According to the current consensus, the guarantee of people's fundamental rights would only be possible through the guarantee of environmental quality. This perspective has been elaborated since the United Nations conception in 1945, the Universal Declaration of Human Rights formulation in 1948, through the elaboration of the report *Our Common Future* in 1987, and attempts to implement programs aimed at carbon emissions reduction, such as in the Treaty of Paris (2015). The unfoldings of this finding have advanced not only among the scientific community. It also reverberated in the legal field, as in ecological justice, as well as in NGOs, social movements, and civil society organizations demands.

This change of perspective on the connection between environment and life quality, and the perception of our unity as living beings that cohabitate in a single planet with diverse non-human species (such as animals, plants, fungi, bacterias, viruses, and insects), although addressed by different fronts, are still based on the idea that human life is more valuable than the others. In this sense, environmental care and preservation have been developed concerning its capacity to provide human well-being and health, bearing in mind food security or for example, its possibilities of being a space for contemplation. Such values that reduce nature to human utilitarianism, almost exclusively conditioned to economic factors, seem insufficient, given the current environmental collapse (Gudynas 2019, Lourenço 2019).

"The classical and traditional environmentalism bears a morality perception that only the human species has intrinsic value. In other words, environmentalism usually represents an anthropocentric – or homocentric – moral perspective. It prioritizes values and practices that promote human interests, needs, and demands to the detriment of other species and nature as a whole, which, in this sense, would only have instrumental value" (Lourenço 2019).

This economic valuation is historic. For example, Latin America has exported its nature since colonial times, through vegetal and mineral extractivism, constantly shaping territories, landscapes, and identities. Nature appropriation and commercialization caused the conflicts (and have been renewing them) for territorial domination, the right of cultural and identity recognition, and, more subjectively, for the territory valuation and protection by other perspectives than those resulting from capitalist utilitarian logic.

The recent tailing dams rupture events, from mining activity in Mariana in 2015 and Brumadinho in 2019, both in Minas Gerais, explained the clashes for territory as well as the social and environmental consequences of the activity. Since before its installation, massive enterprises have had direct and indirect impacts on the environment and the communities in which they operate. The large-scale mining model, pushed in recent years to meet global demand, has intensified territorial conflicts through physical and symbolic violence to populations, adopting practices that violate human rights. This model significantly interferes with local biodiversity, promoting geomorphological changes, waterbody contamination, and vegetation suppression. It also contributes to GHG emissions both in the ore extraction and pelletizing processes. Therefore, nature appropriation reverberates in different ways, interfering in the individuals' experience and sense of belonging toward the place. It also impacts the environmental dynamics of different living beings and favors a capital accumulation mechanism that designates its bonus to a few but divides its onus among many as observed in tailing dams rupture events.

As mentioned before, the link between environmental governance and vulnerability analysis mechanisms and neoliberal values has interfered both in the effectiveness of environmental programs and in the real measurement of the problems, hampering the formulation of concrete alternatives to deal with the urgency imposed by the environmental collapse. It is noticeable that the adopted practices and concepts, such as the term 'sustainable development', combine with keeping a system that anchors in the seek of economic development supported by exploring nature and, consequently, people. Such a discussion is also valid to the meanings attributed to 'development' and 'progress' because, by the same logic, they are used to justify practices such as neo-extractivism[11]. For decades, the established separation between developed and developing countries has supported a standardization sense of the 'quality and way of life' promised by the north developed countries, through a cooperation process with the south countries. However, such collaboration has been articulated unfairly and in an excluding way (Acosta 2016).

> "The ways of organizing society and the economy, the ways of conceiving the world and being in it, and the knowledge and knowledge of a large part of the world population were disqualified and considered poor, outdated, and insufficient, for one reason only: they existed outside the production system and capitalist markets. The "development" purpose is: including territory not thoroughly permeated by capitalist logic and practices to the cycles of capital accumulation; and transforming populations into consumers, subsistence peasants into wage earners or informal workers, natural goods into commodities, collective properties into private and marketable ones" (Lang 2016).

Nature appropriation by neo-extractivist accumulation models, driven by the politics of progressive governments throughout the recent decades, reinforced an environmental, ecological, and political asymmetry between the south and north countries that reconfigures territories, economies, and populations. Resignation supported and naturalized by terms as "curse of the natural resources" or "tropical fatalism"[12] presumes the lack of alternatives to countries, such as Brazil, other than suiting its governance to the neoliberal international market necessities. The paradox of countries rich in natural resources but incapable of economic development is kept by disregarding nature ways of valorizing nature and human labor that are not based on its predatory exploitation (Acosta 2016).

Data on the conflicts related to mining, collected by the Pastoral Land Commission (CPT), from 2004 to 2018, show that more than half of the events were related to water (Wanderley 2019). This finding proves that, beyond ore, mining activity usurp these communities the right and access to water, which is an essential element to life. Due to the waterways appropriation by mining, the rivers become subject to

---

[11] This concept is based on the typologies proposed by Eduardo Gudynas. "Researchers associated with the Social Sciences and environment have critically developed the term to denominate the new extractivist activities scenario in Latin America. This scenario, composed by the relationship between the Mineral Extractive Industry (MNE), the commodities boom, and the Central Government, converges into a high investment project of exploring natural resources to export, against the backdrop of defending national interests and the country's development, in conflict with the interests of traditional communities, social movements, trade unions, and workers." Source: Mining Dictionary.

[12] The term Tropical Fatalismhas already been adopted by the World Bank.

contamination, physical, chemical, and biological alteration. It can adulterate and even extinguish them. In extreme situations, such as tailing dams rupture, the rivers act as chemical tailing dispersers, interfering in numerous nearby communities and their own dynamics. Risk monetization and the following indemnity act as an alternative compensation and way of silencing those directly harmed. Except for the criticism about the effectiveness and fulfillment of the indemnities, which in most cases unfolds in processes that violate individuals' fundamental rights, it is possible to question the existence of viable compensation to be destined to the rivers themselves. During the Fundão tailing dam in Mariana in 2015, the mining tailings reached the Doce River and therefore the Atlantic Ocean[13]. More recently, in Brumadinho, in 2019, the tailings dam from Córrego do Feijão rupture were dispersed by the Paraopeba River, an important tributary of the São Francisco River. From these two extreme scenarios, but inherent to mining practice, two questions arise: is it possible to conceive to the river enough compensation for the tailing 'mud' dumped in its bed? Is the natural recovery time of its dynamics and regenerative cycle respected and guaranteed by current arrangements?

It is known that part of the values and meanings associated with rivers and nature are intimately related to the way people experience them. However, there is also the possibility of recognizing an intrinsic value, proper to nature, independent of anthropocentric values. The independent value proposition is based on the studies of deep ecology that, although it encounters many oppositions, makes it possible to look at nature, disregarding its economic and utilitarian value, thus recognizing the importance of different indigenous world views (Gudynas 2019).

The reflection on environmental ethics offers different contributions, such as ecocentrism[14] and biocentrism. The biocentric perspective recognizes intrinsic life values, whether human or non-human, protecting environmental, landscapes, and ecosystem values. It is imperative to state that there is no denial of human values but its ampliation, contemplating non-human values. Several contributions have been brought to the biocentric discussion, including those related to ecofeminism since much of men's domination over women is built similarly to the men over nature (GUDYNAS 2019).

"Once the recognition of these intrinsic values is achieved, obligations are immediately created, including rights on the environment and living beings, which must be obeyed by people, social groups, companies, the Government, etc. From there, it is possible to start exploring new environmental policies conceived from the biocentric respect perspective" (GUDYNAS 2019).

The new Constitution of Ecuador was the first biocentric document to recognize the rights of nature, which expresses the seek for another possible common goods

---

[13] Oliveira Coimbra, Keyla Thayrinne; Alcantara, Enner; Souza Filho, Carlos Roberto de. An assessment of natural and manmade hazard effects on the underwater light field of the Doce River continental shelf. Science of the Total Environment. Amsterdam: Elsevier, v. 685, p. 1087-1096, 2019. Available at: http://hdl.handle.net/11449/184600.

[14] Ecocentrism perceives animals as having their own cognitive abilities, consequently making it possible to recognize and protect them as a subject of value. Still, some authors argue that animals have intrinsic values and rights (Gudynas, 2019).

valuation. On the probable opposing questions and validations arising from this movement, Gudynas (2019) argues that: "Let us not forget that a Constitution is, in the end, a normative agreement politically built and endorsed by the majority, so there is no need to explain the valorizations origins, just that they are recognized" (Gudynas 2019). That said, it is perceivable that the plurality of cultural and religious values begins to get a room in public policies, ensuring representativeness to different traditions disregarded so far.

There is, therefore, a new view on the environment present in the Constitution of Ecuador: nature's recognition as a subject of rights deepens the debate and influences the decision-making process that directly influences environmental dynamics. Endorsed in 2008, the Ecuadorian Charter contains a session on the rights of nature or Pacha Mama. This session derives from the contemplation of "Well Living" from the original Andean peoples' traditional culture. As Gudynas (2019) warns, the recognition of the term "Pacha Mama," which comes from a world view, alongside "nature", a European cultural definition, is an essential step against centuries of an overlapped homogenizing dominant vision that nullified and repressed different visions and values of the world in their movements about nature and the man himself. This European tradition is represented, for example, by the modern and anthropocentric dualism of nature versus society. Still, "Environmental management or the ecological academy has never substantially incorporated native people's knowledge and their environment interpretation, and there are only a few restrict attempts to recover them" (Gudynas 2019).

> "Se reconece el direcho de la población a vivir en un ambiente sano y ecologicamente equilibrado, que garantice la sostenibilidad y el buen vivir, Sumak Kawasay".

> Se declara de interés público la preservación del ambiente, la conservación de los ecossistemas, la biodiversidad y la integridad del patrimonio genético del país, la prevención del daño ambiental y la recuperación de los espacios naturales degradados".

> "El derecho a vivir en un ambiente sano, ecologicamente equilibrado, libre de contaminación y en armonía con la naturaleza".

> "El derecho a vivir en un ambiente sano, ecologicamente equilibrado, libre de contaminación y en armonía con la naturaliza".

> "El Estado aplicará medidas de precaución y restricción para las actividades que puedam conducir a la extinción de especies, la destruición de ecosistemas o la alteración permanente de los ciclos naturales. Se prohíbe la introducción de organismos y material orgânico e inorgânico que puedan alterar de manera definitiva el patrimonio genético nacional".

> "El Estado promoverá, en el sector público y privado, el uso de tecnologías ambientalmente limpias y de energías alternativas no contaminantes y de bajo impacto. La soberanía energética no se alcanzará en detrimento de la soberania alimentaria, ni afectará el derecho al agua. Se prohíbe el desarrollo, producción, tenencia, comercialización, importación, transporte, almacenamiento y uso de armas químicas, biológicas y nucleares, de contaminantes orgânicos persistentes altamente tóxicos, agroquímicos internacionalmente prohibidos, y las tecnologías y agentes biológicos experimentales nocivos y organismos genéticamente modificados

perjudiciales para la salud humana o que atenten contra la soberanía alimentaria o los ecosistemas, así como la introducción de residuos nucleares y desechos tóxicos al territorio nacional".

"El derecho humano al agua es fundamental e irrenunciable. El agua constituye patrimonio nacional estratégico de uso público, inalienable, imprescriptible, inembargable y esencial para la vida".

"Las personas y colectividades tienen derecho al acceso seguro y permanente a alimentos sanos, suficientes y nutritivos; preferentemente producidos a nível local y en correspondencia con sus diversas identidades y tradiciones culturales. El Estado ecuatoriano promoverá la soberanía alimentaria".

"Las personas tienen derecho al disfrute pleno de la ciudad y de sus espacios públicos, bajo los principios de sustentabilidad, justicia social, respeto a las diferentes culturas urbanas y equilibrio entre lo urbano y lo rural. El ejercicio del derecho a la ciudad se basa en la gestión democrática de ésta, en la función social y ambiental de la propiedad y de la ciudad, y en el ejercicio pleno de la ciudadanía" (Excerpts from the Equatorian Charter 2008, GUDYNAS 2019).

By recognizing the rights of Pacha Mama or Nature, Andean neo-constitutionalism guarantees the respect, existence, and maintenance of its vital regeneration cycles, structures, functions, and evolutionary processes (Article 71). It is worth noting that the right to full restoration is independent of the Government's obligation, or individuals' and legal entities', to indemnify possible affected individuals or communities. That said, it is necessary to provide the restoration of ecosystem biodiversity in case of environmental damage such as the tailings dams rupture events. The constitution also ensures that any citizen, community, or people of any nationality can demand the fulfillment of the rights of nature present in the constitution from public authorities. The document also points out that the Government must promote those actions. According to Gudynas (2019), these factors represent a radical shift in the Latin American constitutional regimes, in which Pacha Mama, or Nature, are now conceived beyond its anthropocentric utilitarian function. The development intended by adopting the Well Living concept is guided by other paradigms that prioritize a more radical structuring on sustainability.

Its implications in politics and environmental management focus on the limits alternative imposed by perspectives based solely on economic valuation imposed on natural goods almost in a totalizing way, beyond conceptual overcoming. Another factor is the complexification of ecosystems' understanding and the subsequent rearrangement of the environmental risk measuring methods and assessments. There are also reflections on legal representation and tutelage changes, which present long analyzes and discussions in the legal field, supported by political ecology considerations.

Therefore, there is evidence that policies are conceived in a participatory, transversal, decentralized, and transparent way. Biodiversity is understood as part of national sovereignty, and the government's property, which has its guaranteed share from the financial benefits that come from possible commercialization processes. The government must avoid negative environmental impacts, especially when damage is inevitable, which goes against current mitigation measures. The liability for damage to nature is objective and attributable to people and institutions. When the risk seems non-existent, its potential onus is transferred to the executive responsible for the enterprise. Citizen participation is expected both in the planning and execution processes with broad dissemination and even prior consultations with communities, actions that point to a nature valuation that goes beyond the merely economic and utilitarian considerations commonly contemplated.

4. Conclusion

As seen in this chapter, due to the environmental collapse we are experiencing, it is urgent to develop alternatives to the systems imposed and conceived as totalizing. The global warming progression, the disasters repetitions, the international environmental governance neo-liberalization, or the limitations and constraints of instruments capable of providing data on the communities and nature vulnerability show that structural changes are urgent, even though there are significant progress around environmental discussions and the life security on the planet. In this context, Latin America perceives itself as a stage for the countless historical violations that have been made in favor of excluding development, resulting even today in conflicts with traditional peoples and environmentalists, silencing communities and suppressing the guarantee of the most fundamental rights to individuals, and as a field of possibilities through, for example, the drafting of the Ecuadorian Constitution.

As demonstrated in this chapter, paradigm changes, such as those established by the Declaration of Human Rights, are formulated collectively when changes are urgent. Whether to safeguard human life itself or the environment, it is necessary to devise alternatives to the current moment that recognize values beyond nature's utilitarian and economic character, given the gaps that this traditional valuation presents. Knowing that it is possible to provide nature a different value, we still need to answer the third question posed at the beginning of this chapter: would we perceive nature as a subject of rights if we reoriented our actions? As stated by Gudynas (2019), "Before 'speaking on behalf of', the effort is in 'knowing how to listen' what emerges from it." It is necessary to listen and learn from the living beings and their environment, genuinely seeking a more harmonious relationship that contemplates a "vision" and an awakening regarding our unity, as a society cannot endure disassociated from nature.

References

Barnett, J. 2020. Global environmental change II: Political economies of vulnerability to climate change. Progress in Human Geography, doi: https://doi.org/10.1177/0309132519898254

Ciplet, D. and J.T. Roberts. 2017. Climate change and the transition to neoliberal environmental governance. Global Environmental Change. 46(2017): 148-156. Available at: https://doi.org/10.1057/s41286-017-0032-z

Giller, K.E., I.M. Drupady, L.B. Fontana and J.A. Oldekop. 2018. Editorial overview:

The SDGs – aspirations or inspirations for global sustainability. Current Opinion in Environmental Sustainability. 34: A1-A2.

Gudynas, Eduardo. 2019. Direitos da natureza: ética biocêntrica e políticas ambientais. São Paulo: Elefante, 340 p.

Hoggett, P. 2019. Introduction. pp. 1-19. *In*: Hoggett, P. [ed.]. *Climate Psychology*, Studies in the Psychosocial. Available at: https://doi.org/10.1007/978-3-030-11741-2_1

Knox, J. 2015. Human rights, environmental protection, and the sustainable development goals. Washington International Law Journal. 24(3): 517-536, Wake Forest Univ. Legal Studies Paper. Available at SSRN: Available at: https://papers.ssrn.com/sol3/papers.cfm?abstract_id=2660392

Lang, M. 2016. Alternativas ao desenvolvimento. *In*: Dilger, G., Lang, M. and Filho, J.P. [eds.]. Descolonizar oimaginário: Debates sobre pós-extrativismo e alternativas a desenvolvimento. São Paulo: Fundação Rosa Luxemburgo.

Lourenço, Daniel B. 2019. Qual o valor da natureza? Uma introdução à ética ambiental. São Paulo: Elefante, 456 p.

Mapp, S. and S.G. Gabel. 2019. The climate crisis is a human rights emergency. Journal of Human Rights and Social Work. Available at: https://doi.org/10.1007/s41134-019-00113-0

Piovesan, F. 1998. A constituição brasileira de 1988 e os tratados internacionais de proteção dos direitos humanos. *In*: Marcílio, M L; Pussoli, L (Coords.). Cultura dos direitos humanos. São Paulo: LTr: 133-151.

Schapper, A. and M. Lederer. 2014. Introduction: human rights and climate change: mapping institutional inter-linkages. Cambridge Review of International Affairs. 27(4): 666-679. Available at: http://dx.doi.org/10.1080/09557571.2014.961806

Schwartz, S.W. 2019. Measuring vulnerability and deferring responsibility: quantifying the anthropocene. Theory, Culture & Society. 36(4): 73-93. Available at: https://dx.doi.org/10.1177/0263276418820961

Svampa, M. 2016. Extrativismo neodesenvolvimentista e movimentos sociais. *In*: Dilger, G., Lang, M., Filho, J.P. [eds.]. Descolonizar o imaginário: Debates sobre pós-extrativismo e alternativas ao desenvolvimento. São Paulo: Fundação Rosa Luxemburgo.

Swim, J.K., T.K. Vescio, J.L. Dahl and S.J. Zawadzki. 2018. Gendered discourse about climate change policies. Global Environmental Change. 48: 216-225.

Tozo, L.S. de O. 2018. Direitos Humanos: o ideal comum a ser atingido por todos os povos e todas as nações. UNICAMP, Direitos Humanos. Jornal da UNICAMP. Available at: https://www.unicamp.br/unicamp/ju/artigos/direitos-humanos/direitoshumanos-o-ideal-comum-ser-atingido-por-todos-os-povos-e-todas

United Nations Human Rights – UNHR. 2015. Understanding Human Rights and Climate Change. Submission of the Office of the High Commissioner for Human Rights to the 21st Conference of the Parties to the United Nations Framework Convention on Climate Change. Available at: https://www.ohchr.org/Documents/Issues/ClimateChange/COP21.pdf

Sustainability and Water Quality Recovery

Almerinda Antonia Barbosa Fadini[1]* and Pedro Sérgio Fadini[2]

[1] São Paulo Federal Institute of Education, Science and Technology – IFSP – Salto, SP, Brazil, Rio Branco Street, 1780, Zip Code 13320-271
[2] Federal University of São Carlos, Chemistry Department, São Carlos, SP, Brazil, P.O. Box 676, Zip Code 13565-905

"I think the waters start the birds I think the waters start the trees and the fish. And I think that the waters initiate men. Water initiates us" – Manoel de Barros (Waters 2001).

1. Water and Pandemic Times

August 2020, this chapter's writing is finishing. The world is amid a pandemic of COVID 19 with a large portion of the world population in complete social isolation, with an astonishing number of infected people and deaths and with a neoliberal capitalist economy in a real shock of reality, the debate related to consumerism and the intensive use of natural resources comes up. According to PNUMA (ONU 2020a), it is evident that the environmental changes and imbalances caused by human actions at local and global scales have led to accelerated transformations and destruction of natural habitats and consequently increasing the risk of epidemic zoonosis.

Globalization is intense and as pointed out by Harvey (2020) in a highly connected world, it becomes much more difficult to avoid a rapid worldwide spread of the virus, from which it can be wrongly inferred that it has a democratic scope since it affects all without distinction, both nations and peoples, regardless of class, gender and race. However, as added by Harvey (2020), there is only a certain truth in this since this global epidemic phenomenon has characterized in practice that the most vulnerable are the most socially and economically affected.

While analyzing data provided by UNICEF (2020), it can be seen that there are significant differences in different countries and social groups in terms of responses to the prevention of this virus. One of the main recommendations to avoid COVID-19 is to intensify hygiene habits, in particular, washing hands with soap. However, only

*Corresponding author: almefadini@ifsp.edu.br

three out of five people worldwide have basic facilities for adopting this practice. Billions of people, mainly residing in less developed countries, live in a situation of sanitary calamity with 40% of the world population, meaning around 3 billion people, have no washbasin with water and soap at home. More than a third of schools around the world and half of the schools in poor countries have no place for children washing hands (ONU 2019, UNICEF 2020).

As the same UNICEF report, at the south of the Sahara, 258 million inhabitants, representing 63% of the population, who live in urban areas, do not have access to water to wash their hands. These numbers are also expressive in East, Central and Southern Asia. Another concerning data is about South Africa, where 47% of the urban population, representing 18 million people, lack facilities to wash hands. In contrast, wealthier urban dwellers have almost 12 times more chance of this access. These data demonstrate that there are relevant differences when comparing the degree of countries' development and social classes to combat the virus spread.

According to ABRASCO (2020), in Brazil the income concentration and structural racism point to the diseases worsening among the poorest social strata, hitting the black population, who mostly live in slums and tenements. Homeless and people detained in the prison system are also affected. These facts reveal social and race differences in the COVID-19 treatment and prevention.

The same occurs when assessing women's situation. According to data from IPEA (2020a, 2020b), there has been a growth in the number of women household heads in Brazil, as they represented 6.3% (9.5 million) in 1995 and 14% (28.6 million) in 2015, being around 13 million white and 16 million black women. Many of these women have an incomplete academic background and low wages. Additionally, they have high charge daily activities requiring them to fulfill a triple journey, that is they are the maintainers of the home which includes also raising and educating their children and home keeping. This social situation directly affects their life quality, directing them to the peripheral areas with lower health care.

It can be seen that there is an inequality of gender, social class, and race in the prevention of this pandemic. This situation triggers an alert and confirms the need to rethink how human beings have been behaving about social inequalities, distribution, and access to drinking water and the sustainability of planet Earth.

2. Drinking Water and Sustainability

The water crisis is not only a concern and a warning for the future but also a current and crucial concern. Census predictions predict that in a few decades the planet will have more than nine billion inhabitants, fading the water resources and reaching a severe shortage. This finding is supported by evidence like population growth, exploitation of natural resources, unrestrained increase in production and consumption, inappropriate land use, the devastation of natural resources, expansion of agricultural borders and pastures, the concentration of people in urban and impermeable areas, waste of water, and the intensive use of irrigated are as among others.

This is a complex reality that is aggravated by the facts that:
- water is not evenly distributed on the planet;

- there is a disproportion in its use with a population minority that presents high consumption, affecting an offer to other people;
- in the same location, not everyone has access to sanitation infrastructure;
- that as a result of exacerbated exploitation there is a searching requirement for increasingly distant sources;
- scarcity often occurs not only due to the lack of the resource but also due to its uneven distribution and contamination;
- strategies for water multiple uses are weakly encouraged by public policies.
- the offer of new sanitation technologies is not adequately social-inclusive.

Among others, these factors have been generating intense conflicts over the use of water in different geographical scales. This water crisis affects all living beings, in different ways and severity, requiring urgent decision-making, based on integrated and participative plans, providing the involvement of different society segments, aiming to achieve environmental sustainability.

This sustainability is presented as a need to preserve the quantity and quality of water as a universal right. In 2010 and 2015, the UN General Assembly passed resolutions that recognize the right to safe and clean water and basic sanitation as human rights. These achievements oblige States to act toward obtaining access to water and sanitation for all, without discrimination, while at the same time giving priority to people living in situations of greater health and social vulnerability (UN 2019).

In 2007, it was enacted in Brazil, Law 11,445 (BRASIL 2007) defining guidelines and the federal basic sanitation policy that, among other achievements, contemplates universal access. For this purpose, in 2013, the National Basic Sanitation Plan was created, considering integrated planning to group up to 2033 with the supply of drinking water, sanitary sewage, solid waste management, and drainage of urban rainwater for the entire population (MDR 2020). Since 2018, the Proposed Amendment to the Constitution (PEC 4/2018) is in evaluation at the Federal Senate, inserting access to drinking water in the list of fundamental rights and guarantees of the Brazilian Federal Constitution (AGÊNCIA SENADO 2020a).

In June 2020, the Brazilian Senate approved the Bill Law - PL 4,162/2019 considered as the new legal framework for basic sanitation which awaits presidential approval (AGÊNCIA SENADO 2020b). This approval has become a controversial issue since, according to Heller (2020), this PL changes existing legislation since 2007 to make changes possible, in particular, the one that increases greater private participation in tenders, putting at risk the service to the most vulnerable areas that would not provide a greater economic return to these private companies. In this sense, Heller (2020) defends the already existing National Basic Sanitation Plan, plans that propose the universalization of sanitation and the establishment of public policies that contemplate the financing, regulation, planning, and participation of society.

This debate is important for strengthening the struggles related to the universal right to drinking water, emphasizing that it cannot be treated separately from other human rights. In this sense, the concept of environmental sustainability must evolve from a single, technical, and economic thinking to a more complex and inclusive look at this topic.

The main objective of this chapter is to deal with sustainability and water quality recovery for which the following scope is defined, which is to present a brief overview of water resources in Brazil and the world, strengthen the right of everyone to access drinking water, demonstrate the relevance of integrated and participative management, emphasizing Brazilian scenario and exposing some innovative technologies of environmental sanitation.

3. A Brief Scenario of Water Resources in Brazil and the World

Water is an essential good for all the living species on Earth having a complex cycle making it a renewable good, but scarce due to uses and quality deterioration. This resource exists in abundance on our planet, occupying approximately 70% of its surface area, but 97% of this percentage stored in seas and oceans of which 29.9% is underground and only 0.3% is present in rivers and lakes (MMA 2020).

It should also be considered that due to the different geographical and climatic conditions of each location, this fraction of 0.3% is not evenly distributed over the planet, as denoted by the fact that Brazil has 13.7% of the entire world freshwater, while countries in the Middle East and Africa have a great shortage of this resource. Also noteworthy is the heterogeneity of regional availability;taking into account Brazil as a case example, the country has an average annual rainfall of 1,760 mm but due to its continental dimensions, the annual total of rainfall varies from less than 500 mm in the semi-arid region of the Northeast to more than 3,000 mm in the Amazon region (ANA 2019a).

Further evidence of regional differences according to MMA (2020) is provided by the comparison between the water supply in the Paraná River Basin, which has 6% of the national resource and the Amazon River Basin with 73% of the country's fresh waters. In the first basin, densely populated states of the country are located, among them, São Paulo stands out with 21.9% of the Brazilian inhabitants according to IBGE (2020), while the Amazon River Basin covers the regions with the lowest demographic density in Brazil.

As pointed by IBGE (2010, 2020) and at UN (2019) that has been occurring in Brazil and the World (Table 1), a concerning growth in the number of inhabitants, overloading and threatening the world's natural resources.

Table 1 values point out a relationship between the economy and population growth, allowing highlighting that the demographic transition had an impulse from the middle of the twentieth century. Since then, the number of world inhabitants has intensely grown as evidenced by data referring to the period from 1970 to 2019 were in 49 years there was a world demographic increase of 208% and 222% in Brazil. As in other developing countries, also in Brazil, the number of inhabitants' growth was stimulated in the mid-1950s due to advances in preventive medicine and hygiene and basic sanitation policies adoption, in addition to the industrialization process and consequent urbanization, which got stronger from the 1970s.

Table 1: Brazilian and World Population Growth

World population			Brazilian population		
Period of time (year)	*Population growth (billion)*	*Population increase (%)*	*Period of time (year)*	*Population growth (billion)*	*Population increase (%)*
1950-1970	2.5-3.7	148	1950-1970	52-94.5	181
1970-2019	3.7-7.7	208	1970-2019	94.5-210	222
2019-2050	7.7-9.7	126	2019-2050	210-233	111
1950-2050	2.5-9.7	388	1950-2050	52-233	488

Source: IBGE 2010, IBGE 2020, UN 2019.

A projection made for 100 years between 1950 and 2050 presents an expectation of population growth of 388% for the world and 448% for Brazil. However, it is necessary to highlight that in certain developed countries, a negative growth rate has been occurring due to lower birth rates than those of mortality. This is a result of social policies adopted in these countries, such as educational, work, and quality of life improvements which, even though not accessible to all, enable a reduction in fertility rates that supplant the increase in longevity.

However, world projections raise concerns, considering that the wealthiest countries, even with a demographic drop, have a greater consumption power and in the case of several poor or developing countries, a more intense population increase occurs in urban areas, driven by the process of rural exodus. This is a phenomenon that is aggravated when a more affluent social stratum seeks to replicate the highly consumerist way of life of rich nations, demanding an increased use for natural resources, especially the water one.

In this context, it is worth mentioning that population growth in cities further deepens the water crisis since urban areas occupy, according to Leite (2009), is less than 1% of the world surface but concentrates more than half of the world population, consuming 75% of the energy and generating 2/3 of the waste. Forecasts, for the coming decades, indicate that developing countries' cities will concentrate 80% of the urban population, especially in megacities.

As pointed out by Gaete (2015) and Simão (2017), the UN initially designated New York and Tokyo as the first two megacities on the planet, i.e., cities with more than 10 million inhabitants. In the year 2000, the megacities were already eighteen and in 2015 this number reached twenty-two. In 2030, there will be 41 megacities, 29 of which in Asia. The major concern is that most megacities will be located in less developed countries with the aggravation that in several of these large population agglomerations will be concentrated high levels of social problems, in an absence of urban infrastructure, with precarious sanitation systems, and critical drinking water availability.

According to the UN, a person needs a water amount of 110 liters/day to meet daily consumption and hygiene needs; however, in countries on the African continent, this availability is between ten and fifteen liters/person/day. In Brazil, this value is 150 liters/person/day and in the United States, the average consumption per capita is 300 liters/day, while in New York the expenditure is 2,000 liters/person/ day (CETESB 2020, ANA 2020). These data demonstrate the discrepancy among countries, concerning meeting sanitary needs.

According to the National Water Agency, ANA (2019a), the water demand in Brazil is growing with an estimated increase of around 80% of the total withdrawn in the last two decades and with a 2030 forecast withdrawn 26% higher. It is noteworthy that the most intense water uses in the country are irrigation, which due to the high demand for monoculture and export agriculture was responsible in 2018 for the withdrawal of 49.8% followed by human supply with 24.4% and industry with 9.6%.

This general picture leads to an increase in conflicts over water use, both international and national. Considering that water does not obey the border limits among countries and according to Zuffo and Zuffo (2016), laws that contemplate transnational waters tend to be fragile and are difficult in solving problems in a friendly way as is the case in the regions of the Middle East and North Africa that have severe problems of water scarcity and extreme dependence for the production of food in irrigated areas. Conflicts arise and are accentuated by water control with the military, political and economic clashes as the main categories.

In Brazil, several of the conflicts related to water resources are located in areas with high demographic density and intense industrial activities, such as in the Southeast and South regions. In these areas, due to water pollution, there is an increase in water treatment costs. Another problem is related to the high withdrawal costs because due to scarcity, water is sought in places increasingly distant from urban centers (Castro 2012).

Associated with these factors, there are also problems related to the Permanent Preservation Areas (PPA's), which should have vegetation cover, aiming at the preservation and maintenance of water resources. According to the MMA (2011), forested areas make water shading possible, minimizing temperatures, increasing the habitat quality, allowing an adequate nutrients source for aquatic and wild organisms.

These riparian areas also have an important role as natural filters for pesticides and agricultural fertilizers, preserving surface and underground water resources. Although PPA's are legally protected, this does not mean that this happens fully, as improper uses such as crops, livestock, mining, and urban occupation take place over these areas, compromising their environmental function. For this reason, it is important the planning the land use, protecting the PPA's, aiming for a healthy ecosystem, resource availability, and water quality, even allowing the use for human supply at good purity levels.

The population growth requires a supply of food from the agricultural and livestock activities, which in many cases progressively occupies the Permanent Preservation Areas, causing deforestation and transport of fertilizers and pesticides that end up contaminating the watercourses.

According to Lopes and Albuquerque (2018), the National Program for Agricultural Pesticides (PNDA) was implemented in Brazil in 1970, granted agricultural credits with the State as one of the main sponsors of this practice, allowing a low cost for products register and exemption for several States. These actions boosted the pesticide market, placing Brazil at the top of the world consumer ranking since 2008. Even with the laws that regulate the production, use, and trade of these products, inspection compliance is still very fragile.

The impact on water resources in rural areas associated with the incipience of urban basic sanitation puts even more at risk the preservation of water quality. Recent data point out that Brazil has almost 35 million people without access to treated water, approximately 100 million inhabitants without sewage collection, representing 47.6% of the population and that only 46% of the sewage generated in the country is treated (Tratabrasil 2019).

Taking into account that in 2015, 84.7% of the Brazilian population was living in urban areas and 15.3% in rural areas, it is clear that much remains to be done in terms of more equitable sanitary improvements for the population and that care with drinking water becomes an essential element for maintenance and even for an increase in water supply for localities (IBGE 2016).

These data denoting the need for public and social policies that allow access to water for all, both in quantity as in quality, regardless of the degree of development of the countries. In this sense, water is seen as a universal right that must be institutionally guaranteed by all representatives of nations to minimize segregation and social injustices arising, among other problems, from the poor distribution of this natural resource.

4. The Universal Right to Drinking Water

In 1948, the United Nations General Assembly adopted and proclaimed the Universal Declaration of Human Rights which considers "that the recognition of the inherent dignity of all members of the human family and their equal and inalienable rights is the foundation of freedom, justice and peace in the world" (ONU 2018). Since then, commissions, studies and standards have been created to reinforce universal and international rights, such as the adoption in 2002 of the General Comment No. 15 on the right to water by the United Nations Committee on Economic, Social and Cultural defining that:

> "The human right to water provides that everyone has sufficient, safe, acceptable, physically accessible, and reasonably priced water for personal and domestic uses. Universal access to sanitation is not only fundamental to human dignity and privacy, but also one of the main mechanisms for protecting the quality of water resources" (ONU 2020b).

In 2011, the UN Human Rights Council adopted Resolution 16/2 that everyone should have access to safe drinking water and sanitation as a human right (ONU 2020c). In 2015, during the Sustainable Development Summit, a new agenda was sought, including an agreement on global climate change. On this occasion, through the joint contributions of the UN Member States and civil society, were formulated the 17 Sustainable Development Goals (SDGs) to be reached by the year 2030. Among the 17 objectives, this chapter highlights the number 6 "Drinking Water and Sanitation: Ensuring availability and sustainable management of water and sanitation for all" (ONU 2015).

Studies about the objectives of the UN and the goals established for the year 2030 are important elements for the creation and improvement of indicators in the national, state, and municipal spheres aimed at solving social and environmental problems. When examining each objective, it can be seen that these unfold into several variables, requiring technical views and social representatives to establish public policies that meet the interests of all.

In this sense, the relevance and the need to formulate and improve laws, environmental plans, and integrated, systemic, and participatory management of water resources are increasing.

5. Integrated and Participative Management of Water Resources in Brazil

In Brazil, the concern with the use of water has as its landmark, the 1934 Water Code, which presented rules linking the resource use to the public authority control, prioritizing the public health interest and water security. From these new laws, regulatory decrees, the creation of secretariats, and ministries related to the management of water resources were created.

The Stockholm Conference took place in 1972, as a forum in which the economic issue had a lot of weight in decisions, but which nevertheless provided positive reflexes for the environmental area and in Brazil, it stimulated the formulation of laws and the creation of national and international control bodies and also the Special Secretariat for the Environment – SEMA (Fadini 2005). In 1988, with the promulgation of the Constitution of Brazil, a framework was established for the water issue, where the Union started to legislate on the management of water resources (Zuffo and Zuffo 2016).

Law 9433 was created in 1997, instituting the National Water Resources Policy and creating the National Water Resources Management System (BRASIL 1997). This law defined aspects related to the framing of water bodies in quality and use classes, the charging and granting of water use rights and the implementation of decentralized and participative management integrating Union and States and contemplating the installation of hydrographic basin committees constituted by representatives of public authorities, users and civil society.

The National Water Agency (ANA) was created in 2000 by law 9984 (BRASIL 2000) as a regulatory agency, that requires compliance with the objectives and guidelines of the Water Law 9433/97 (ANA 2020). The National Water Agency has four action lines related to water resources, which are being Regulation, Monitoring, Law Enforcement, and Planning.

In Brazil, one of the greatest advances in the field of water resources policies observed in recent decades concerns the systemic and participatory character of management, through the Hydrographic Basin Committees. This model is not immune to conflicts of interest, but it has evolved as an innovative alternative under the impetus and protagonism of the global and national social movements that took place since the 1960s and also from the strengthening of the influence of the World Environment Conferences held since the 1970s.

In this process, environmental education (EE) has played a prominent role in the consolidation of social and ecological achievements, through participatory and inclusive strategies (Hoeffel et al. 2002, 2004). EE plays a role in stimulating greater reflection on the role of the citizen in the construction and management of public policies aimed at achieving quality of life and preserving natural resources. The same search for sustainable alternatives together for the localities, contributing to a transformative pedagogical praxis, using the method of reflection followed by action and object of new reflection, according to Freire (1975).

Environmental education as a transversal and interdisciplinary practice seek, through different areas of knowledge, the formation of social agents that reflect critically and ethically about the problems of an economic development model that is based on privileging a minority to appropriate and exploit in a predatory way, the natural resources. It also seeks to sensitize and encourage these agents to engage in the construction of inclusive and supportive alternatives that meet the principles of sustainability.

Another highlight is the role of scientific research that has enabled advances in the understanding and solution of problems related to sanitation and quality recovery of water resources, presenting new technologies and innovative practices that assist in public policies and the well-being of society.

6. Water Quality Recovery: Research, Innovation, and Technologies

"Water is not a commercial product like any other but, rather, a heritage which must be protected, defended, and treated as such" (European Commission 2000).

The history of surface water contamination goes back to the existence of human agglomerates. Until just over two centuries ago, water contamination was mainly due to organic matter from human waste as well as associated pathogens. The excess of such organic matter in the resources generally caused only one type of contamination that was readily perceived since the organic molecules consume dissolved oxygen, causing a shortage of this, which is the main oxidizing agent of the aquatic environment. The fish found dead on the water surface were already evidence of this type of kind of contamination.

In the scarcity of dissolved oxygen, a change in the processes of biological succession occurs, where other electron receptors begin to act, such as Mn (IV), Nitrate, Fe (III), and Sulfate, after which it comes to the production of methane and carbon dioxide. Carbon in the anaerobic environment (Froelich et al. 1979). The action of sulfate as an oxidizing agent leads to the formation of reduced sulfur compounds, such as hydrogen sulfide, which has a characteristic odor, easily associated with impacted environments.

With the evolution of the industrial means of production, the transformation industry gained strength, establishing a productive cycle with an impact on the environment, where the raw material is sought in nature, subjected to transformations that consume energy often from fossil fuels, generating waste and a finished product with added value that moves the economy. Agriculture has also intensified, requiring greater production in smaller areas, requiring water consumption for irrigation and artificial soil fertilization to increase production.

When mentioning the agricultural production of food, biofuels, and plant inputs, there is a reaction known as primary productivity, which occurs when carbon dioxide with water leads to the formation of organic matter and gaseous oxygen, that consumes energy from solar radiation, transforming inorganic carbon into organic. For it to occur on a large scale, nutrients such as forms of nitrogen, phosphorus, and potassium, among others, are incorporated into the soils, as a means of fertilizing them and plant nutrition.

Such demand stimulated intense manufacture of ammonia, through artificial processes, like the Harber-Bosch one (Zhao et al. 2020) and also phosphoric acid, which is obtained from the mining of phosphate rocks that are later attacked with sulfuric acid in an environmentally harmful process that releases toxic chemical elements and waste into the environment (Pérez-López et al. 2007, Guerrero et al. 2020). Phosphorus compounds are also used in industrial processes for food preservation, manufacture of toothpaste and soaps, metallic surface treatment, and others.

These kinds of phosphorus use, led to a wide scenario of enrichment of aquatic bodies in phosphorus, which is the main responsible for the so-called eutrophication process, causing an increase in the primary productivity and consequent proliferation of algae and aquatic macrophytes.

As a consequence of the higher organic carbon content in aquatic bodies, there is a less light entry, death of living biomass, degradation of organic matter, consumption of dissolved oxygen, and associated environmental deleterious consequences. Besides that, some species of algae can produce toxins, exhibiting a harmful effect on human health when water is ingested.

6.1. The Twentieth-Century Economic Growth and Water Pollution

The twentieth-century saw a transition from the knowledge of the existence of acute effects of pollution, to the perception of chronic ones, from the understanding of bioaccumulation processes in living organisms, characterized by a higher rate of acquisition of contaminants than the speed of elimination of these.

Characterized by the transfer of contaminants along the food chain, causing high levels of contamination at the top of the food chain, the biomagnification process was also verified. A landmark in the history of this understanding was the publication of the book The Silent Spring (Carson 1962), referring to the bioaccumulation and biomagnification of organochlorine compounds, which ended up having a strong impact on the lives of wild birds.

Still, around the middle of the twentieth century, industry and construction started to stimulate mining activity. The exploration of metals from natural reservoirs was intensified as well as their use, in the form of plates, bars, and other parts that required chemical treatments and consequent discharge of metal ions into the environment or even the direct release from mining waste. This period represented an intensification of production, associated with low energy efficiency and lack of knowledge about complex biogeochemical processes of great importance for the protection of aquatic life and human health. Implications of an ecotoxicological nature were slowly being understood by the scientific community and at an even lower speed, this understanding was translated into environmental legislation.

In the field of toxic chemical elements biogeochemistry, only publications from the 1970s-1980s led to a full understanding that total concentrations of metals explained the toxicity of aquatic environments less successfully than the concentration of their different chemical species with distinct bioavailability and consequent differentiated toxicities (Florence et al. 1982). This perception was also very sedimented from the understanding of the case of mercury contamination in the Minamata Bay, where Chisso Co between the 1920s and 1960s, dumped methyl-Hg into ocean waters, a mercurial species among the most toxic known to mankind, causing the death of 1,043 people (Harada 1995).

As global industrialization has progressed, the impacts on water resources have also intensified. Aquatic bodies around the world, receiving discharges of treated or untreated wastewater, got both water and bottom sediments contaminated. Sediments keep a history of anthropic contamination and also can act as internal sources of contaminants in aquatic ecosystems, continuing to export pollutants to the water column, even after several years after external sources break off (Mozeto et al. 2012).

6.2. Synthetic Organic Molecules and Their Impacts

Until the twentieth century middle, water resource contamination was mainly originated by sanitary sewage and also by inorganic salts emission from industrial wastewaters, this started to change dramatically with the use of synthetic organic molecules for a wide range of purposes. An executive summary published in the Pure and Applied Chemistry Journal in 1998, stated that at the turn of the century, a human being carried in his body at least 500 molecules, in measurable concentrations, that did not even exist in 1920 and also that no human being was born without carrying acquired contamination still in the mother's womb (Jost 1998). Exposure to synthetic molecules was firstly very much associated with pesticides. Some of them are already abandoned and others are more recently introduced and still widely used, such as in countries like Brazil that have an agricultural economy.

In addition to these synthetic molecules are others linked to drugs, personal hygiene products, and countless industrial additives, such as plasticizers, preservatives, and flame retardants, which are widely used and end up reaching water resources.

More recently, the presence of microplastics in the environment has occupied a prominent place in the environmental scientific literature, primarily in ocean waters, but today also in sanitary sewers, where microfibers from the washing of fabrics are discharged in large quantities in water resources. Many of these contaminants represented by relatively new molecules are called emerging contaminants and are not part of environmental legislation, which are still adapting to this new class of molecules.

The adverse effects of several of these contaminants are poorly known as their harmful power on human health with little understanding as to their action on the quality of aquatic life. In addition to emerging contaminants, Persistent Organic Pollutants (POP) are added, which combine toxic harmful power with persistence in terms of degradation in the environment and long-distance transport capacity (Wang et al. 2020). Among them, it is worth mentioning the organo-chlorinated pesticides, industrial additives, such as polychlorinated biphenyls, and also dioxins.

In the face of water scarcity, attention has been paid to the possibility of capturing water from the urban environment. In an evaluation carried out by Nino et al. (2012), Polycyclic Aromatic Hydrocarbons – HPA were monitored in an aquatic body in the urban environment, intended to partially supply the population of a city in the State of São Paulo, Southeast Brazil. The authors detected high concentrations of HPA in these waters, mainly from residues from the incomplete burning of fossil fuels deposited on the urban surface, after the occurrence of rains and consequent urban runoff in succession to periods of drought.

Synthetic molecules, such as medicines, drugs of abuse, and personal care products, have been found in studies conducted in different places around the world, presenting a generalized scenario of the occurrence of these molecules. In Brazil, the same happens, and such contaminants have been detected in aquatic bodies in the urban environment (Campanha et al. 2012) and more polluted rivers, such as the Jundiaí River, in the State of São Paulo, a water supply source, which receives treated and untreated sewage. In this river, a set of molecules with resistance to degradation was identified and are candidates to indicators of the existence of this new category of contaminants (de Sousa et al. 2014).

Machado et al. (2016), pointed out that such contaminants may not be removed at conventional water treatment plants and end up reaching homes, such as urban water supply. This large set of synthetic organic molecules existing in the environment can is a consequence of the high number of substances listed by the Chemical Abstracts Service (ACS 2020), that today is around 68 million of which only about 400 thousand have their trade-in somehow regulated. Alternatives treatment techniques are today necessary for this kind of removal.

6.3.   Advances in Water Treatment Technologies

Classically, for sanitary sewage, the following treatments are performed: primers that aim at the removal of solid materials, in sieves and grids; secondary ones that seek the removal of dissolved organic matter, being generally both aerobic and anaerobic biological steps, or else a combination of both; tertiary that aims to remove nitrogen and phosphorus and; quaternary (this is still in process) which aims to remove organic micropollutants. Such strategies already allow the direct reuse of sewage, through its potabilization (Liu et al. 2020, Scruggs et al. 2019).

Indirect reuse is a common practice in Brazil, characterized by the release of sewage into rivers, which do not take long to have their water intake for water treatment plants. This procedure is not forbidden, but there is still no legislation that allows direct reuse, that is the transformation of sewage into the water supply. Treatment strategies have experienced advances, such as the Membrane Bioreactors (MBR) technologies, where after biological treatment, the effluent is filtering through PVDF (Polyvinylidene Difluoride) membranes. Conventional methods of treatment down to the second level are efficient for removing dissolved organic matter, while

MBR reactors, which combine activated sludge reactors with separation of biomass by membranes, even provide the removal of viruses (O'Brien and Xagoraraki 2020).

MBR technology involves a final stage of ultrafiltration (UF), after which it is still possible to implement devices for performing Reverse Osmosis (OR) and Nanofiltration (NF). OR is strategic for removing dissolved salts, when water flow by successive cycles of use and reuse and continuous salt enrichment.

NF systems can be suitable for removing water pathogens being that a quality that meets drinking water standards can reach when the disinfection step is also adopted. Furthermore, oxidative treatments can also act, such as combinations of UV ultraviolet radiation, hydrogen peroxide (H_2O_2), and ozone (O_3). Such processes are efficient in removals such as pharmaceuticals, hormones, personal hygiene products, drugs of abuse, additives from contamination by microplastics, sweeteners, and molecules from numerous types of industrial applications (Hespanhol et al. 2019).

Adsorption technologies have also found application in water purification since the removal of drugs, through the use of traditional methods using activated carbon (reference) or innovative materials such as zeolites (de Sousa et al. 2018). Adsorbent materials use can also provide the removal of phosphorus, a nutrient present in wastewater. This nutrient occurs in relatively high concentrations in sewers, not being significantly in Sewage Treatment Plants that operate up to the secondary level, which can lead to eutrophication of aquatic bodies.

Among materials used for phosphorus removal, maybe highlighted those that lead the formation of the insoluble salt on Lanthanum modified clay (Kuroki et al. 2014) or even biosorption in natural materials, like sawdust, where retention is associated with the interaction of chemical phosphorus species with biofilms containing Fe (III) oxyhydroxides (Pantano et al. 2016).

Adequate treatment of wastewater, both industrial and sanitary, is of fundamental importance for the maintenance of aquatic life, preservation of the water quality resources that can have multiple uses such as recreation, supply, navigation, irrigation, and electricity generation. Besides that, the existence of a good quality source reduces operating costs related to the water treatment process contributing to the maintenance of the public health quality.

As regards phosphorus, in particular, it is a strategic nutrient for soil fertilization and plant nutrition that is on a depletion route, as it is a non-renewable resource. Studies indicate that between 5 and 10 decades (Cordell et al. 2009), the world will know the depletion of phosphate rocks, which poses the possibility of recovering the nutrient present in sewers and also in impacted water bodies, as a strategy that reconciles water and food security. Various strategies that insert technologies and a variety of unit operations in sewage treatment systems can generate a treated effluent that even meets a standard of potability, allowing its direct reuse if local legislation permits (Liu et al. 2020).

Steps that require pumping, pressure increase, filtration, stirring, and incidence of ultraviolet radiation are energy-consuming. Costs can be minimized through the use of photovoltaic energy from solar panels (Zhang et al. 2020, Ganiyu et al. 2019) or biogas gas energy (Liu et al. 2020), taking into account that anaerobic systems, responsible for methane production can be promising in the micropollutants treatment (Granato et al. 2020).

Water reuse is an option to promote water resources and aquatic life preservation. It may mean less need to search and pump water over long distances or lower water tables. One of the difficulties of using various operations to purify sewage aiming reuse is energy consumption. For this, it is necessary to promote an advance in the search for alternative sources of energy in association with strategies for the reuse of sewage.

Technologies aimed at water reuse are still far from the world reality, as already demonstrated in the course of this chapter, many areas of the planet do not even have access to the basic sanitation system. For this reason, the role of scientific knowledge in seeking alternatives to achieve higher water purity, reinforcing the public policies need aiming access to good quality water to all, regardless of purchasing power or place, is extremely important for the planet where they live.

7. Short Remarks

This chapter brings an overview of water resources availability and quality in the world and Brazil. In most of the world's areas, there is no efficient and comprehensive basic sanitation program. As reported by the World Health Organization (WHO) and UNICEF (2017), 30% of people worldwide, around 2.1 billion people, do not have safe drinking water access at home, while 6 out of 10 people, or 4.4 billion people, do not has adequate managed sanitation facilities (UNICEF 2017).

These data claims for an urgent rethink regarding the adopting world development model. Territories organizations carried out privileging to industrialization, urbanization, agriculture, and services have been shown harmfully predatory for all the living beings. The world economy obeys the capitalist system where the growth of GDP is taking into account as a development sign.

For this reason, the search for environmental sustainability permeates new socioeconomic and philosophical perspectives that prioritize the balance between society and nature. In this context, the water quality recovery must be in line with the achievement of good life quality to everyone and not just to a privileged minority. Understanding how works the natural mechanisms of water resources are essential for making decisions aimed at preserving them, contemplating the distribution, and access to this relevant resource. In this scenario, it is worth point out that the historically conquered environmental laws have complied, investments in sanitation technologies encouraged and implemented, and environmental education projects stimulated.

Valuing and encouraging the engagement of civil society in struggles in defense of the environment, and especially of water, is paramount to face the world scenario. The same is valid about the search for integrated and participatory water management, totally based on scientific studies contemplating new technologies with universal coverage, regardless of the economic power of the attempted societies.

Policymakers must consider universal rights to drinking water and sanitation so that they are fulfilling comprehensive and equitable instruments that provide water and health guarantees to the nations and societies that reside in them.

Acknowledgments

Authors are grateful to Professor Dr. João Luiz de Moraes Hoefel for the draft review and also to CNPq (grant 305627/2018-0), INCTAA (grants CNPq 465768/2018-8; FAPESP 2014/50951-4) and CAPES – Finance Code 001.

References

ABRASCO. 2020. População negra e Covid-19: desigualdades sociais e raciais ainda mais expostas. Viewed May 15th, 2020. https://www.abrasco.org.br/site/noticias/sistemas-de-saude/populacao-negra-e-covid-19-desigualdades-sociais-e-raciais-ainda-mais-expostas/46338/.

AGÊNCIA SENADO. 2020a. Projeto institui acesso à água potável como direito fundamental na Constituição. Viewed April 14th, 2020. https://www12.senado.leg.br/noticias/materias/2020/01/21/projeto-institui-acesso-a-agua-potavel-como-direito-fundamental-na-constituicao.

AGÊNCIA SENADO. 2020b. Senado aprova novo marco legal do saneamento básico. Viewed June 26th, 2020. https://www12.senado.leg.br/noticias/materias/2020/06/24/senado-aprova-novo-marco-legal-do-saneamento-basico.

ANA – Agência Nacional de Águas. 2019a. Conjuntura dos recursos hídricos no Brasil: informe anual/Agência Nacional de Águas. Brasília: ANA. Viewed April 04th, 2020. http://conjuntura.ana.gov.br/static/media/conjuntura-completo.bb39ac07.pdf.

ANA – Agência Nacional de Águas. 2020. Dia Mundial da Água incentiva cooperação entre os países. Viewed March 23th, 2020. https://www.ana.gov.br/noticias-antigas/dia-mundial-da-agua-incentiva-cooperaassapso-entre.2019-03-15.6545387713.

ANA – Agência Nacional de Águas. Lei nº 9.984 de 2000. Viewed March 23th, 2020. https://www.ana.gov.br/acesso-a-informacao/institucional/sobre-a-ana.

BRASIL. Lei nº 9.433, de 8 de Janeiro de 1997. DOU. Viewed May, 23th, 2020. http://www.planalto.gov.br/ccivil_03/leis/l9433.htm.

BRASIL. Lei nº 9.984, de 17 de Julho de 2000. DOU. Viewed May, 23th, 2020. http://www.planalto.gov.br/ccivil_03/LEIS/L9984.htm.

BRASIL. Lei nº 11.445, de 5 de Janeiro de 2007. DOU. Viewed April, 14th, 2020. http://www.planalto.gov.br/ccivil_03/_ato2007-2010/2007/lei/l11445.htm.

Campanha, M.B., A.T. Awan, D.N.R. deSousa, G.M. Grosseli, A.A. Mozeto and P.S. Fadini. 2015. A 3-year study on occurrence of emerging contaminants in an urban stream of São Paulo State of Southeast Brazil. Environmental Science and Pollution Research. 22: 7936-7940.

Carson, R. 1962. Silent Spring. Houghton Mifflin Co. Fortieth Anniversary Edition, New York.

CAS. 2020. http://web.cas.org/cgi-bin/regreport.pl. Viewed, July, 30th, 2020.

Castro, C.N. 2012. Gestão das Águas: experiências internacional e brasileira. Texto para discussão 1744. IPEA – Instituto de Pesquisa Econômica Aplicada, Brasília.

CETESB – Companhia Ambiental do Estado de São Paulo. 2020. O problema da escassez de água no mundo. Viewed April, 08th, 2020. https://cetesb.sp.gov.br/aguas-interiores/informacoes-basicas/tpos-de-agua/o-problema-da-escasez-de-agua-no-mundo/.

Cordell, D., J.-O. Drangert and S. White. 2009. The story of phosphorus: global food security and food for thought. Global Environmental Change. 19: 292-305.

deSousa, D.N.R., S. Insa, A.A. Mozeto, M. Petrovic, T.F. Chaves and P.S. Fadini. 2018. Equilibrium and kinetic studies of the adsorption of antibiotics from aqueous solutions onto powdered zeolites. Chemosphere. 205: 137-146.

deSousa, D.N.R., A.A. Mozeto, R.L. Carneiro and P.S. Fadini. 2014. Electrical conductivity and emerging contaminant as markers of surface freshwater contamination by wastewater. Science of the Total Environment. 484: 19-26.

Fadini, A.A.B. 2005. Sustentabilidade e identidade local: pauta para um planejamento ambiental participativo em sub-bacias hidrográficas da região Bragantina. Dissertation, UNESP, Rio Claro - SP, Brasil.

Florence, T.M. 1982. The speciation of trace elements in waters. Talanta, 29: 345-364.

Freire, P. 1975. Pedagogia do Oprimido. Paz e Terra, Rio de Janeiro.

Froelich, P.N., G.P. Klinkhammer, M.L. Bender, G.R. Luedtke, D. Cullen, P. Dauphin, D. Hammond, B. Hartman and V. Maynard. 1979. Early oxidation of organic matter in pelagic sediments of the eastern equatorial Atlantic: suboxic diagenesis. Geochimica et Cosmochimica Acta. 43: 1075-1090.

Gaete, C.M. 2015. Mapa da urbanização no mundo entre 1950 e 2030. Arch Daily Brasil. (Trad. Julia Brant) Plataforma Urbana: 2015. Viewed April, 12th, 2020. https://www.archdaily.com.br/br/763172/mapas-a-urbanizacao-no-mundo-entre-1950-e-2030.

Ganiyu, S.O., L.R.D. Brito, E.C.T.A. Costa, E.V. Santos and C.A. Martínez-Huitle. 2019. Solar photovoltaic-battery system as a green energy for driven electrochemical wastewater treatment technologies: application to elimination of Brilliant Blue FCF dye solution. Journal of Environmental Chemical Engineering. 7: 102924.

Granatto, C.F., G.M. Grosseli, I.K. Sakamoto, P.S. Fadini and M.B.A. Varesche. 2020. Methanogenic potential of diclofenac and ibuprofen in sanitary sewage using metabolic cosubstrates. Science of the Total Environment. 742: 140530.

Guerrero, J.L., S.M. Pérez-Moreno, F. Mosqueda, M.J. Gázquez and J.P. Bolívar. 2020. Radiological and physico-chemical characterization of materials from phosphoric acid production plant to assess the workers radiological risks. Chemosphere. 253: 126682.

Harvey, David. 2020. Política anticapitalista em tempos de coronavírus. Viewed May, 17th, 2020. https://blogdaboitempo.com.br/2020/03/24/david-harvey-politica-anticapitalista-em-tempos-de-coronavirus/.

Harada, M. 1995. Minamata disease: methylmercury poisoning in Japan caused by environmental pollution. Critical Reviews in Toxicology. 25: 1-24.

Heller, L. 2020. Saneamento básico é um direito humano universal. O silêncio proposital da mudança do novo marco legal. Interview with Léo Heller. Viewed July, 26th, 2020. http://www.ihu.unisinos.br/601005-o-silencio-absoluto-acerca-dos-direitos-humanos-na-mudanca-do-marco-legal-do-saneamento-basico-nao-foi-um-esquecimento-foi-proposital-entrevista-especial-com-leo-heller.

Hespanhol, I., R. Rodrigues and J.C. Mierzwa. 2019. Reúso potável direto - estudo de viabilidade técnica em unidade piloto. Revista DAE. 67: 103-115.

Hoeffel, J.L., A.A.B. Fadini and C.F.S. Suarez. 2002. Environment, sustainable tourism and academic responsability. pp. 415-427. *In*: W. Leal Filho [ed.]. Teaching Sustainability at Universities. Peter Lang, Frankfurt, Germany.

Hoeffel, J.L., A.A.B. Fadini, F.B. de Lima and M.K. Machado. 2004. Moinhod'água: rural communities and environment – environmental education activities in an environmentally protected area. pp. 247-258. *In*: W. Leal Filho and M. Littledyke [eds.]. International Perspectives in Environmental Education. Peter Lang, Frankfurt, Germany.

IBGE – Instituto Brasileiro de Geografia e Estatística. 2010. Sinopse do Censo Demográfico 2010: Brasil. Viewed Mach 11th, 2020. https://censo2010.ibge.gov.br/sinopse/index.php?dados=4&uf=00.

IBGE – Instituto Brasileiro de Geografia e Estatística. 2016. Pesquisa nacional por amostra de domicílios: síntese de indicadores 2015 / IBGE, Coordenação de Trabalho e Rendimento. Rio de Janeiro.

IBGE – Instituto Brasileiro de Geografia e Estatística. 2020. População: Projeção da população do Brasil e das Unidades da Federação. Viewed March, 25th, 2020. https://www.ibge.gov. br/apps/populacao/projecao/.

IPEA – Instituto de Pesquisa Econômica Aplicada. 2020a. Retratos das desigualdades – Gênero e Raça: Chefe de família. Viewed May, 13th, 2020. https://www.ipea.gov.br/ retrato/indicadores_chefia_familia.html.

IPEA – Instituto de Pesquisa Econômica Aplicada. 2020b. Retratos das desigualdades – Gênero e Raça: População. Viewed May, 13th, 2020. https://www.ipea.gov.br/retrato/ indicadores_populacao.html.

Jost, J.W. 1998. Executive Summary. Pure and Applied Chemistry. 70: v-vii.

Kuroki, V., G.E. Bosco, P.S. Fadini, A.A. Mozeto, A.R. Cestari and W.A. Carvalho. 2014. Use of a La(III)-modified bentonite for effective phosphate removal from aqueous media. Journal of Hazardous Materials. 274: 124-131.

Leite, Carlos. 2009. As megacidades e o desenvolvimento sustentável: concentração de desafios e oportunidades no mundo cada vez mais urbano. *In*: Cidadania planetária: dimensões globais, sociais e subjetivas. n. 9. FDC, jul./out. pp. 57-63. Viewed April, 10th, 2020. http://acervo.ci.fdc.org.br/AcervoDigital/Artigos%20FDC/Artigos%20DOM%20 09/CIDADES%203%20Cidadania%20planet%C3%A1ria.pdf.

Liu, L., E. Lopez, L. Dueñas-Osorio, L. Stadler, Y. Xie, P.J.J. Alvarez and Q. Liu. 2020. The importance of system configuration for distributed direct potable water reuse. Nature Sustainability. 3: 548-555.

Liu, T., J. Li, Z.K. Lim, H. Chen, S. Hu, Z. Yuan and J. Guo. 2020. Simultaneous removal of dissolved methane and nitrogen from synthetic mainstream anaerobic effluent. Environmental Science & Technology. 54: 7629-7638.

Lopes, C.V.A. and G.S.C. Albuquerque. 2018. Agrotóxicos e seus impactos na saúde humana e ambiental: uma revisão sistemática. Saúde Debate. 42: 518-534.

Machado, K.C., M.T. Grassi, C. Vidal, I.C. Pescara, W.F. Jardim, A.N. Fernandes, J.S. Santana, M.C. Canela, C.R.O. Nunes, K.M. Bichinho and F.J.R. Severo. 2016. A preliminary nationwide survey of the presence of emerging contaminants in drinking and source waters in Brazil. Science of the Total Environment. 572: 138-146.

MDR – Ministério do Desenvolvimento Regional. 2020. Plano Nacional de Saneamento Básico – PLANSAB, mais saúde, qualidade de vida e cidadania. Viewed June, 20th, 2020. https://www.mdr.gov.br/saneamento/plansab.

MMA – Ministério do Meio Ambiente. 2011. Áreas de Preservação Permanente e Unidades de Conservação & Áreas de Risco. O que uma coisa tem a ver com a outra? Relatório de Inspeção da área atingida pela tragédia das chuvas na Região Serrana do Rio de Janeiro. Wigold Bertoldo Schäffer [et al.]. MMA Brasília. Viewed March 24th, 2020. https://www. mma.gov.br/estruturas/202/_publicacao/202_publicacao01082011112029.pdf.

MMA – Ministério do Meio Ambiente. 2020. Água: Um recurso cada vez mais ameaçado. Viewed March 24th, 2020. https://www.mma.gov.br/estruturas/secex_consumo/_ arquivos/3%20-%20mcs_agua.pdf.

Mozeto, A.A., M. Montini, S.A. Braz, F.G. Martins, A. Soares, M.R.L. Nascimento, F.A.R. Barbosa, P.S. Fadini and B.M. Faria. 2012. External versus internal loads of nutrients of an urban eutrophic tropical reservoir (Southeastern Brazil). Journal of Environmental Science and Engineering A. 1: 598-610.

Niño, L.R., R.J. Torres, A.A. Mozeto and P.S. Fadini. 2014. Using urban streams as drinking water: the potential risk in respect to polycyclic aromatic hydrocarbons (PAHs) content in sediments. Polycyclic Aromatic Compounds. 34: 518-531.

O' Brien, E. and I. Xagoraraki. 2020. Removal of viruses in membrane bioreactors. Journal of Environmental Engineering. 146: 03120007.

ONU – Organização das Nações Unidas. 2015. Transformando Nosso Mundo: A Agenda 2030 para o Desenvolvimento Sustentável. Viewed April 03rd, 2020. https://nacoesunidas.org/pos2015/agenda2030/.

ONU – Organização das Nações Unidas. 2018. Declaração Universal dos Direitos Humanos. Viewed April 02th, 2020. https://nacoesunidas.org/wp-content/uploads/2018/10/DUDH.pdf.

ONU – Organização das Nações Unidas. 2019. Mais de 2 bilhões de pessoas no mundo são privadas do direito à água. Viewed April 02nd, 2020. https://nacoesunidas.org/mais-de-2-bilhoes-de-pessoas-no-mundo-sao-privadas-do-direito-a-agua/.

ONU – Organização das Nações Unidas. 2020a. Surto de coronavírus é reflexo da degradação ambiental, afirma PNUMA. Viewed July 10th, 2020. https://nacoesunidas.org/surto-de-coronavirus-e-reflexo-da-degradacao-ambiental-afirma-pnuma/

ONU – Organização das Nações Unidas. 2020b. O Direito Humano à Água e Saneamento. Viewed April 03th, 2020b. https://www.un.org/waterforlifedecade/pdf/human_right_to_water_and_sanitation_media_brief_por.pdf.

ONU – Organização das Nações Unidas. 2020c. O Direito Humano à Água e Saneamento. Viewed April 03th, 2020c. https://www.un.org/waterforlifedecade/pdf/human_right_to_water_and_sanitation_milestones_por.pdf.

Pantano, G., J.S. Ferreira, F.W.B. Aquino, E.R. Pereira-Filho, A.A. Mozeto and P.S. Fadini. 2016. Biosorbent, a promising material for remediation of eutrophic environments: studies in microcosm. Environmental Science and Pollution Research. 24: 2685-2696.

Perez-Lopes, R., A.M. Alvarez-Valero and J.M. Nieto. 2007. Changes in mobility of toxic elements during the production of phosphoric acid in the fertilizer industry of Huelva (SW Spain) and environmental impact of phosphogypsum wastes. Journal of Hazardous Materials. 148: 745-750.

Scruggs, C.E., C.B. Pratesi and J.R. Fleck. 2020. Direct potable water reuse in five arid inland communities: an analysis of factors influencing public acceptance. Journal of Environmental Planning and Management. 63: 1470-1500.

Simão, Jorge R. 2017. Megacidades, urbanização e desenvolvimento sustentável. Hoje Macau: Perspectivas. Viewed April 02nd, 2020. https://hojemacau.com.mo/2017/12/20/megacidades-urbanizacao-e-desenvolvimento-sustentavel/.

TRATABRASIL. 2019. Novo Ranking do Saneamento Básico evidencia: melhores cidades em saneamento investem 4 vezes mais que as piores cidades no Brasil. Viewed June, 23th, 2020. http://www.tratabrasil.org.br/images/estudos/itb/ranking2019/PRESS_RELEASE___Ranking_do_Saneamento___NOVO.pdf.

UN – United Nations. 2019. World Population Prospects. Viewed March, 11th, 2020. https://population.un.org/wpp2019/.

UNICEF – Fundo das Nações Unidas para a Infância. 2017. 2,1 bilhões de pessoas não têm acesso a água potável em casa, e mais do dobro de pessoas não tem acesso a saneamento seguro. Viewed April, 10th, 2020. https://www.unicef.org/angola/comunicados-de-imprensa/21-bilh%C3%B5es-de-pessoas-n%C3%A3o-t%C3%AAm-acesso-%C3%A1gua-pot%C3%A1vel-em-casa-e-mais-do-dobro.

UNICEF – Fundo das Nações Unidas para a Infância. 2020. Viewed May, 22th, 2020. https://www.unicef.org/brazil/comunicados-de-imprensa/lavar-maos-com-sabao-fundamental-contra-coronavirus-fora-de-alcance-de-bilhoes.

Wang, J., R.P.J. Hoondert, N.W. Thunnissen, D. van de Meent and A.J. Hendriks. 2020. Chemical fate of persistent organic pollutants in the arctic: evaluation of simple box. Science of the Total Environment. 720: 137579.

Zhang, S., P. Xiaofeng, Z. Yue, Y. Zhou, H. Duan, W. Shen, J. Li, Y. Liu and Q. Cheng. 2020. Sulfonamides removed from simulated livestock and poultry breeding wastewater using an in-situ electro-Fenton process powered by photovoltaic energy. Chemical Engineering Journal. 397: 125466.

Zhao, L., J. Zhao, J. Zhao, L. Zhang, D. Wu, H. Wang, J. Li, X. Ren and Q. Wei. 2020. Artificial N_2 fixation to NH_3 by electrocatalytic Ru NPs at low overpotential. Nanotechnology. 31: 29LT01.

Zuffo, A.C. and M.S.R. Zuffo. 2016. Gerenciamento de recursos hídricos: conceituação e contextualização. Elsevier, Rio de Janeiro.

Environmental Perception of Students at General Secondary School Located in Niassa Special Reserve

Francisco Gonçalves Nhachungue[1]*, Sônia Regina da Cal Seixas[2] and Benjamim Olinda Bandeira[3]

[1] Teacher at UniRovuma-Niassa. PhD Student in Energy and Environment - FNCM - UPMaputo/Mozambique

[2] PhD in Social Sciences IFCCH/UNICAMP. Professor at NEPAM and PSE/UNICAMP (Brazil) and Co-supervisor of Francisco's doctoral thesis

[3] PhD in Population Biology and Ecology Sea. Professor in the Department of Biology/UP Maputo (Mozambique) and Supervisor of Francisco's doctoral thesis

1. Introduction

The present study had its spatial focus on Niassa Special Reserve (NSR), and it sought to discuss the environmental perceptions of a group of students at 16th June General Secondary School in Mecula in relation to its insertion in the area of biodiversity conservation.

Since the beginning of 1990s, after the Signature of General Peace Agreements, Mozambique started the restoration process of biodiversity conservation areas and the creation of new areas for the purposes of protection of species and special biomes in extinction (Muhale 2020). This desideratum is based on the Principles of Stockholm Declarations, according to which, "The natural resources must be preserved; the wild fauna and flora must be preserved" (NEBBIA and UNEP s/d.; Gurski et al. s/d.) and infulfilment of different international agreements on this matter.

Mozambique ratified, under Resolution No. 2/94 of 24th August, the United Nations Convention on Biologic Diversity, established in June 1992 that proves the need to raise awareness and take immediate actions to assure correct and sustainable management of biologic diversity with the aim to construct a source of livelihood of present and future generations (Republic of Mozambique 1994, Kotsakis 2009).

The 'Biological Diversity' is understood as variability between living beings of all origins, including lacustrine, marine ecosystems and other aquatic ecosystems

*Corresponding author: fnhachungue@yahoo.com.br

and, ecological compounds that make part of it (Republic of Mozambique 1994). The conservation of biological diversity or simply biodiversity assumes two perspectives: *in-situ* and *ex-situ*. *In-situ* conservation is the conservation of ecosystems and natural habitats, as well as maintenance and recovery of a viable population of species in its natural environment, whereas *ex-situ* conservation refers to the conservation of components of biological diversity out of their natural habitats.

The issues of *in-situ* biodiversity conservation, the case of Niassa Special Reserve, constitute an enormous challenge to the government authorities, especially to the managers of such areas. This challenge has to do with the necessity of implementing mechanisms that actively involve the local communities in the shared use of biodiversity. The issue of shared management of subsistence resources between humans and wild fauna in conservation areas, such as the land, springs and other environmental resources, leads to reciprocal conflicts among the stakeholders. On the one hand, the local communities feel isolated from the management process of fauna and flora conservation from which derives the feeling of not taking protective measures against the attacks by wild animals to human integrity to agricultural fields, etc. On the other, the managers complain about the negative attitudes of the local communities, allegedly by breaking the national legislation on biodiversity conservation and triggering undesirable practices, such as poaching, agricultural practices under uncontrolled fires, illegal mining and others.

The manifestation of these socio-ecological phenomena is considered as a problem of shared use of common resources (Vieira et al. 2005, Hardin 1968). According to the authors, this type of resources comprises two basic characteristics:

(i)   The problem of restricted access or access control of potential users;
(ii) The problem of shared use, that is, each user is able to subtract the earnings that also belong to all other users.

These two characteristics wind up the environmental degradation in biodiversity conservation areas. The limited access to the use of environmental resources (land, springs, flora, fauna and underground resources) imposed by the conservation law, as well as restricted participation of local communities in the decision-making process on management, lead to the aggression of the environment with common loss. Therefore, there is a need to adopt the coexistence mechanisms in the use of natural resources in conservation areas.

According to Salvemini and Remple (2014), the range of success in biodiversity conservation demands the implementation of mechanisms that aim at engaging the local communities in various spheres of social, economic, cultural and environmental life. It requires the use of participatory methods in the evaluation of local problems and their causes through discussions of focal groups among community members. In these processes, the interesting parts identify the characteristics, resources, challenges and opportunities that their landscape offers and how to maximize their sustainable use. The discussions and maps created by the community are a material reference for future analysis and planning for they will provide relevant information about specific socio-ecological characteristics of the determined conservation areas.

For Vieira et al. (2005), there are four mechanisms of using resources: free access, communal, government and private properties. The free access testifies the "tragedy of commons" (Hardin 1968), that is quick exhaustion of rare and special environmental resources and the restriction to their access aims at preserving such resources in a long term. The mechanism of communal properties implies the restriction capacity to the access of environmental resources to a determined group. That could not be applicable in conservation areas like NSR. In the same way, the government property in isolation could not be viable if it does not allow to grant certain areas to the private sector, and it would be difficult to ensure effective control of use, integral oversight and protection of resources under study due to lack of financial resources.

The setting of private property mechanism allows the institutional arrangement that influences the stakeholders, including the local communities in compliance with the legislation and norms related to biodiversity conservation. However, it is necessary that the concessionaries integrate the local communities in the process of management of privatized areas, establishing access to the resources of common use and indispensable for life. This position is supported by Chardonnet (2019) when considering that no matter the type of governance, funding, political support and skilful techniques of management, the support to the development of local communities ensures effective conservation of biodiversity.

It is clear that the development of local communities is a social aspect that manifests through training assurance of individuals residing within as well as the outskirt of determining areas of biodiversity conservation.

The school is a social space in which learners, teachers, educational managers and other education stakeholders share different bits of knowledge, values, life experiences and competencies defined in the curricula for different school subjects. The training of individuals based on the local curricula approach has its focus on the valuation of environments in which the educational institutions are located.

Thus, the local or realistic curriculum (INDE 2007) is a component of the national curriculum that corresponds to 20% of the total time predicted for teaching each subject and is made of contents defined locally as being relevant for the learner integration in his community (INDE 2003).

The local curriculum was at the first stage implemented at Primary School Education and gradually, it became an innovative aspect at Secondary School Education (SSE) as an intermediate level between High Education (HE) and Primary School Education (PSE) (INDE 2007). This curriculum envisages the involvement of different social sectors, as active forces of the society, to give their maximum contribution in the development of competencies (cognitive, affective and psychomotor) of their learners, for a good integration in the communities (Castiano 2006). In addition, the Curricular Plan itself at the General Secondary School Education (CPGSS) recognises the weak horizontal and vertical articulation among the programs and subjects at General Secondary Schools (GSS) due to the absence of common goals between the various subjects. In this sense, the CPGSS recommends the diversification of educative offers and topics through the extra curriculum activities, creation of groups or circle of interest, youth association, pair work, debates, brainstorming, role-plays and others (INDE 2007) in order to stimulate abilities of opinion confrontation and question about the reality and propose alternative solutions to current problems in their respective environments.

The operationalisation of these inclusive strategies, integrated and centered on learners is assured by the teacher who plays the role of mediator between the community competencies and those established by the official curriculum. All this should be based on the UNESCO strategic guidance to GSS, according to which it should develop teaching methods, innovative school organisation chart and services adapted to the teaching of necessary competency and abilities for life and behavioural education (INDE 2007).

Concentrating on aspects of behavioural competencies development, we next, analyse the CPGSS in terms of its more varied objectives to local environmental issues taken as a whole among all subjects at GSS (1st and 2nd learning cycles).

The first objective of GSS: "Provide young learners with an integral and harmonious development through a set of competencies: knowledge, abilities, attitudes and values articulated in all areas of learning" (INDE 2007) and comprises all aspects related to competencies for life, recognition of the environment in which the learner lives; the territorial potentials in different scales (local, regional and global) in order to allow them to act consciously in determined realities.

In this way, we express the objectives of some of the subjects in both learning cycles at GSS with a focus on the subjects of Geography, Biology and Agro-Livestock, by scientific inherence of their intervention areas (Table 1). These subjects deal with environmental issues, although it is known that the subjects have an interdisciplinary and cross-cutting character. Such character is emphasised in the CPGSS itself when stating that "the development of competencies is an exercise that will be present in all moments of learner's life, either within or out of classroom. For such reason, its approach should be done in a cross-cutting manner" (INDE 2007).

The selection of such subjects has to do with the fact that the authors have scientific, professional affinities, academic and life experiences gathered in contact with different professionals in such subjects.

After the discussions about biodiversity conservation, the school curriculum and its contribution to the development of varied competencies, biodiversity conservation and presentation of the objectives of the three selected subjects from the CPGSS, we next express the objectives of this research.

The general objective consists of evaluating learners' perception of the environment in relation to their insertion in the biodiversity conservation area.

In more specific terms, the work sought to characterise learners' socio-demographic variables; identify the subjects at GSS focused on the study of environmental aspects; analyse the relationship between learners' environmental knowledge and their insertion in the area of biodiversity conservation; and capture learners' perceptions about socio-environmental conflicts.

The study and perception of the relationship between human beings and the environment in schools favour the conception of mechanisms of sustainable use of resources (Santos et al. 2008, Malafaia and Rodrigues 2009, Almeida et al. 2012, Fogaça e Limberger 2014).

Learners are a social group whose environmental perceptions deserve appropriate attention to understand how the awareness of biodiversity conservation develops at school, in the community and their families and likewise understand the extent to which such contents influence their daily practices in relation to the surrounding

Table 1: Prescribed Objectives in CPGSS in Mozambique

Subject	Prescribed objectives in CPGSS	
	1° cycle	*2° cycle*
Geography	Expand and consolidate competencies (...) knowledge, abilities and correct attitudes towards nature and society.	Contribute to the understanding of the interdependency relationship between nature and human activity (...) and the changes in the environment.
	-	Develop, in learners, abilities that allow them to use knowledge from Geography to understand the factors that influence (...) the well being of Man and of nature.
Biology	Contribute to the protection, conservation and sustainable use of natural resources especially those of biological diversity of our country, for the benefit of the current and future society.	Valuate the importance of protection and conservation of the school environment through individual and collective responsibility.
	Instil in learners the love for nature, establishing an effective relationship with organisms.	Publish and apply techniques of conservation of environment in the community.
Agro-Livestock	Contribute to the application of new production techniques in families and in the community as a way of increasing production and productivity, improve the diet and ensure food security.	Develop abilities for the application of new production techniques in the family and in the community, as a way of increasing production and productivity, improve diet and ensure food security.

Source: Authors (2020), adapted from CPGSS (INDE 2007)

environment. Chierrito-Arruda et al. (2018) integrate the physical, cultural, social and historical dimensions as an object of analysis of environmental perception in such a way that people should experience its surrounding with emphasis on the dimensions.

The environmental perception is understood as the branch of psychology whose object is the study of interrelations between man and his actions on the environment (Álvaro and Bassani s/d). Likewise, Gifford (2014) considers that understanding the environment is to examine the transactions between individuals and their environments built or natural, including the investigation of behaviours that inhibit or promote sustainable changes for the increasing of pro-environmental awareness. In this sense, Bassani et al. (2007) point out the reciprocity of the relationship between people and the environment, that is a person modifies the environment and this, in turn, acts and modifies the person. These inter-relationships stir the life and culture, cognitive, affective and psychomotor aspects of the stakeholders. Gifford et al. (2011) show the relevance of the commitment that people have to the physical environment: "people use physical mean among them, according to complex rules and strong preferences, although these aspects are not always conscious. The importance is clear when they are committed". Therefore, among these complex interrelationships, as people feel committed, for example, with the process of conservation of environmental resources through mechanisms of active participation, they valuate their perceptions. Gifford (2014) considers it as 'reciprocal transactions' between humans and their natural or built environments with consequences within the same environments.

"Many environmental problems are rooted in human behavior and, therefore, they can be solved through behavior comprehension. The pro-environmental behavior is important. Influences in the pro-environmental behavior includes childhood experience; knowledge and education; personality; perceived behavioral control; values, attitudes and world views of various types; assumed responsibility and moral commitment; place, norms and habits; goals; effectivity and many demographic factors. These influences (...) interact among them" (p. 544).

Thus, the author sets three important challenges:
(i) Learn more about how these influences moderate and mediate one another;
(ii) Discover which domains of behaviour related to the environment are more influential in which social domains and;
(iii) Expand and deepen the knowledge of functionalities of the society, in terms of production and consumption of goods and services and how wide the influences of socio-political aspects and their contribution are to values, attitudes and behaviour.

These three aspects (challenges) guide the choice of method and respective instruments for the collection of environmental perceptions of learners in the area of study.

The study seeks to answer the questions that follow:
• How are the research learners' socio-demographic variables characterised?
• What are the subjects at GSS that deal with issues of the environment?
• What is the relationship between learners' environmental bits of knowledge and their insertion in the biodiversity conservation area and;
• Which perceptions do learners have about socio-environmental conflict?
• The discussions about these issues are taken in an integrated manner and are based on materials and methods that are presented next.

2. Materials and Methods

2.1. Characterisation of Niassa Special Reserve (NSR)

The NSR connects itself to the National Reserve of Selous in Tanzania with a total area of 50.000 km^2 (Booth and Duham 2014). The NSR covers the totality of Mecula District and it extends through more than seven districts across Niassa and Cabo Delgado Provinces in the North of Mozambique. It was created in the colonial period as a Game Reserve by the Legal Act No. 10.578 on October 9, 1954 (Republic of Mozambique 1954, SRN 2006) and its limits have been changing due to conservation policies, as well as the need for expanding its area. From Game Reserve it passed to

Partial Game Reserve through the Legislative Diploma 1997 on June 23, 1960 and in 1999 it ascended to the category of National Reserve[1] (Republic of Mozambique 1999), occupying 42.000 km[2] (Booth and Dunham 2014, Begg and Begg 2009) and exhibiting new limits (SRN 2006).

According to the chart below, 'A' represents the map of NSR and the location of Mecula head-office in the central zone of the Reserve. Figure 1 ('B' and 'C') illustrates the geographical location of NSR in Niassa Province and the African context.

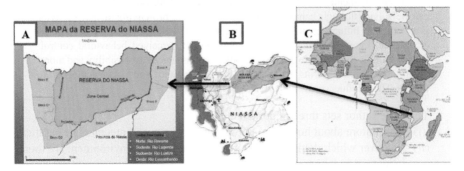

Figure 1: n.o 1 (A, B e C): Geographical Location of Niassa Special Reserve
Source: http://cuamba.blogspot.com/2009/06/reserva-natural-do-niassa.html

NSR keeps the biomes and ecosystems with special ecological, social, economic, and environmental importance with emphasis on miombo forest (*Brachystegia*)[2]. Its exposition to extinction risk due to the manner people explore the subsistence resources and for commercial purposes generates the need to keep this area.

2.2.   Population and Socio-Economic Activities

The reserve is inhabited by a population estimated at more than 60,000 inhabitants, distributed in 40 villages of which 21.342 inhabitants reside in Mecula District (INE 2019). These inhabitants dedicate themselves to agriculture, fishing, hunting, mining and generally they enter into conflict with the implementation of conservation Law and surviving mechanisms based on extractivism of environmental resources.

2.3.   School Network in the Area of NSR (Mecula District)

The school network of Mecula District is composed of 26 schools of which 17 are Primary School Education (PSE) that comprise only the two first learning cycles (from 1st to 5th standards), eight is Complete Primary Schools (CPS) that comprise all three learning cycles (from 1st to 7th standards) and one is General Secondary School (GSS). These schools are distributed through Locality Areas and Administrative Posts of the District, as Table 2 illustrates data based on statistics of Mecula DSEYT[3].

---

[1] Decree No. 81/99 of 16 November
[2] Covers an area of 95% of the total surface of the Reserve (Branch et al. 2005).
[3] District Services of Education Youth and Technology.

Table 2: Mecula District School Network

Locality/Administrative post	Nº/Type of School (2019/2020)		
	GSS	Basic education	
		PSE	CSE
Mecula Head Office	1	0	0
	0	0	4
	0	6	0
Lugenda	0	0	1
	0	7	0
Naulala	0	0	1
	0	2	0
Mbamba	0	0	1
	0	0	0
Administrative Post of Gomba	0	0	0
	0	1	0
Administrative Post of Matondovela	0	0	1
	0	1	0
Subtotal	1	17	8
Total			26

Source: Authors (2020). Based on statistic figures of Mecula DSEYT (District Services of Education Youth and Technology)

The 16th June General Secondary School of Mecula, the research object, is situated in the Locality of Mecula head office in Mecula District. Apart from being located in a rural area, the school is within the central block of biodiversity conservation in Niassa Special Reserve (NSR).

The school receives learners who finish the Primary School Education (PSE) from different Administrative Posts and Localities of Mecula Districts and from other Districts of Niassa Province who come to continue with their studies. Being a General Secondary School (GSS), it comprises the 1st learning cycle (Grade 8 to 10) and the 2nd learning cycle (Grade 11 to 12) (Figure 2).

2.4. Universe and Research Sample

According to statistic figures from Mecula District Services of Education, Youth and Technology, in the academic year of 2019, the school enrolled 501 learners distributed as follows: 210 females and 291 males.

From this universe, 19 learners, corresponding to 3% of the total universe number, were involved in the research. Of 19 learners, two were females and 17 were males. The size of this sample, associated with the type of research study does not allow the generalisation of perceptions (Trivinos 1987). However, it gives an idea of what might be happening in the environmental perceptive field of learners from the researched school.

Figure 2: Frontal View of 16th June General Secondary School of Mecula

This sample is a subset of people and entities involved in the research field, in March 2019, in the context of our thesis for the doctorate in course.

The learners were selected randomly with the collaboration of the School Directorate, teachers and observing the three criteria: (i) include learners from 1st and 2nd Cycles of learning (from Grade 8 to 10 and from Grade 11 to 12, respectively); (ii) select the students from schoolyard during break; and (iii) involve learners available to cooperate with the researcher in the response to the questionnaire (Figure 3).

2.5. Research Type, Techniques and Instruments

The type of research, techniques and instruments used are similar to those applied by Chierrito-Arruda et al. (2018) and Malafaia and Rodrigues (2009).

Figure 3: Part of the Group of Learners involved in the Research

In relation to the approach, the research is qualitative based on ecological-naturalistic assumptions, as Trivinos (1987) states:

The research type, techniques and instruments used are similar to those applied by Chierrito-Arruda et al. (2018) and Malafaia and Rodrigues (2009).

The research is based on a qualitative approach, focusing on ecologico-naturalist assumptions. According to Triviños (1987):

> "The environment, with its physic and social characteristics, transmits to subjects peculiar features that are unavailed in the light of understanding of meanings that it establishes. Because of that, as an attempt to understand the human conduct, isolated from the context in which it manifests, creates artificial situations that falsify the reality, take to mistake and elaboration of inadequate assumptions and, to ambiguous interpretations" (p. 122).

The techniques used for comprehension and description of this influence exerted by the environment on the behavioural and habits manifestation, socio-cultural and environmental practices of the researched individuals were bibliographic consultation, document analysis and fieldwork. The fieldwork consisted of direct observation through which the researcher had to take part in the community's daily life.

The instruments used were semi-structured interviews and questionnaires. The questionnaire contained five groups of perceptive aspects, that involved open-ended questions, closed questions and multiple choice on the Likert scale (Alexander et al. 2003). The first group of questions characterises the socio-demographic variables of learners involved in the study; the second group identifies the subjects at General Secondary School that deal with the issues of the environment; the third group of questions elicits learners' knowledge about the environment; the fourth group deals with socio-environmental conflicts.

2.6. Data Interpretation

The data were interpreted through content analysis, defined by Chizzotti (2000), as a method of processing and analysing information gathered through techniques of data collection expressed in a document. Therefore, the methods are applied to written texts, gestural, visual and verbal information with the aim to understand critically the meaning of information, its expressed and latent content, covert and overt meaning.

Bard in (1977) defines content analysis as a set of methodological instruments applicable to extremely diversified discourses, that is it consists of interpretation of written and discursive content of a determined group of interviewed people.

The content analysis was based on 5 (five) stages, as Table 3 illustrates.

2.7. Data Analysis

Presentation of results consisted of methodological triangulation, subdivided into three integrated moments: the first being dedicated only to the results, the second reserved for discussion and the third moment corresponded to conclusions.

We inspired ourselves in the format of data analysis adopted by Chierrito-Arruda et al. (2018) that beyond the similarities in the research type, techniques and

Table 3: Stages of Content Analysis and Its Characteristics

Ord. Nº	Stage designation	Characteristics
1	Analysis Organisation	Pre-analysis, physic and questionnaires' content evaluation, material conservation conditions, gender and school level order, the way instruments were filled in.
2	Codification	Allocation of identification code to each questionnaire, corresponding to each learner: L_1 MHO, L_2 MHO, L_3 MHO, L_4 MHO, etc, meaning respectively: Learner$_1$ Mecula Head Office; Leraner$_2$ Mecula Head Ofiice; Learner$_3$ Mecula Head Office, Learner$_4$ Mecula Head Office, etc.
3	Categorisation	Classification and grouping of constituent elements of a set of aspects contained in a questionnaire, done through the analogy of similarity and difference.
4	Inference	It allowed the understanding of tendencies and the sense of responses to questionnaires and interviews carried out.
5	Computer processing	Use of computer statistic programs *IBM SPSS – Version 20* (Pilati & Porto Unpublished) and Excel. Apart from the cross-reading of variables of analysis, there was carried out the calculation of learners' age Mean and interviewed learners' household, reading of results through tables and graph.

Source: Authors (2020), adapted from Bardin (1977), Mozzato and Grzybovski (2011).

instruments used, it offers advantages in the coherence and comprehension of result discussion and post the presentation of conclusions. The analysis is made according to the sequence of the research objectives, as suggested by the diagram below:

2.7.1. Characterization of Socio-Demographic Variables

The socio-demographic characteristics (age, sex, origin and the school level attended) allowed an integrated knowledge of the group of learners involved. For such a purpose, two different variables were associated, forming two groupings. The analysis was made looking at the results of each grouping, that is, the 'provenience' was associated with 'school level' and 'age' was associated with learners' gender. Then, it was carried out the crossed analysis, whose results were systematised and illustrated in bar graphs. Next, the school subjects that deal with the environment were registered and analysed.

2.7.2. Subjects at GSS Dealing with Environmental Aspects

In this aspect, learners' perception about subjects dealing with environmental aspect was identified through these questions: if 'they have ever heard about environment', 'the circumstances and place' and 'name only two subjects at GSS that deal with such topics'? The results of this item were presented in the form of interview quotations and synthesise by a pie chart. After the analysis of subjects dealing with the environment, learners' knowledge about the environment was analysed.

2.7.3. Learners' Knowledge About Environment

Learners' environmental knowledge was spotted from questions that included the 'concept of environment', 'components of environment', 'importance of environment preservation' and 'manners of environment preservation'. Thus, for the analysis of the concept of environment, there were four established subcategories of aspects which included in the concepts of the environment from learners' point of view, for inference under Lucena and Freire researches (2015) and Mattos and Andrade (2018): (i) subcategory of abiotic/inert aspects (springs, rocks, soils, minerals, etc.); (ii) biotic aspects (flora and fauna, without including Man); (iii) biophysics aspects (biotic and abiotic); and (iv) cross-cutting aspects (abiotic, biotic and socio-environmental, including the Man). The results were presented in quotations of some learners' stretches.

For the analysis of environmental components, there used the same subcategories applied to the concept of the environment (Mattos and Andrade 2018).Abiotic and biotic aspects were gathered, making up the biophysics aspects. Thus, this variable was analysed under biotic, biophysics and wide-ranging/socio-environmental aspects. The results referred to two questions are synthesised in a histogram/column graphs.

The importance of environmental preservation in the local context was analysed through five subcategories established by Malafaia and Rodrigues (2009) and Tamaio (2000):

(i) Romantic – It establishes a vision of super-nature or provider mother-nature;
(ii) Utilitarian – It presents an ecocentric and determinist interpretation, that is it considers nature as the source of all and without it, the man no longer exists;
(iii) Wide-ranging – It involves the totality of natural aspects (biotic, abiotic, socio-economic, cultural and environmental);
(iv) Reductionist – It postulates that the environment is strictly related to physico-natural aspects (water, air, soil, rock, fauna and flora), excluding the human being and all his production;
(v) Socio-environmental – It presents the man and the built environment in an historico-cultural process as constitutive elements of nature.

The results of this variable 'learners' environmental knowledge' were presented in bar-graphs, while the results about the manners of environment preservation were presented through some quotations. After the collection of environmental knowledge, perceptions about socio-environmental conflicts were analysed.

2.7.4. Socio-Environmental Conflicts

For the collection of learners' perceptions about socio-environmental conflicts in NSR, it was analysed, on the one hand, the variable 'attacks by wild animals', using the Likert scale ('much frequent, frequent and non-frequent'), confronted with the variable 'learners' provenience'. The option 'there is no attack' was not included because, during the exploratory study, it was observed that there was a generalized feeling of the local communities in relation to the problem of attacks by wild animals. The results were synthesised and presented in a table with the percentual standard.

On the other, it was tested the variable 'participation in community queries', given its importance in mechanisms of involvement in biodiversity conservation. This variable allowed only two options for closed responses (YES/NO).

These results were also presented in a table with percentual standard, a product of confrontation with the variable 'learners' provenience'.

3. Results

This subsection is reserved for the presentation of research results, according to the sequence of specific objectives.

3.1. Learners' Socio-Demographic Characterisation

Characterisation of socio-demographic variables (age, sex, origin and school level) corresponds to the first specific objective of this research and constitutes the starting point for environmental perception analysis of the learners involved in the research.

The results reveal that from 19 learners, 44.4% are from Mecula head office. The 33.3% are attending the 1st Cycle at General Secondary School and 11.1% are attending the 2nd Cycle. The 27.8% are from out of Mecula District of which 16.7% belong to 1st Cycle and 11.1% are attending the 2nd Cycle. The 10.2% of learners come from Macalange Locality of which 5.6% are in the 1st Cycle and the equal percentage of learners are attending the 2nd Cycle.

The rest of the learners are from Naulala and Mbamba localities and Matondovela Administrative Post. These represent, in total, 16.8% of all interviewed, being 5.6% each, and all are attending the 1st Cycle at GSS (Figure 4).

In relation to the age, 89.4% of learners involved in the study are male and 10.6% are female in an age range of 16 to 24 years old. The female learners are in the age range of 19-21 years old.

The male learners aged 18 are the majority in the study, that is 26.3%. The interval between 5.3 to 10.8% represents learners of the age range between 16 to 20 years, while 15.9% corresponds to learners in the age range of 21 to 24, of which 5.3% correspond to each age of the range, except the age of 22 (Figure 5).

The socio-demographic variables lead us to understand the age range characteristics, gender, provenience and the school level attended by the interviewed learners. As the research being conducted in the school context, we next present the subjects dealing with environmental aspects.

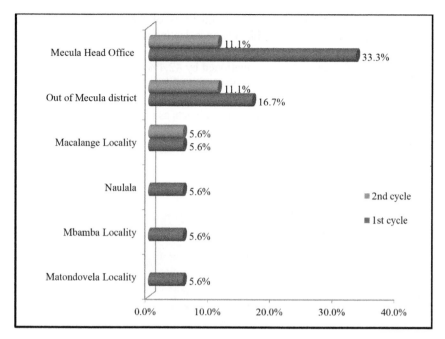

Figure 4: Origin and School Level
(*Source*: The authors, 2020)

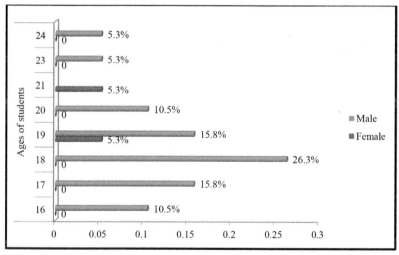

Figure 5: Age and Sex of Interviewed Learners
(*Source*: The authors, 2020)

3.2. Subjects at GSS Dealing with Environmental Aspects

The totality of (100%) of learners stated that they have heard about the environment precisely at school. According to their responses, environmental issues at GSS are taught in the subjects of Geography, Agro-Livestock and Biology.

Figure 6 illustrates the percentual frequencies of the indicated subjects by learners from both learning cycles, like the ones dealing with environmental aspects at GSS.

The subject of Geography was mentioned by 42.4% of learners as the one that deals more with environmental aspects. Next, Agro-Livestock and Biology with 21.2 and 15.2%, respectively, were mentioned.

Also, 21.2% of learners indicated other subjects as not applicable (N/A) at GSS (Figure 6).

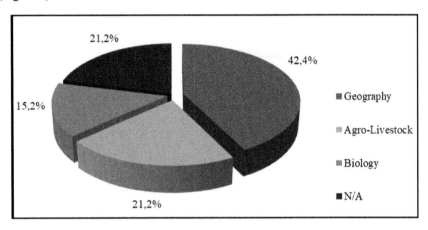

Figure 6: Subjects that Deal with Environment in Both Cycles at GSS
(*Source*: the authors, 2020)

3.3.   Learners' Environmental Knowledge

In relation to the concept of environment, 100% of the interviewed defined environment under biophysics perspective: L_1 MHO, L_2 MHO, L_3 MHO, L_4 MHO.

> ... Environment is everything surrounding us not made by Man (L_1 MHO, personal communication)
>
> ... Environment is everything surrounding us (L_7 MHO, personal communication)
>
> ... Environment is everything surrounding us (L_8 MHO, personal communication)
>
> ... Environment is everything surrounding us that we live on (L_9 MHO, personal communication)
>
> ... Environment is everything surrounding us (L_{10} MHO, personal communication)

The environmental components presented mostly by 52.6% of learners, independently of their cycle, are linked to biophysics aspects. These concepts and environmental components exclude the Man (Figure 7).

In relation to the importance of environment preservation, 100% of learners stated that it is crucial to preserve the environment. The importance of preservation is centred on the utilitarian subcategory (68.4% of the interviewed learners) (Figure 8).

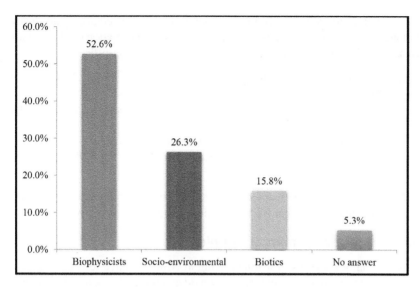

Figure 7: Concepts and Components of the Environment
(*Source*: The authors, 2020).

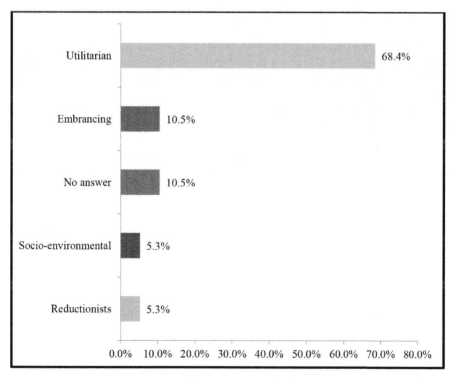

Figure 8: Importance of Environment Preservation
(*Source*: The authors, 2020).

In relation to concrete actions on environment preservation, all learners mentioned the need of not practising illegal acts to the conservation of biodiversities, such as poaching through traps or poison and uncontrolled bush-fire.

> ... we can preserve animals in specie and as in the jungle [sic] (L_9 MHO, personal communication)
> ... don't use poison, don't leave the poachers to be try [sic] (L_{10} MHO, personal communication)
> .... avoid poaching (L_{11} MHO, personal communication)
> ... avoid poaching and uncontrolled bush-fire (L_{13} MHO, personal communication).

Although underlying learner's awareness of environmental preservation measures, it is understood that it is only under a utilitarian perspective.

3.4. Socio-Environmental Conflicts

One of the major concerns of learners living in NSR is related to attacks to people and the destruction of agricultural crops by wild animals and lack of social inclusion in community queries.

> "More value is given to animals than to people ... When the animals kill us, damage our farms ... no measure is taken ... but when it happens the other way round ... we are taken to jail in Marrupa" (António, personal communication).

In general, the perceptions vary from 'much frequent' to 'frequent'. According to the table below, 27.8% of learners from Mecula head office confirmed the existence of frequent attacks by wild animals and 16.7% said that the attacks do not happen frequently.

Learners coming out of Mecula District also assume the variation between 'much frequent' to 'frequent' in a percentage of 22.2% of responses (Table 4).

Table 4: Perceptions About the Attacks by Wild Animals

Learners' provenience	*Perception of attacks by wild animals*			*Total*
	Much frequent	*Frequent*	*Non-frequent*	
Mecula Head Office	11.1%	16.7%	16.7%	44.4%
Macalange Locality	5.6%	5.6%	0.0%	11.1%
Matondovela Administrative Post	5.6%	0.0%	0.0%	5.6%
Mbamba Locality	5.6%	0.0%	0.0%	5.6%
Naulala Locality	5.6%	0.0%	0.0%	5.6%
Out of Mecula District	11.1%	11.1%	5.6%	27.8%
Total	44.4%	33.3%	22.2%	100.0%

Source: The Authors 2020

In relation to the participation in community queries, it was noticed that 58% of learners never participated in any community query, especially those from the localities inwards. No learner from Macalange, Matondovela, Mbamba and Naulala confirmed to have participated in any community query. Only learners from Mecula head office (29.4%) and those from out of the District (11.8%) confirmed to have participated once (Table 5).

4. Discussion

The present study focuses on learners' perception of the environmental world at 16th June General Secondary School in relation to their insertion in an area of biodiversity conservation, case of Niassa Special Reserve. Data discussion is sequenced according to the specific objectives of the research.

4.1. Learners' Socio-Demographic Characterisation

The majority of learners involved in the research (77.8%), no matter their provenience, attend the 1st Cycle at GSS. However, the majority of this group (50%) is constituted by learners from District head office and those from out of Mecula District. Although the great number of participants were male learners (89.4%), comparatively to female learners, the socio-demographic characteristics allowed to gather learners' conscious perceptions, given that a representative number (73.7%) have the minimum age of 18 years.

According to MINED (2007), school education is a basic right and a fundamental instrument for the development of human capital in Mozambique.

According to data about the distribution of school network obtained from District Services of Education, Youth and Technology in Mecula, in Localities out of Mecula head office, there are only Complete Primary School (CPS), that is learners from those regions only attend the Primary School Education (PSE) and after finishing it, they move to the District head office to continue with their studies at the unique General Secondary School. The difficulties in their permanence at the District head office; the insecurity to move long distances in the jungles and their insertion in the production of family economies are factors that interfere negatively in the continuation of learners' studies at GSS in those distant localities from Mecula head office.

4.2. Subjects at GSS Dealing with Environmental Aspects

Learners mentioned the subjects of Geography, Agro-Livestock and Biology as the ones that deal more with environmental aspects. In fact, the contents of these subjects contribute to the development of cognitive, affective and psychomotor competencies in learners at GSS in relation to environmental issues, with a focus on biodiversity conservation.

According to learners' responses, the subject of geography is at the top position in dealing with environmental issues.

The prescribed objectives in the CPGSS assert that geography envisages to "amplify and consolidate competencies (...) knowledge, abilities and correct attitudes

Table 5: Participation in Community Queries

Community queries	Learners' provenience						Total
	Mecula H. O.	Macalange L.	Matondovela A. P.	Mbamba L.	Naulala L.	Out of Mecula	
Yes	11.8%	11.8%	5.9%	5.9%	5.9%	17.6%	58.8%
No	29.4%	-	-	-	-	11.8%	41.2%
Total	41.2%	11.8%	5.9%	5.9%	5.9%	29.4%	100.0%

Source: The Authors, 2020

towards the nature and society. Contribute to the comprehension of interdependency relationship between the nature and human activity (...) and the changes in the environment" (INDE 2007).

This perception equates with the researches by Freia and Mahumane Junior (2014) that underline the importance of School Geography in the process of abilities formation and interventive capacities in learners' local reality.

Being an empirical subject, Agro-Livestock deals with theoretical and practical issues, with more focus on the last ones. Its objectives at GSS level are to "Contribute to the application of new production techniques in a family and in a community, as a way of increasing production and productivity, improve diet and assure food security".

This subject discusses new production techniques, aiming at socio-environmental sustainability.

The subject of biology in both cycles, according to the prescribed objectives in the CPGSS, deals specifically with issues of biodiversity conservation: "Contribute to the protection, conservation and sustainable use of natural resources specifically of biological diversity in our country, for the benefit of current and future societies. Valuate the importance of protection and conservation of the school environment through collective and individual responsibility."

Moser (2005) and INDE (2007) emphasise the importance of an integrated, cross-cutting and interdisciplinary approach to environmental contents. On the same platform, Freia and Mahumane Júnior (2014), in reference to the interdisciplinary approach of environmental issues, say that 'issues of defence and conservation of Nature (...) should never be left aside of any learning level...".

The contents of subjects at GSS, in general, when giving special treatment to the study of the environment in which the learner lives (Freia and Mahumane Júnior, 2014), will contribute to the development of environmental competencies in learners, with effects on their families. Teachers can create opportunities for a permanent alliance of theory and practice, be it in a local curriculum approach or in interdisciplinary excursions (Nhachungue 2014, Nhachungue and Rafael 2018, Langa 2014) with the aim to develop knowledge, attitudes, and sustainable actions to the environment.

The socio-environmental approach (Tamaio 2000) or a wide-range approach (Malafaia and Rodrigues 2009) appears to be important in subjects at GSS in Mozambique.

4.3. Learners' Environmental Knowledge

Learners demonstrate a more varied environmental knowledge to biophysics aspects, from concept and components to manners of environment preservation. This conception is analysed under the following dichotomic perspectives: biophysics also observed by Mattos and Andrade (2018) in the research results about the environmental perception of a group of learners and, other which is determinist/ exclusive provider of resources for the survival of Man, also noticed with evidence in Malafaia and Rodrigues (2009) research results.

(i) The biophysics perspective excludes the Man in the context of the environment. It is our understanding that the expression 'everything surrounding us' reproduces a concept of 'human centrality', that is the Man is at the centre (although it is understood that He is excluded) and surrounded by something named environment. In addition, Tamaio (2000) research attributes the generalising character to the definitions of 'everything surrounding is nature', presented by learners involved in the research. There is a lack of harmony in understanding this definition because, taking into account our perception, the individual at the centre of something is part of it. But in learners' perception, the Man surrounded by something is not mentioned as one of the elements of the environment set (see Figure 7). This definition isolates the Man from the eco-systemic view of his permanent interaction with the environment. Therefore, he is seen as a static element at the centre.

Malafaia and Rodrigues (2009) research that involved a group of teens and adults of a school concluded that the majority (81.8%) of learners presented concepts of the environment with a reductionist character, that is with focus on physico-natural aspects, excluding the Man. Therefore, no matter the used subcategories, these results are in agreement with our findings in the learners' group of Mecula, which takes us to infer that it is a generalised conception.

Learners' environmental conceptions are still grounded on the definition of 'Nature' turned to the concept of 'everything that surrounds us, except the Man'. Gurski et al. (n/d) consider that the term 'Nature' was substituted for the term 'Environment'. We consider that this vision came to have a 'new and wide-reaching approach' in which the Man is integrated with the biosphere set, taking the term 'Environment', in the second half of the past century, at the conference about the biosphere, organised by UNESCO in Paris, in 1986 (Gurski et al. n/d).

Mozambican Environment Legislation as:

> "The Mean in which the Man and other Living beings live and interact among them and with the Environment itself. It includes the air, light, land and water; the ecosystems, biodiversity and its ecological relationships; all organic and inorganic matter; all socio-cultural and economic conditions that affect the lives of communities" (Republic of Mozambique 1997).

This definition is in agreement with the approaches at the United Nation Conference about Man and the Environment (Stockholm Conference) and with the United Nation Convention about Biological Diversity. Taking this definition as a reference, the Man and his socio-cultural life experiences is an active being in the environment.

(ii) The determinist perspective, or the one stating that the environment is the exclusive provider of Man's survival resources, defends that the Man is only at the centre and without carrying out any interactive action with the environment. Because of that, there is a generalised thought in the researched area, according to which, 'fetch resources from nature for our survival without which, our life is impossible'. This conception of 'fetching resources' leads to self-exclusion of individuals in the process of biodiversity conservation, as well as aggravates the socio-environmental conflicts in Niassa Special Reserve.

4.4. Socio-Environmental Conflicts

In relation to perceptions of conflicts, wild animals' attacks (Jorge, 2012) and exclusions in community queries' programs in Niassa Special Reserve were emphasised by learners in their responses. The conclusions reached in the relationship between 'learners' provenience' and 'the perception of attacks by wild animals' indicate that there is a direct relationship between learners' provenience and the perception of attacks by wild animals. The more we move to regions out of District head office, the less are the monitoring activities to damages and the chasing away of animals, and this is the reason why a great number of learners from these regions (Macalange, Matondovela, Mbamaba and Naulala) understand the level of attacks as 'much frequent' in relation to those from the District head office.

Lack of involvement of communities away from the District head office takes us to the perception that these programs are focused only on the region where the political and administrative powers are more influential, that is Mecula District head office.

The community queries are one of the mechanisms of participatory involvement of communities in biodiversity conservation actions (Kotsakis 2009), in order to avoid the imposition of global conservationist policies locally. The queries underline the relevance of collecting local community's sensibilities (debates and discussions) about the current state and future perspectives of biodiversity conservation with the aim to trace adequate policies to local communities.

Moeller (2010) advocates that the sharing of knowledge through community queries and the use of participatory methods increase often the group cohesion and can instil in people the feeling of co-ownership of the group process. Beyond the public policies, there are stimulating mechanisms for local socio-economic development through micro-credit, scholarships and other forms (BirdLife International 2010). This view is shared by Salvemini and Remple (2014), who consider that the achievement of success in biodiversity conservation demands the implementation of mechanisms that envisage the engagement and support local communities in various spheres of social, economic, cultural and environmental life.

If the criteria of participation in community queries on community projects is adulthood, then the learners from the school involved in the research have the requirements for such purpose, no matter their provenience.

5. Conclusions

The present study aimed at evaluating learners' perceptions about the conservation of biodiversity in Niassa Special Reserve. The same carried out a theoretical analysis about men-environment perceptions with more focus on biodiversity conservation, having also analysed pedagogical-methodological issues which helped in the understanding of learners' perception at 16th June General Secondary School as an institution inserted in a conservation area.

In relation to the characteristics of socio-demographic variables, our conclusions are centred on 'age' and 'provenience' variables given that they end up influencing the understanding of others.

Concerning age, it has been concluded that the researched learners' adulthood contributed to a conscious collection of their environmental perceptions related to biodiversity conservation in Niassa Special Reserve. In relation to the reduced number of learners from localities distant from the District Head Office (provenience variable), it has been concluded that the difficulties of their maintenance at the District Head Office, insecurity in the mobility through the jungles, their early involvement in the production of subsistence domestic economies, the long distances between their localities and the perimeter of the Secondary School are the factors hindering learners' continuation of their studies at GSS, as well as their delay in entering such level, mainly those from out of Mecula District Head Office.

From environmental educative potentials offered by different school subjects at GSS, it has been concluded that beyond the subjects of Geography, Agro-Livestock and Biology, there is a space for environmental contents to be dealt with in forms of local curriculum, cross-cutting topics, groups of interest, group studies, environmental clubs under a cross-cutting and interdisciplinary perspective and feel like the environmental agent that can help to solve local socio-environmental conflicts.

In relation to learners' knowledge about the environment, the study concluded that there is a need to improve the approach to the concept of environment in classes, not only in the subjects mentioned by learners but also in other curricular subjects at GSS.

The environmental conception should consider the Man as an active being and always in permanent interaction with the so-called 'nature' in the production of his own space and other survival resources. With this conception, the school will be developing cognitive and effective psychomotor competencies that will influence the construction of 'new environmental conceptions' turned to biodiversity conservation in Niassa Special Reserve.

Concerned with socio-environmental conflicts, it has been concluded that learners have a perception of phobias in relation to wild animals that terrorise the communities.

The need to ensure cognitive and effective psychomotor evolution based on the 'sense of belonging to the environment' has been found to be important in the creation of environmental awareness turned to biodiversity conservation in learners. Such awareness revealed itself to be necessary through the adoption of participatory mechanisms in community queries, extra-school activities such as lectures and community debates about topics in socio-environmental daily life.

Acknowledgements

We thank first to God and everyone who supported the field surveys: Mecula's educational entities, students and teachers.

We extend our thanks to MSc. Arlindo Malize for the translation of the text from Portuguese to English.

References

Alexandre, J.W.C., D.F. Andrade, A.P. Vasconcelos, A.M.S. Araujo e M.J. Batista. 2003. Análise do número de categorias da escala de Likert aplicada à gestão pela qualidade total através da teoria da resposta ao item. XXIII Encontro Nacional de Eng. de produção, ENEGEP-ABEPRO.

Almeida, G.S., P.O. Sousa, A.M. Souza, C.R. Souza and K.A. Oliveira. 2012. Percepção das populações do interior e do entorno do Parque Nacional Serra do Divisor-Acre sobre a caça cinergética e seus efeitos sobre a abundância dos recursos faunísticos. Enciclopédia Biosfera-Centro Científico Conhecer, Goiânia. 8(15): 1902.

Alves, M.C.L. e M.A. Bassani. s/d. A psicologia ambiental como área de investigação da inter-relação pessoa-ambiente.

Bardin, L. 1977. Análise de Conteúdo. São Paulo.

Bassani, M.A., J.M.G. Ferraz e M.A. Silveira. 2007. Percepção ambiental e agroecologia: considerações metodológicas em Psicologia Ambiental. Resumos do II Congresso Brasileiro de Agroecologia, Revista Brasileira Agroecologia. 2: 1.

Begg, C.M. and K.S. Begg. 2009. Mitigation of negative human impacts on large carnivore populations: Niassa National Reserve. Mozambique – Annual progresso report.

BirdLife. 2010. Biodiversity Conservation and Local Communities.

Booth, V.R. and K.M. Dunham. 2014. Elephant poaching in Niassa Reserve. Mozambique: population impact revealed by combined survey trends for live elephants and carcasses. Fauna and Flora international. Oryx. 1-10.

Branch, W.R., M.O. Rodel e J. Marais. 2005. Herpetological survey of the Niassa Game Reserve, northern Mozambique – Part I: Reptiles. 2005.

Craig, G.C. 2009. Aerial survey of wildlife in the Niassa Reserve and adjacent areas. Mozambique.

Craig, G.C. 2012. Aerial survey of wildlife in the Niassa Game Reserve. Mozambique.

Castiano, J.P. 2006. Currículo Local do Ensino Básico em Moçambique: finalmente os saberes locais vão entrar oficialmente na Escola? In: Revista Síntese, n.º 6, Ano II, FCS, Maputo.

Chardonnet, B. 2019. Africa is changing: should its protected areas evolve? Reconfiguring the protected areas in Africa.

Chizzotti, A. 2000. Pesquisa em Ciências Humanas e Sociais. 4.ed., Cortez, Brasil.

Chierrito-Arruda, Yaegashi, Paccola e Grossi-Milani. 2018. Percepção Ambiental e Afectividade: Vivências em uma horta comunitária. In: Ambiente and Sociedade, Vol. 1, São Paulo.

D´Amorim, M.A. 1997. Estereótipos de género e atitudes acerca da sexualidade em estudos sobre jovens Brasileiros. Temas em psicologia.

Fogaça, T.K. e L. Limberger. 2014. Percepção ambiental e climática: estudo de caso em colégios públicos do meio urbano e rural de Toledo-PR. Revista do Departamento de Geografia-USP. 28: 134-156.

Freia, A.C.B., A. Mahumane Júnior e B.J. Bernardo. A contribuição da Geografia na solução dos problemas locais: Que estratégias? In: Duarte, S.M. e Raimundo, I.M. (Orgs.). 2014. Geografia em Moçambique: Passado, Presente e Futuro. Educar, Maputo.

Gifford, R. 2014. Environmental Psychology matters.

Gifford, R., L. Steg and J.P. Reser. 2011. Environmental Psychology. Blackwell Publishing Ltd.

Gurski, B., R. Gonzaga and P. Tendolini. s/d. Conferência de Estocolmo: um marco na questão ambiental.

Hardin, G. 1968. The Tragedy of the Commons. Science, New Series. 162: 1243-1248.

INDE. 2003. Plano Curricular do Ensino Básico: Objectivos, Política, Estrutura, Plano de Estudos e Estratégias de implementação. UNESCO.

INDE. 2007. Plano Curricular do Ensino Secundário Geral (PCESG). Documento Orientador: Objectivos, Política, Estrutura, Plano de Estudos e Estratégia de Implementação. Imprensa Universitária UEM.

Instituto Nacional de Estatística [INE]. 2012. Estatísticas do Distrito Mecula. Maputo.

Instituto Nacional de Estatística [INE]. 2019. IV Recenseamento Geral da População e habitação 2017. Maputo.

Jorge, A.A. 2012. The sustainability of leopard *Panthera pardus* sport hunting in Niassa National Reserve, Mozambique. Thesis submitted in fulfillment of the academic requirements for the degree of Master of Science, School of Life Sciences. University of KwaZulu-Natal, South Africa.

Kotsakis, A. 2009. Community Participation in Biodiversity Conservation: Emerging Localities of Tension. Published in Amanda Perry Kessaris (ed.). Law in the Pursuit of Development: Principles into Practice.

Langa, A.A. 2014. A integração de conteúdos de Geografia local na disciplina de Ciências Sociais. Caso de Escolas do Ensino Básico de Namaacha. *In*: Revista de Pós-Graduação-Educação, Pesquisa e Sociedade- Colecção cadernos de pesquisa Geografia, Maputo.

Lucena, M.M.A. e E.M.X. Freire. 2011. Percepção ambiental sobre uma Reserva particular do Património Natural (RPPN), pela comunidade rural do entorno, Semiárido brasileiro. Educação ambiental em acção.

Malafaia, G. e A.S.L. Rodrigues. 2009. Percepção ambiental de jovens e adultos de uma escola municipal de ensino fundamental. Revista Brasileira de Geociências, Porto-Alegre. 7(3): 266-274.

Mattos, D.M. e L.N.P.S. Andrade. 2018. Percepção ambiental dos alunos do 8° ano da Escola Estadual Coronel Antônio Paes de Barros no Município de Colider/MT. Revista Brasileira de Educação em Geografia. Campinas. 8(16): 167-191.

Moeller, N.I. 2010. The protection of traditional knowledge in the Ecuadorian Amazon: A critical ethnography of capital expansion. Submitted for the degree of Doctor of Philosophy - Department of Sociology, Lancaster.

Moser, G. 2005. Psicologia Ambiental e estudos pessoa-ambiente: Que tipo de colaboração multidisciplinar? Universidade Paris V. Psicologia USP. 16(1/2): 131-140.

Mozzato, A.R. e D. Grzybovski. 2011. Análise de Conteúdo como técnica de análise de dados qualitativos no campo da Administração: potencia e desafios. RAC Curitiba. 15(4): 731-747.

Muhale, I.J. 2020. O papel da produção científica na institucionalização de políticas ambientais em Moçambique 1980-2014. Campinas.

Nebbia, T. UNEP. s/d. Integração entre o meio ambiente e o desenvolvimento: 1972-2002.

Nhachungue, F.G. 2014. O currículo local na cidade de Lichinga: estudo comparativo entre as Escolas Primárias Completas "A" e "B" (2007-2012). *In*: Revista de Pós-Graduação-Educação, Pesquisa e Sociedade - Colecção cadernos de pesquisa Geografia, Maputo.

Nhachungue, F.G. e R.S. Rafael. 2018. Excursão Geográfica: potencialidades da prática educativa. Revista Brasileira de Educação em Geografia. Campinas. 8(16): 348-359.

Nhachungue, F.G., R.L. Carmo e S.R.C. Seixas. 2019. Reflexões sobre a protecção da biodiversidade em áreas habitadas pelo Homem- Modelos e teorias demográficas: o caso da Reserva Nacional de Niassa em Moçambique. In: Textos NEPO- Explosão demográfica 50 anos depois de *"The Population Bomb"*, Campinas.

Nhachungue, F.G., S.R.C. Seixas e B.O. Bandeira. 2019. A dinâmica das áreas de conservação do Brasil e Moçambique: Estudo comparativo da Serra do Japi, Mata Santa Genebra e Reserva Nacional do Niassa. Momentum, Atibaia. 1(17): 1-20.

Pilati, R. e J.B. Porto. s/d. Apostila para tratamento de dados via SPSS.

Santos, B.R., N.D. Santos, A.C.M. Ayres, L.F.M. Dorvillé, L.F. Rodrigues e R.C.B. Travassos.

2008. Representações sociais de meio ambiente e qualidade de vida no ensino médio. Revista vozes em diálogo (CEH/UERJ).

Salvemini, D. and N. Remple. 2014. Community-based approaches to landscape management. *In*: Chavez-Tafur, Jorge and Roderick J. Zagt (eds.). Towards productive landscapes. Tropenbos international, ETFRN NEWS 56, The Netherlands. Xx+224.

Seixas, S.R.C., J.L.M. Höeffel e M. Bianchi. 2010. Qualidade de Vida, Ambiente e Subjetividade na APA Cantareira. *In*: Höeffel, J.L.M., Fadini, A.A.B. e Seixas, S.R.C. (Orgs.). 2010. Sustentabilidade, Qualidade de Vida e Identidade Local: Olhares sobre as APAs Cantareira (SP) e Fernão Dias (MG). RiMa, São Carlos.

SRN. 2006. Plano de Maneio da Reserva Nacional do Niassa (2007-2012). s/l.

Tamaio, I. 2000. A mediação do professor na construção do conceito de Natureza: uma experiência de educação ambiental na Serra da Cantareira e Favela do Flamengo- São Paulo/SP. Dissertação de Mestrado em Educação aplicada às Geociências. Instituto de Geociências-Universidade Estadual de Campinas, São Paulo.

Triviños, A. N.S. 1987. Introdução á Pesquisa em Ciências Sociais - a Pesquisa qualitativa em Educação: o Positivismo, a Fenomenologia, o Marxismo. Atlas S. A., São Paulo.

Vieira, P.F., F. Berkes e C.S. Seixas. 2005. Gestão integrada e participativa de recursos naturais: conceitos, métodos e experiências. APED-Secco, Florianópolis.

Legislation

Moçambique. 1954. Boletim Oficial, III Série Número 13, Portaria n.º 10.578- Extingue a "Coutada do Niassa" estabelecida no aviso da Comissão de caça publicado no B.O. n.º 13, 3ª Série de 27 de Março último e cria a "Reserva de Caça do Niassa".

República de Moçambique. 1994. Boletim da República I Série número 34: Resolução n.2/94 de 24 de Agosto- ratifica a Convenção das Nações Unidas sobre diversidade biológica.

_____. Lei n.º 20/97 de 1 de Outubro. Lei do Ambiente.

_____. 1999. Boletim da República I Série número 45: Decreto n.º 81/99 de 16 de Novembro-altera os limites da Reserva Nacional do Niassa e revoga o Diploma Legislativo n.º 2884 de 24 de Maio de 1969.

Part III
Environmental Risks and Sustainability

Mining and Sustainability

Maria José Mesquita[1]*, Rosana Icassatti Corazza[1], Maria Cristina O. Souza[1], Guilherme Nascimento Gomes[1], Isabela Noronha[2] and Dione Macedo[3]

[1] Institute of Geosciences, University of Campinas – UNICAMP, Brazil
[2] Centre for Environmental Studies and Research, University of Campinas – UNICAMP, Brazil
[3] Ministry of Mines and Energy – MME, Brazil

1. Introduction

From lithic to lithium, ancient and modern societies are geographically built up in the surroundings of geological resources.

Mineral-based products that enable everyday life cannot be taken for granted, as the modern economy is inextricably linked to infrastructure or artefacts dependent on mining and they mediate contemporary lifestyles. Mining activities are not limited to the extraction of metallic minerals; the so-called industrial minerals are essential today and for the future, from 'social minerals' such as sand and clay to those used in high-tech industries, such as cobalt, nickel and lithium.

The critical character of mining should not, however, obscure the need for it to be shaped in an economically, socially, and environmentally sustainable way. To begin with, mining is an earthmoving activity that alters topography and local dynamics, accumulates large volumes of waste, interferes with ecological processes, and requires a constant influx of water and energy (Bridge 2004). It is therefore imperative to think about how mining is performed and at what rates and scales.

According to the International Geosphere-Biosphere Programme (IGBP), the demand for and extraction of natural resources, including minerals, has increased exponentially since the post-World War II period, driven by population and industrial and technological goods production growth. This process has been called 'The Great Acceleration', as it has boosted the demand for energy, the emission of pollutants and the biodiversity loss (Steffen et al. 2015). The authors argue that the rapid increase in waste and pollutants puts pressure on the planet's ecosystems, taking them beyond their limits of resilience and bringing about profound environmental changes, sometimes irreversible with catastrophic consequences for humankind itself. So, in the proposition of Anthropocene as a new geological period, anthropogenic actions

*Corresponding author: majo@unicamp.br

and activities become an important geological agent, raising new questions about the role of humanity in the future of the planet (Picanço and Mesquita 2018).

Attention is drawn to the fact that anthropic activities are not homogeneous and, despite the risks that fall upon all living beings on the planet, not all forms of social organisations of human life are to be blamed (Svampa 2019). The concern about achieving more sustainable practices has been increasing every year since the proposition of the Eco-development notion by Maurice Strong, the former Secretary-General of The United Nations Conference on the Human Environment (also known as the Stockholm Conference), in 1972, when UN created its Environmental Programme (UNEP) with the mission of supporting developing countries in the implementation of environmentally sound policies and practices.

It initiated a long and yet unconcluded discussion about the pressures of economic growth and the depletion of crucial natural resources, as well as the role of technological development in the control of the undesired backlashes of the advancement of societies (Brüseke 1995). Herrera et al. (1976) discussed the possibilities of societies in dealing with scarce resources and environmental crises by admitting a greater degree of freedom to forms of social organisation based on citizenship participation and knowledge-based decision-making. The approach to basic needs proposed in this work seems to have influenced the vision of 'needs' set out in the 1987 Report of the United Nations (UN) World Commission on Environment and Development, which adopted the concept of 'sustainable development', meaning the kind of "development that meets the needs of the present without compromising the ability of future generations to meet their own needs" (Brundtland 1987).

The UN Development Agenda incorporated Amartya Sen's capacity approach and Mahbub Ul-Aq's human development and implemented this vision within the United Nations Development Programme (UNDP) in the early 1990s. In 2000, UNDP went further, institutionalising and helping nations implement the Millennium Development Goals, which gave ground to the Sustainable Development Goals (SDGs) as part of Agenda 2030 in 2015. The SDGs are a compendium of 17 goals that countries should follow and plan to achieve by 2030 with the universal objective of "end poverty, protect the planet and ensure that all people enjoy peace and prosperity" (UN 2015).

Considering the definition of the Brundtland Report, mining, as it means the extraction of mineral and non-renewable resources, could not be properly sustainable. At the same time, mining enterprises involve many processes other than the mineral extraction itself that can be potentially harmful to the environment and society. Assuring the safety of tailing dams, water quality and availability, controlling air pollution, and especially considering the needs and sovereignty of the people who are affected by these processes over their territory are among the issues to be faced. Those are issues that require advances to render mining as sustainable as possible.

In line with those considerations, in 2017, UNDP, in cooperation with Columbia University and the World Economic Forum, introduced the 'Atlas for SDGs to the mining sector' (WEF 2016). That can be considered a further step, including SDGs as guiding principles to help mining enterprises to promote changes and become more sustainable.

It is also appropriate to highlight some criticism of the UN's approaches to sustainable development. Several authors have argued that threats to nature and people originate not only from population growth but most notably from the development model conventionally sought by nations (Svampa 2019, Acosta 2016, Lander 2016). The idea that development based on economic growth could be the only path to the well-being of society conflicts with the physical limits of the planet, be it the availability of resources or its entropic capacity, that is its capacity to support mineral exploitation by the products of technological development. In this regard, if there will be no questioning about the wasteful, conventional development pattern, sustainable development will remain an oxymoron: either it is development, or it is sustainable.

Even with the discourse on sustainable development, the needs and consumption of minerals and rock have tended to increase intensively and steadily as the world's population grows. Just like the most developed countries, developing countries have also industrialised within the same trend, reinforcing the "historical human tendency to translate economic success and well-being into material consumption" (Guilbert and Park 2007). It remains mandatory to earnestly question the relation between well-being and consumption and conventional wisdom about development.

The present chapter aims to discuss the challenges and possibilities that arise when one brings together mining and sustainability, especially in the light of SDGs. Mining is as ancient as humankind and is not going to disappear in the times to come. This chapter points out urgent issues to face if mining is required to be, in some way, more sustainable. Meeting the increasing demand for mining products, or mitigating this demand, while dealing with its alarming social and environmental impacts challenges researchers, industry, government, and society at large to pave a safer road for mining to achieve sustainability. To assist in facing this challenge, the contributions of this chapter is organised in four sections: the role of mining in the societies through time; a bibliometric analysis on mining and sustainability; mining and the SDG; and the conclusions. The conclusions summarise the previous discussions and present some recommendations to develop further this critical research, policy, and entrepreneurial field.

2. Mining and the Role of Minerals in Society

Since its dawn, humankind has been relying on one or another type of mineral good. With modern industrialisation, countries strive to obtain mineral raw materials to spur development, reach autonomous growth and gain international competitiveness. The more economically developed society is, the more it uses mineral inputs with ever more technological and sophisticated uses.

Imagine one's own house, how much it uses bricks, cement, hardware, glass, and plastics, all inputs from the mineral origin. Also, agriculture has become increasingly dependent on mineral fertilisers; in the years 2018-2019, Brazil has imported 44% of potassium, sulphur, and phosphatic rock for agriculture (IBRAM 2019). Automobiles, buses and aircraft, in addition to being mainly powered with fossil fuels, are now built with lighter and more sophisticated metal alloys.

Since 1950, the world population has grown by factor 2.7 while the global material consumption by 3.7, and from 1950 to 2010, the global average of material used by people increased from 5.0 to 10.3 tons per capita a year (Schaffartzik et al. 2014). Therefore, as world population growth accelerates along with a general but uneven improvement in living standards, one observes an enormous boost on the global demand for mineral resources of all kinds.

Furthermore, sustainability transitions to a low carbon economy are going to require electric and electronic devices and energy infrastructure which are intensive in the use of mineral resources (Smil 2010, Sovacool et al. 2020). Besides, the geographic concentration of mineral deposits in a few countries is expected to pose further sustainability trade-offs (Månberger and Stenqvist 2018). Hence, as one expects global continued growth of mineral resources demand in the years to come, it would be inescapable to unveil the sustainability challenges the mining industry and society are going to face.

2.1. A Concise History of Mining

Through the Paleolithic period, human groups used lithic tools for hunting and fishing. Later on, in the Mesolithic, pigments were employed in funerary rituals. The oldest known mine is a hematite's one, over 40,000 years old, and located on Bomvu Ridge (Swaziland), in southern Africa (Mutunhu 1981). The Neolithic period was marked by the transition from predation to agriculture, and outstanding achievements, such as the polished stone technology, ceramics and metallurgy, and the beginning of livestock and agriculture. Salt became an important commodity, including as a currency (Shepherd 1993). Native copper, probably the first used metal, for its ductility and was easier to shape than the chipped stones.

The discovery of new techniques for the obtention of metallic alloys has enabled the advent of bronze, around 3000 B.C. in today's Iran (Matthews and Fazeli 2004). Obtained from a mixture of tin oxide (cassiterite) with native copper, bronze is more rigid and yet more suitable to shape than copper. Different cultures were actively involved in copper and tin extraction, and bronze production and commercial networks flourished in the Mediterranean and Near East (Shepherd 1993).

Developed by the Hittites around 1385 B.C. in the Anatolian Peninsula, the iron technology, for its turn, quickly spread to other regions of the known world (Sommer 2007). In the Iron Age, the use of metals as currency came up around 700 B.C., in Lidia kingdom, and initially, the electrum, a natural alloy of gold and silver, was used for minting the first known coins, later replaced by gold and silver ones (Wood et al. 2019). The wealth of Athens included numerous silver occurrences, the most famous of which were the Laurion mines (Hopper 1968). In Western Europe, the Romans dominated the extraction and production of metals and alloys used in plumbing and the damp-proofing of houses and buildings, as well as for the manufacturing of domestic utensils such as bowls and dining appliances (Shepherd 1993).

The low middle age European commercial revival increased the demand for metals. There were, at that time, problems involving excavation, ventilation, and lighting problems emerging from the reliance on lower ore grade (Gregory 1984). New technologies, new machines, water tanks and more efficient smelting furnaces have been developed to cope with them. The most influential book in the history of mining, De Re Metallica, written by Agricola in 1556, accurately describes the bulk of these techniques (Agricola 1950).

During Mercantilism, the colonial usurpation of the Americas by Iberian nations was the cornerstone of the colonial empire's wealth, sustained by the extraction of precious metals in mines in Mexico and Potosi, Peru (currently Bolivia). The amalgamation method, or the 'Metodo de Patio', developed by Bartolomé de Medina in Pachuca, Mexico, allowed a proverbial silver production of Spanish America in the modern era (Martos 2006). Portuguese America, for its part, with its rich alluvial deposits in Minas Gerais, contributed with about two-thirds of world gold production in the eighteenth century (Boxer 1969). The mineral production had in both Portuguese and Spanish Americas substantial social costs. Millions of indigenous people were subjected to slavery and compulsory labour practices, such as the Encomienda and the Mita (Garner 1988). The black slave trade intensified through the XVIII century when three million people were captured in Africa to work as slaves in the gold Brazilian mines and plantations (Alencastro 2000).

The industrial revolution, partly financed by American gold and silver, was established thanks to new iron metallurgy techniques based on mineral coal energy (Clark and Jacks 2007). Coal mining has raised numerous occupational safety hazards in underground galleries, such as inadequate drainage, ventilation and lighting, coal lift and explosions caused by gas leakage, predominantly methane (Gregory 1984). Slavery, female, and child labour were usual in English coal mining during the British industrial revolution, and, according to John (2013), the country attended until 1841 to ban female and child labour. Clark and Jacks (2007) point intensive mechanisation as a drive towards slavery overcoming in the mining industry in the nineteenth century. It is noteworthy that iron and coal miners played a political role in working classes' manifestations during the nineteenth and twentieth centuries. Technological advances in the nineteenth century, such as the Cornwall water pump, the discovery of nitro-glycerine and the invention of dynamite, and the development of pneumatic drilling tools, first used in 1900, increased mining productivity. The Davy's cold flame lamp started to be used in the coal mines in the nineteenth century, and electric power began to be implemented in 1920 (Betrán 2005). At that time, industrialisation gave rise to the use of various inputs such as, among others, phosphate, borate, potassium salts, and sulphur.

Through the mining history, one has seen technology development allowing the augmentation in scale, intensity and speed of mineral extraction and use. Steffen et al. (2005) stress the great acceleration of society-led changes which poses a progressive strain on natural systems since the second post-war period since 1945. An example of this is the global fertiliser (phosphates, nitrogenous, potash) and other minerals production and use that increased ten times from 1950 to 2000, putting biogeochemical cycles at or beyond their resilience limits. As far as one may see, the pressures underlying these strains are going to continue, given a global acceleration of total and per capita demands for Earth's resources of all kinds, unless one urgently changes our societies' route towards a proper sustainable paradigm of organising the socioeconomic processes.

2.2. Key Features of the Mining Industry

The mining industry is formed by the business segments of exploration, mining and mineral transformation. Mineral production's critical point is related to the infrastructure viability; mining does not choose the place where it will be implemented, it has locational rigidity, that is mineral resources only occur where geological processes allow them to.

There are many classifications for mining activities in the mining industry. According to Moon (2006), the mining industry can be grouped according to their production, mainly into major and junior. The junior companies are smaller and flexible than majors, they could be involved only in the exploration of new deposits and negotiate the discovery with major companies, or they can share the discovery and production.

Artisanal and Small-Scale Mining (ASM) does not appear within the hierarchical and regulated global mining system. According to IGF (2017), "ASM is a complex and diversified sector across much of the developing world. It ranges from informal individual miners seeking a subsistence livelihood to small-scale formal commercial mining entities responsibly producing minerals". Many legal mining cooperatives are also gaining space. In the years 2004, 15 million individuals were working in the gold ASGM (Veiga and Baker 2004), and it was responsible for 20 to 30% of the total global production of gold in 2004 (Swain et al. 2007). The efficient recovery of gold and the drastic reduction of mercury use depends on the technological sophistication of processes, which is at a higher level in South America, intermediate in Asia and Central America, and lower in Africa (Seccatore et al. 2014).

2.3. Geological Resources for Modern Life

Mining is part of a production chain made up of mineral-based industries. This economic activity produces and disseminates a multitude of products that are related to the quality of life of the populations (IBRAM 2019).

The geological resources are broadly subdivided into metallic minerals, industrial rocks and minerals (or simply industrial minerals), energetic resources, and water (potable and aquifers). Any metallic object, from a simple pot to the most complicated scientific instrument, is made from a variety of mineral products, including the electrical energy transmitted cables, car, refrigerator, brick, tile or the coating of metallurgical furnaces. Even modern food production relies on mineral fertilisers and soil conditioners (IBRAM 2019).

Ore is the mineral goods with economic value. The metallic ore is composed of minerals made by metallic elements as ferrous (iron, manganese, chromium, molybdenum, nickel, cobalt, tungsten, vanadium), and non-ferrous divided into base-metals (copper, lead, zinc, tin), light metals (aluminium, magnesium, titanium, beryllium), and rare-metals (lithium, beryllium, rare earth elements as caesium, indium, and platinum group metals as platinum, rhodium, iridium) (Manning 1995).

Industrial minerals are natural substances applied to products and processes such as raw materials, inputs, and additives, in the most diverse industrial segments. They are essential in the manufacture of products demanded by the post-industrial and urban society, like plastics, optical fibre, electronic components, pharmaceutical, and space industry (Manning 1995). Another group of industrial minerals is the so-called social minerals since they have traditional uses in the implantation of urban infrastructures such as aggregates (gravel and sand), lime, cement, paints, glass, ceramics, and paper.

The urban infrastructure development could be measured by the consumption of social minerals. In Brazil, from 2007 to 2014 gravel and sand production has risen 170%, and cement production 150% as a result of the Brazilian Growth Acceleration Programme (PAC), a public policy to develop infrastructure, sanitation, housing, transport, and other sectors in the country, according to the Mineral Summary 2007-2014 (ANM official site). Lebdioui (2020) also pointed out the importance of linking these social or low-value minerals to the beneficiation chain to input other economic sectors in Africa and elsewhere. In 2007, when the Algerian Mediterranean Float Glass (MFG) was created, all its domestic glass demand was imported. Nowadays, MFG produces all the float glass demand (30% of the production) and exports 70%. Most of the glass raw materials (silica sand, dolomite, soda ash, feldspar) were first imported, but the industry has progressively developed their silica sand and dolomite mining and production.

The global mining industry is challenged to supply the ore materials for the next decades, significantly to mitigate climate change. According to Månberger and Stenqvist (2018) projections, from 2015 to 2060, sees increased demand for materials in the electric vehicle batteries, which is 87,000%, 1,000% in the wind power devices, and 3,000% in solar and photovoltaic cells production. Projections by Sovacool et al. (2017) for global crucial ore material demand from 2015 to 2050 show a spectacular growth, as are the cases of lithium (956%), cobalt (585%), indium (241%), vanadium (173%), nickel (108%), and silver (60%). Therefore, any national and international policies must account for where, how and with what intensity those strategic minerals are going to be mined.

The World Bank Group also reported its projections in the publication called Minerals for Climate Action. The Mineral Intensity of the Clean Energy Transition, according to which, meeting the growing demand for clean energy technologies are going to require increased production of minerals, such as graphite, lithium and cobalt, estimated in nearly 500% by 2050. It also estimates that over 3 billion tons of minerals and metals are going to be needed to deploy wind, solar and geothermal power, as well as energy storage, required to limit temperature rise below 2°C. In other words, the clean energy transition will be significantly mineral intensive. Furthermore, while the growing demand for minerals and metals provides economic opportunities for resource-rich developing countries and private sector entities alike, significant challenges will likely emerge if the climate-driven clean energy transition is not managed responsibly and sustainably (World Bank 2020).

Another key piece to fit the high-technology demand puzzle is the fact that most of these metals, companion metals, are recovered only as a by-product, in other words, they are rare metals (less than 0.1%) inside the structure of ore minerals and rarely formed viable deposits (Nassar et al. 2015). Examples include gallium and indium, core components in the electronic and solar energy applications, the cobalt used in high-temperature alloys, or several rare-earth elements employed in offshore wind technology (Nassar et al. 2015).

Once no resource is renewable in the human time-scale, meeting the demand of all referred materials requires the sustainable extraction and processing of minerals urgently throughout the mining chain to deal with social (especially for the vulnerable communities), economic and environmental trade-offs while searching for synergies. Growing demand for minerals and metals provides opportunities but also significant challenges that will emerge if the transition is not managed within a socially responsible, fair and sustainable way. United Nations' Sustainable Development Goals may represent a way to build upon these opportunities and overcome challenges. In the next section, one proposes to briefly review the academic literature on the subject.

3. A Brief Bibliometric Account of Mining and Sustainability as an Emerging Interdisciplinary Field

As demand for minerals rapidly grows worldwide and the global and local concerns about its impacts and prospects increases, attention on these issues have become an up-to-date matter to States, corporations and society at large. Research advances and a massive literature corpus emerges. Since this is a methodological approach that applies when a new research topic attracts the attention of academic societies giving rise to a huge body of literature, a bibliometric study offers the opportunity to map these efforts on the revised subject.

This section provides a brief bibliometric study that may prompt further reviewing exercises to understand the growing literature focusing on sustainability in the mining industry. A properly designed bibliometric study allows mapping an emerging research field, identifying not only the chief published papers and correspondent indexed journals but also the leading researchers and their respective fields of research, their academic areas, their institutions, the funding agencies and the like.

The exercises presented below are not intended to provide an exhaustive state of the art of published research on sustainability and the mining industry. Based on a preliminary and thus far from comprehensive search strategy, it seeks to render a glimpse of what one may discern as an emerging field of interdisciplinary research.

The bibliographic search methodology was set out as follow. The Web of Science database was used as a source for bibliometric research. The search strategy was refined using a combination of keywords on the theme reviewed, resulting in two research equations: the first is intended to outline the state-of-the-art of 'mining' and 'sustainability' studies, while the second is proposed to refine preceding data to specifically target research on 'SDGs' and 'mining' (Table 1). It is worth mentioning that the search strategy was limited to identifying only scientific papers published in all languages available in the indexed databases.

The result of the first search equation comprised 3,344 papers between 1990 and 2020[1] (Figure 1). It can be seen that there is a prominent increase in the number

---

[1]   Research was done on September 4, 2020.

Table 1: Search Equation and Filters Used in the Web of Science

Search Equation 1	Topic: ('mining') AND Topic: ('sustainable*') NOT Topic: ('data mining') NOT Topic: ('text mining') AND Types of documents: (Article)
Search Equation 2	Topic: ('mining') AND Topic: ('sustainable development goal*') NOT Topic: ('data mining') NOT Topic: ('text mining') AND Types of documents: (Article)
Indexed Base Types	SCI-EXPANDED, SSCI, A&HCI, CPCI-S, CPCI-SSH, ESCI.
Time Interval	All years
Languages	All languages

Note: The research done on topics is the same as searching for the frequency of words contained in the title, keywords and abstract.

Source: Elaborated by the authors based on Web of Science on September 4, 2020

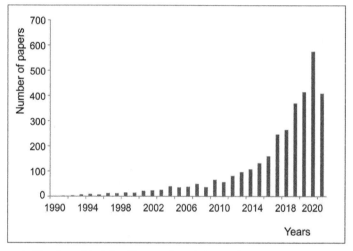

Figure 1: Number of Articles, Keywords 'Mining' and 'Sustainable*', 1990-2020
Source: By the authors based on Web of Science on September 4, 2020

of articles on the subject with a growth of 133% between 2015 and 2019, including 576 papers only in 2019. The first paper, entitled "Regulation of Large Rivers – Problems and Possibilities for Environmentally-Sound River Development in South-America", was published in 1990 in a conference in Venezuela (Petts 1990). This one was not a paper directly related to the mining industry, but it does concern with specific environmental and social impacts produced by mining activities.[2]

---

[2] One observes that sustainability, which was a growing concern in the event of Rio-92, had at least one previous awakening. In Stockholm in 1972, the United Nations Conference on Human Environment has introduced the term ecodevelopment, preceding in twenty years the Second Earth Summit in laying the foundations of a new conception of development. In 1992, in Rio de Janeiro, Brazil, in the context of a thriving environmental multilateralism, this Second Earth Summit gave rise to ambitious global conventions such as Biological Diversity and Climate Change. Rio-92 was a landmark in urging society's perception of the complex interdependencies among environmental, social, cultural, and economic dimensions of development (Guimarães and Fontoura 2012).

The Rio-92 context may explain, in broad terms, the publication looming on 'sustainability' and 'mining' at that time. To begin to map the research field, one observes the institution that published most during the period on the subject being the University of Queensland with 115 articles (Table 2). The University of São Paulo (USP), the prestigious Brazilian institution, appears in the seventh position with 28 papers.

Under the second equation exercise, the survey for mining and SDG concerning the period 2009-2020 has resulted in just 50 papers (Figure 2). The first available paper was entitled "Towards a Sustainability Criteria and Indicators Framework for Legacy Mining Land", published in 2009 in the Journal of Cleaner Production, the authors being from the University of Queensland (Brisbane, Australia) (Worrall et al. 2009). Within the period, the Australian university also stands out in the first position. In these terms, it is also possible to observe the pioneering in the area of mining and sustainability from the University of Queensland (Table 2).

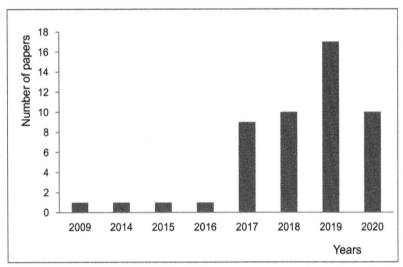

Figure 2: Number of Papers, Keywords 'Mining' and 'SDG', from 2009 to 2020
Source: By the authors based on Web of Science on September 4, 2020

Considering the subject of research funding, from the first search equation, the agencies that most financed research on the topics of mining and sustainability were the National Natural Science Foundation of China (NSFC) with 269 articles, followed by the European Union (59 articles), the Research Funds Fundamentals of China for Central Universities (56 articles) and the Brazilian Institution National Council for Scientific and Technological Development (CNPq) (47 articles). CAPES, another Brazilian institution, appears in the seventh position among the leading research funding agencies. It deserves mentioning that the mining and SDGs research (search Equation 2) revealed an important role of CNPq in supporting the research underlying three out of 50 papers. Despite the low funding amounts, it is noteworthy that a developing country's research funding agency has stood out on the publications' account of the reviewed subject under that period.

Table 2: Top 10 Institutions with the Largest Number of Publications, for Both Search Strategies

Keywords 'Mining' and 'Sustainable*'			Keywords 'Mining' and 'SDG'		
Institutions	*Countries*	*Number of papers*	*Institutions*	*Countries*	*Number of papers*
University of Queensland	Australia	115	Univ Queensland	Australia	6
Chinese Acad Sci	China	74	Geol Soc London	United Kingdom	2
China Univ Min Technol	China	73	Leiden Univ	Netherland	2
Univ British Columbia	Canada	42	Newcastle Univ	United Kingdom	2
Univ Witwatersrand	South Africa	40	Queens Univ	Canada	2
China Univ Geosci	China	39	Univ British Columbia	Canada	2
Univ Sao Paulo	Brazil	28	Univ Technol Sydney	Australia	2
Univ Chinese Acad Sci	China	26	Univ Waterloo	Canada	2
Monash Univ	Australia	24	Univ Western Australia	Australia	2
Pontificia Univ Catolica Chile	Chile	24	Univ Wollongong	Australia	2

Source: by the authors based on Web of Science on September 4, 2020

In summary, the results of the two surveys allow us to point out a multidisciplinary nature of research in mining, sustainability, and SDGs. Therefore, it is appropriate to emphasise the main areas of research involving these themes. The first search has shown a large concentration of papers related to the area of environmental sciences and ecology (1,510), about 45% of the total, although it is possible to find publications in other areas of knowledge (Table 3).

When analysing the 50 papers that deal specifically with topics regarding 'mining' and 'SDGs' (second search), there is also a corresponding concentration in the same research area, about 54% of the total. Also, other areas stand out, such as technological sciences, engineering, business economics, and geology, among others that indicate multidisciplinary (Table 3).

Still, the second search allows identification of the 17 SDGs mentioned by the publications, as well as the quantification of their frequency within the literature reviewed (Figure 3). Out of the second search results, 18 papers did not mention any specific SDGs. In this case, to measure SDGs in this equation, a citation for each SDGs was added to these 18 papers. Besides, a paper written in Russian was found that could not be analysed due to the language barrier.

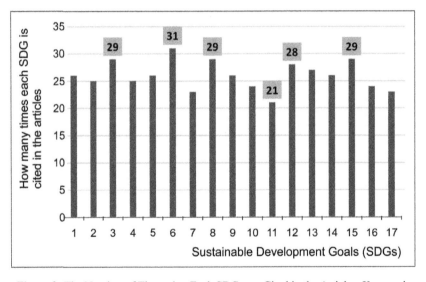

Figure 3: The Number of Times that Each SDG was Cited in the Articles, Keywords 'Mining' and 'SDG'. *Source*: By the authors based on Web of Science data 2020

It was observed that all 17 mining-related SDGs are mentioned similarly. SDG 6 (Clean Water and Sanitation) was the most cited with presence in 31 papers, followed by SDG 3 (Good Health and Well-Being), SDG 8 (Decent Work and Economic Growth), SDG 15 (Life on Earth) all of which have been highlighted in 29 papers. It is noted that the mentioned SDGs are related to direct impacts on society, such as the concern with the quality of sanitation and the quality of employability. However, the least mentioned was the SDG 11 (Sustainable Cities and Communities) with only 21 mentions.

Table 3: Main Research Areas for Both Search Strategies

Keywords 'Mining' and 'Sustainable*'		Keywords 'Mining' and 'SDG'	
Research areas	Number of papers	Research areas	Number of papers
Environmental Sciences Ecology	1,510	Environmental Sciences Ecology	27
Engineering	691	Science Technology Other Topics	11
Science Technology Other Topics	605	Engineering	9
Mining Mineral Processing	258	Business Economics	5
Geology	253	Geology	3
Business Economics	245	Public Environmental Occupational Health	3
Water Resources	238	Water Resources	3
Agriculture	153	Biodiversity Conservation	2
Metallurgy Metallurgical Engineering	138	Marine Freshwater Biology	2
Energy Fuels	129	Chemistry	1
Mineralogy	97	Development Studies	1
Geography	82	Energy Fuels	1
Materials Science	74	Geography	1
Chemistry	71	International Relations	1
Development Studies	63	Linguistics	1
Public Administration	63	Meteorology Atmospheric Sciences	1
Social Sciences Other Topics	59	Mining Mineral Processing	1
Biodiversity Conservation	48	Oceanography	1
Computer Science	47	Public Administration	1
Meteorology Atmospheric Sciences	43	Remote Sensing	1
Physical Geography	42	Social Sciences Other Topics	1
Government Law	37	Thermodynamics	1

According to the two presented search equations and the corresponding analytical procedures, the intellectual production by researchers seems to allow one to hypothesise the emergence of an interdisciplinary field of research in mining and sustainability. Within the limits of this preliminary bibliometric study, this field would be characterised by the prevalence of research by leading institutions and scholars in mining research and sustainability research groups, financed by leading national and international agencies, representing a broad range of areas of knowledge, published in an identified group of academic journals with their research focuses distributed more or less evenly across the seventeen SDGs. Such bibliometric exercises, by their incipient character, invites further research on the topic.

4. Sustainable Development Goals and Mining

Despite the industry advances in the management of impacts and risks and improving efficiency, safety and work conditions, severe social, economic, and environmental problems still arise from mining hitherto.

Fraser (2019) points out that the current reputation of mining is a story that has been highlighted in conventional and social media, worsening the already bad reputation of the mining industry. The author refers to "tailing storage facility collapses; competition for scarce resources such as water; labour issues; environmental impacts; court cases and legal challenges for failure to respect indigenous rights and titles" (Fraser 2019). UNDP recognises that, for a long time, issues like violation of human rights, violence, including gender-based, and armed conflicts, traditional communities and indigenous peoples displacement, inequality exacerbation, ecological deterioration, water contamination, soil erosion, manifold risks acceleration to the environment, health, dignity and people's lives, alongside with fiscal evasion and corruption have been associated with mining sector activities, especially in less developed countries (UNDP 2016).

Social and environmental bad records are not exclusive to the mining industry. Academy, policymakers, and organised civil society have pointed out the urge to change the whole production and consumption dominant paradigms to face major environmental and social crises that have led our civilisation to the brink of collapse.

Consistently with a systemic, science-based turn of global environmental governance, on the occasion of the seventieth anniversary of the United Nations, in 2015, all the 193 UN member states signed the Resolution 70/1 or 'Transforming our World: the 2030 Agenda for Sustainable Development', also known simply as 2030 Agenda. Member states agreed to intend to achieve the 169 targets of the 17 SDGs, devoted to reaching economic development while safeguarding human rights, social equality and fairness and the environment protection (UN Resolution 70/1 2015)[3].

Still, as underscored in the previous section, products from mining are indispensable components of goods and services demanded by our society. It is far from evident how the mining industry could cope with SDGs. It is certain that SDGs pose challenges that could bring about the opportunity to make decisive contributions to the transition to a more sustainable future.

To unveil these challenges, UNDP along with the UN Sustainable Development Solutions Network (SDSN), the Columbia Centre on Sustainable Investment (CCSI, which is a joint centre of Columbia Law School and the Earth Institute at Columbia University), and the World Economic Forum (hereafter WEF), prepared in 2016 an Atlas for Sustainable Development Goals (SDGs) to the mining sector, mapping SDGs challenges to be faced by mining (WEF 2016).

---

[3] To find more, access to the website https://sdgs.un.org/goals

Thus, one identifies a few illustrative cases from WEF (2016) and other sources to grasp ways by which the mining industry may be challenged but also contribute to SDGs. Some of these challenges and contributions can be understood in the following paragraphs.

A significant contribution for the SDG1—end poverty in all its forms—should be the public release of information about the tax and royalty payments industry makes to governments, preferentially on a project-basis. Tax and royalties constitute the bulk of resources for local governments' budgets and should be spent on local social priorities, like health, education, and infrastructure, which are pivotal for poverty reduction. This is a measure that is requested by The Extractive Industries Transparency Initiative (EITI)[4] and voluntary industry policies, but it should be integrated into national mining regulatory frameworks and mandatory for industry.

In the case of SDG2—zero hunger, achieve food security and improved nutrition and promote sustainable agriculture—a fine example can be seen in the contribution of Finnish Geological Survey (GTK) to farmers in Ethiopia. GTK has acted as an information and technical knowledge provider, examining the soil and helping the surrounding agricultural community to understand soil features and advising them for the substitution of costly and ineffective industrial fertilisers for local lime, resulting in increased crop yields.

A useful input from the mining industry to SDG3—ensure healthy lives and promote well-being for all at all ages—would be, for instance, due to their impacts of high stakes, the tailing ponds safety and management. After the tragic of Vale-BHP Billiton joint-venture tailing collapse at Brumadinho, Brazil, on January 25, 2019, one could mention three initiatives that come to attention: the communities' mobilisation and articulations with Academia and other social actors to fight for local peoples' rights (Milanez and dos Santos 2020); the movement of institutional investors, concerned with social and environmental issues related with their investment portfolios; and a global initiative under United Nations Environmental Programme (UNEP) towards the creation of a Global Industry Standard on Tailings Management (UNEP 2020).

To help to achieve SDG4—ensure inclusive and equitable quality education and promote lifelong learning opportunities for all—there is a room for the mining industry action, as one may recognise the up-to-date debate on knowledge development, technical training and local communities and global social awareness for sustainable livelihood opportunities beyond mining. The involvement of the mining industry with education projects in local communities does exist, mainly integrating Corporate Social Responsibility (CSR) strategies with examples worldwide, and here one briefly mentions the Brazilian case of Fundação José Carvalho, in the State of Bahia, in Northeastern Brazil. Created in 1975 by the late owner of Ferbasa, one of the top 10 mining and siderurgy companies of that State, this is a foundation that offers educational opportunities mostly for poor and neglected children. The foundation owns 90% of Ferbasa's voting shares and 51% of the company's capital. Since its creation, the initiative helps thousands of youngsters with education, industrial and agricultural technical training, and artistic and athletic activities. Initiatives alike

---

[4] https://eiti.org/

may help young people in impoverished communities that usually surround mining sites, especially in the Least Developed Countries (LDC), and their contribution may not be underestimated within local, specific plots. Meanwhile, one has two considerations here. First, it is not possible to dismiss the fact that education has been a major State obligation with its duties defined by the 1988 Constitution in Brazil, and CSR, locally efficient as it may be, cannot be a substitute for this mandate. Secondly, one should note with Hoi et al. (2013) that CSR social investments may be associated with aggressive tax avoidance practices, which, in spite of being legal, has been considered in the definition of illicit financial flows (IFF)[5] within the ODS framing by UNCTAD and Forstater (2018).

Challenges of SDG5—achieve gender equality and empower all women and girls—to the mining industry, local communities and governments and society at large are complex and cannot be underestimated. A practical example of the ways to meet them is reported as the global partnership between Barrick Gold Corporation (Barrick) and White Ribbon (WR). This partnership involves host communities since 2012 to prevent gender-based violence and to promote a positive role for men in ending violence against women in three locations – north-eastern Nevada, United States; Lumwana, Zambia; and Porgera, Papua New Guinea. The foremost challenge of bringing people together to face this critical problem relied on the conduction of a previous phase, including workshops, focus groups, consultations and interviews with a broad range of stakeholders at the national, regional and local levels working together to identify and address issues.

As mineral production chain and tailing ponds are water-intensive and polluting, there is plenty of room for improving efficiency and directing innovation towards SDG6—ensure availability and sustainable management of water and sanitation for all—in the mining industry. An illustration of coping with this challenge is provided by a Chinese case, at Jiangxi Copper Company's Dexing mine that, with a partnership with BioTeQ Environmental Technologies, has built a water treatment plant that treats the wastewater while recovers copper from it. During its early six months of activity, the plant treated 3 billion litres of wastewater, recovering 700,000 pounds of copper, fully covering costs of treatment with recovered copper revenues, and providing water reuse in the site.

Recognising that mining is energy-intensive, WEF (2016) envisages many challenges posed by SDG7—ensure access to affordable, reliable, sustainable, and modern energy for all—for the industry. Most of the examples that illustrate how to cope with this challenge have to do with measures to boost energy efficiency and incorporate renewable energy into mine power supplies and facilities. That is the case of using locally available resources as, amongst other cases, the Sunshine for Mines, an initiative of Rocky Mountain Institute and Carbon War Room, whose first project is a 40 MW power plant in South Africa with a partnership Gold Fields; or the Glencore's Raglan mine, in northern Canada, which is replacing its diesel fuel with wind power. Those are projects aligned with a decarbonising perspective of growth, and one must remind that the intensive nature of renewable energy technologies in terms of rare minerals is already posing other challenges to sustainable energy (Smil 2010).

---

[5] https://sdgpulse.unctad.org/illicit-financial-flows/

When it comes to SDG8—promote sustained, inclusive and sustainable economic growth, full and productive employment and decent work for all—there is an implicit supposition in our view that social problems could be faced mainly by growth and employment. Work beyond straightforward, conventional salaried employment or small business should have the attention of public and private decision-makers since there will be no place for everyone in the marketplace and the mining industry has long been recognised as a labour-saving one. Nevertheless, in the view of prevailing unacceptable, outrageous slur on Human Rights in mining practices such as in Congo, it is very knowledgeable that efforts must be made towards eliminating child labour and any form of slavery-like conditions of work. In line with this fundamental assumption, one can point out the Cable News Network (CNN) initiative of investigative journalism on cobalt production in the Democratic Republic of Congo, which has contributed to raise public concern and awareness and to claim more accountability and transparency in corporate practices on the matter. To account for the heavy dependence of high-tech industries on mineral resources, like cobalt, there is a long way ahead to face this vital problem[6].

Innovation has been considered a pivotal issue to foster transitions towards sustainability, and technological transformations have also been seen for a long time to develop economies beyond their resource's dependence. SDG9—build resilient infrastructure, promote inclusive and sustainable industrialisation and foster innovation—in this sense, renews the challenge, and puts it into an urgent time frame; moreover, one considers the time requirements for investments on infrastructure and industry to mature. Beyond incremental changes in mining sites, operations and processes, what is needed is more enduring commitments amongst the several actors in the whole society with disruptive changes in the way industry and consumption rely on non-renewable minerals. The way ahead has to do with once more, a circular economy that helps to recover useful material from waste, from a better design – 'recyclable by design' products, integration of processes towards a proper industry ecology (Ghisellini et al. 2016).

As pointed out by WEF (2016), mining can exacerbate inequalities in host places and countries. SDG10—reduce inequality within and among countries—brings to the front of the debate the need to face this secular problem. Interestingly, two ways to manage the problem, according to WEF (2016) include, on the one hand, targeting diverse groups with social investments (education, health, infrastructure, respecting local culture, traditional livelihood and costumes) and, on the other hand, stimulating the involvement of communities in the budgetary monitoring and planning of revenues from mining in the local level.

One of the most evident, yet challenging tasks of SDG11—make cities and human settlements inclusive, safe, resilient and sustainable—for the mining industry might be linked to the fact that its principal impacts are precisely on the operations' sites. Water, soil and air contamination tailing pond risks, biodiversity loss and so on health risks for workers and communities are strongly connected with operation and processing sites, surrounding territories and mining-impacted areas, which are not

---

6   https://edition.cnn.com/interactive/2018/05/africa/congo-cobalt-dirty-energy-intl/

always correctly dimensioned by environmental impact risks assessments. Making these territories and communities safe, resilient and sustainable is, by all means, an explicit condition for SDG11 to be met by the industry. Another challenge has to do with the end of the life cycle of mines. In this case, WEF (2016) mentioned several examples of alternatives for them. Beginning with Eden Project in Cornwall, United Kingdom, which converted former mines into parks and educational facilities as the largest underground bike park in the planet and an underground data centre in the USA (both built from former limestone mines), a physics laboratory also in USA (a former gold mine), and a museum in Norway (former zinc mine). Alternative mine reclaiming to former salt mines are known in the form of a subterranean theme park in Romania, asthma therapy spas in Ukraine, and a cathedral in Poland and Colombia.Meeting SDG12—ensure sustainable consumption and production patterns—poses, among other paths, clear opportunities to close mineral resources production and consumption cycles. To cope with this, WEF (2016) highlights Canada, USA, and European initiatives towards zero-waste and ambitious recycling in the mining industry. The Canadian program called Towards Zero Waste, for example, is a partnership involving industry, Academia, and government and aims at minimising waste, implementing closed-system processing, and refining tailings to recover marketable resources.

Mining operations may be an additional risk factor to communities in the face of climate change. So, SDG13—take urgent action to combat climate change and its impacts— invites to think about threats and measures to overcome them. In Peru, as noted by WEF (2016), a project known as La Granja, implemented by Rio Tinto, International Council on Mining and Metals (ICMM) and the local community identified vital climate risks, as drought, cold snaps, floods and landslides, enabling a better understanding of the problems and the resilience improvement.

As one assists a global call for broad and in-depth energy/sustainability transitions and to embrace the Industry 4.0, a mineral course is underway to dig rare metals and other minerals from bed sea, switching on an alert light about links between the mining industry and SDG14—conserve and sustainably use the oceans, seas and marine resources for sustainable development. In this sense, it is noticeable that the study published by World Resources Institute (Haugan et al. 2019), under the initiative of High-Level Panel for a Sustainable Ocean Economy (hereafter Ocean Panel), identifying ways to alleviate such pressures on the ocean by seabed mining. This initiative includes, for example, fostering research and implementing practices on the circular economy in order to reduce the mineral intensity of electronics and energy systems, to spur ocean governance at global and local levels, and to improve efficiency and sustainability in land-based mining sites and operations.

Amongst the many ways, scales and processes mining activities impact biodiversity reviewed by Sonter et al. (2018), one may stress here the case of environment and human contamination by mercury and cyanide in the Amazon Basin with menaces to its biodiversity and human populations. This case exemplifies one of many challenges SDG15—protect, restore, and promote sustainable use of terrestrial ecosystems, sustainably manage forests, combat desertification, and halt and reverse land degradation and halt biodiversity loss—poses to mining activities.

The difficulties faced by the initiative of the International Council on Mining and Metals (ICMM) to stop mining in World Heritage Sites (WHS)[7] in 2003 and renewed in 2016, cited by the authors, denotes the dramatic nature of this challenge. ICMM initiative is not being monitored and has not its results evaluated, while the UNESCO's State of Conservation Reports depicted in 2015, more than 80% of the WHS properties endangered by mining, oil and gas activities.

The literature on resource curse has long diagnosed unscrupulous practices historically associated with mining, especially in the Least Developed Countries (LDC), such as bribery, tax evasion, corruption, human rights violations, and social and environmental conflicts (Ross 2015). Having such diagnostics in mind, meeting SDG16—promote peaceful and inclusive societies for sustainable development, provide access to justice for all and build effective, accountable and inclusive institutions at all levels—may defy old mining practices and their relations to host governments and communities. Better knowing conflicts, mapping, examining, and diffusing the information related to them can contribute to raising public awareness, communities' empowerment, and better solutions. Such are the cases of several initiatives that map socio-environmental conflicts all around the world, including cases involving mining operations, as are the global initiative The Environmental Justice Atlas[8] (Temper et al. 2015), and the Brazilian-based initiative 'Observatório dos Conflitos Ambientais de Minas Gerais'[9] for the State of Minas Gerais (Cf. Zhouri 2018).

SDG17 points to the need to strengthen the means of implementation and revitalise the global partnership for sustainable development. Among its targets highlighted by WEF (2016) as relevant to the mining industry, one underlines the first one, which stands for the need to 'Strengthen domestic resource mobilisation, including through international support to developing countries, to improve domestic capacity for tax and other revenue collection'. To illustrate how the industry can meet this target, WEF (2016) have chosen the EITI which aims at establishing, for the host countries involved, a global standard for oil, gas and mineral resources governance. The host countries which implement EITI must disclose tax payments, licences, contracts, production, and other information on resource extraction so that public awareness is raised and a broader debate on the use of revenues is spurred. As useful as this initiative may be, it is necessary to recognise some drawbacks. For instance, in the case of VALE S/A, a Brazilian mining major, one can observe its commitment to the EITI abroad—in the United States of America, Mozambique, Peru, Philippines, Zambia and the United Kingdom—while Brazil itself does not take part on it. Furthermore, if Morlin (2017) data and analysis prove to be correct, the same VALE S/A has performing strategies associated with tax avoidance, fiscal evasion, and trade misinvoicing. This seems to be the case of the creation and use of a business office in Switzerland in 2006 to triangulate the iron exports between Brazil and major importers.

---

[7] https://whc.unesco.org/en/extractive-industries/

[8] https://ejatlas.org/commodity

[9] https://conflitosambientaismg.lcc.ufmg.br/observatorio-de-conflitos-ambientais/
mapa-dos-conflitos-ambientais/

From the preceding discussion, stemming from local government and communities collaborations, mining should, more obviously, provide decent employment, promote water and energy efficiency, take care of residues of production and toxic waste, implement best practices, adopt adequate mitigation and compensation measures, cultivate transparent relationships with local government, and collect legal taxes. However, having in mind the complex, integrated nature of the issues and the SDGs themselves, one must pay attention to the possible trade-offs to be managed and synergies to be achieved. In this sense, within the ten years yet left to the horizon of 2030 Agenda, it remains a broad range of tasks to be done to establish a minimal set of accomplishments for SDGs and the mining industry.

5. Concluding Remarks

This chapter has brought to the attention of readers a contemporary narrative whose axis lies in the challenges of sustainability for the mining industry. It sought to emphasise, with the support of specialised literature, the fact that mining concerns a very ancient way in which humankind exploits natural resources in order to produce instruments, utilities and goods that, with an ever-advancing degree of technological sophistication, contribute to the construction of civilisation as we know it.

Throughout this long process, mining products' demand accelerated, deepening the dependence of our society and our ways of life on the activities of this industry.

Today, when scholars, some of leading progressive governments, innovative corporations and relevant segments of organised civil society seem to broadly recognise the need for a profound, sustainable transition, both in terms of production and consumption, mining activities and mining industry have a long path ahead, still to be paved. From the high-tech industry to renewable energies, infrastructure, and the retrofitting of domestic, industrial, and service facilities, all sectors will demand even more and specific geological resources.

Based on a bibliographic and documentary survey, supported by two brief bibliometric exercises, the contribution of this chapter is pointing to urgent issues to be addressed and improved if one wants mining to be somehow more sustainable. Given the growing demand and supply of mining products, as well as their alarming social and environmental impacts, challenges are outlined to understand the relationship between mining and sustainability. Some of these crucial challenges have been pointed out specifically in the light of the SDG.

The bibliometric analysis on mining and sustainability has enabled an initial mapping of an emerging, interdisciplinary, and perhaps fragmentary research field. Further research is needed to refine this mapping and confirm, or reject, this hypothesis and advance knowledge on the subject.

The recommendations that follow this diagnosis aim at steering the continued development of this challenging field of research have to do with the requirements for a sustainable mining industry transformation with research, policy and entrepreneurship implications and are presented below.

First of all, social (especially for the vulnerable communities), cultural, economic, and environmental trade-offs solving and active searching for synergies amidst those dimensions should be a first rang priority in private and public decision-making in all decision scales, local, national and internationally.

Second, a significant broadening of the of decision-making focus affecting mining sector, from the industry to the whole production and product life cycle, including the global product and value chain (from the exploration, mining, processing, to the market, waste disposal, including capital and financial fluxes), allowing a transition towards a circular economy.

Third, an ample transition of sectors using mineral resources towards mandatory requisites on recyclability by design, stepping through precise information, material and energy cross-sector, or cross-industry integration, to achieve a broader circular economy.

Fourth, any international agreement to reduce carbon emissions required to limit temperature rise below 2°C shall consider the enormous increase in production of the so-called green elements as lithium, cobalt, nickel, silver, indium among others. One must account for where, how and with what intensity these strategic minerals are going to be mined. Investments in mineral exploration must increase to support long-term sustainable practices.

Fifth, SDGs may have a crucial role to play in the mining transition to sustainability. Nevertheless, the fragmented character one observes through a bibliometric analysis of the literature body concerned to this subject and the examples that emerge when one focuses on SDGs and mining literature, call for a more integrated and universal perception of the challenges SDGs poses to the humanity's relations with the natural world, and to the ways humans act and interact with each other.

The sixth and final recommendation concerns with governance aspects. Self-regulatory initiatives have been identified in the mining industry as well as in other sectors, including the declaration of principles, adherence to good practice, sector and intersectoral agreements, internal sustainability policies, and other voluntary forms of governance. Based on the premise of the modern State, despite its current crisis, as the locus of caring for the common interest, one argues for the leadership of public power in making the principles, practices, objectives, and goals of sustainability supported by the SDGs mandatory. The governance of resources implies a central role for the State, especially in ensuring spaces for participation and articulation among the multiple involved actors, such as organised civil society, local communities, the industry itself, health, education, sanitation and environmental protection agents, and the academic community. In addition to being 'multi-actor', governance has to be multi-level or multi-scale: from local government to multilateral organisations and initiatives, and national regulations and policies. This governance has to face the complexity of dealing with issues that concern many sectors of the economy (sectors that produce and use mineral resources). Finally, it needs to be open to advancing knowledge of the problems: it needs to be flexible to adapt to the new inputs that science is making all the time. In other words: governance must be multi-actor, multi-level, adaptive, and deal with complexity.

In sum, SDGs and sustainability issues in the mining sector, as this chapter has sought to put in evidence, require a more systemic view of the embedding of business, or production, in society and the environment. Requiring exploration of resources that are, by their very nature, exhaustible, mining has clear implications for future generations ability to face their needs. Communities which are depending on the mining industry, directly or indirectly, are coming to face the need to move to ways of life beyond mining eventually.

Tasks to build a sustainable transition comprehends but not limits to foster non-mining business development, contribute tocitizenship education and qualification for a beyond- and post-mining local development, ecosystems and landscape restoration, increased tax revenues and fiscal justice, and resilient infrastructure synergies. Furthermore, conflicts associated with land use and resources exploitation remain a crucial challenge in the unrelenting defence and full observation of human rights.

SDGs challenges mining businesses, social organisations, and local government not only to adapt and develop new ways of solving economic, social and environmental issues by internalising new rationalities for licensing, investing, operating, and decommissioning. Moreover, they challenge this plethora of social actors to face cross-cutting issues that unveil more profound considerations about local development needs and the very concepts and beliefs about conventional wisdom on development and progress.

Ultimately, one must radically change the way one understands and promotes industrial development, by stimulating a circular economy in corporations, cross-sectors, embedded in a more conscious societal frame within a transboundary panorama. Through these transformations, deeply embedded in social practices that also needs transformation, demand for mineral extraction shall fall to rates that do not exceed capabilities of the circular economy to be built and the planetary safe operating space for humanity so that it does not compromise the needs of the future generations.

References

Acosta, A. 2016. O bem viver: uma oportunidade para imaginar outros mundos. Editora Elefante, São Paulo, Brasil.

Agricola, G. 1950. De Re Metallica [1555]. H.K. Hoover and L.H. Hoover [translated]. Dover Publications, Inc. New York.

Alencastro, L.F. 2000. O trato dos viventes: formação do Brasil no Atlântico Sul. Companhia das Letras, São Paulo.

ANM (National Mining Agency). Mineral Summary from 2007 to 2014. Available from: https://www.gov.br/anm/pt-br/centrais-de-conteudo/publicacoes/serie-estatisticas-e-economia-mineral/sumario-mineral. Retrieved 2020-09-12.

Betrán, C. 2005. Natural resources, electrification and economic growth from the end of the nineteenth century until World War II. Revista de Historia Económica. XXIII: 47-81.

Boxer, C.R. 1969. A idade de ouro do Brasil (dores de crescimento de uma sociedade colonial). Brasiliana, São Paulo.

Bridge, G. 2004. Contested terrain: mining and the environment. Annual Review of Environment and Resources. 29: 205-259.

Brundtland, G.H. (ed.). 1987. Our Common Future: The World Commission on Environment and Development. Oxford University Press, Oxford.

Brüseke, J.F. 1995. O problema do desenvolvimento sustentável. *In*: C. Cavalcanti [ed.]. Desenvolvimento e natureza: estudos para uma sociedade sustentável. Cortez, São Paulo.

Cavalcanti, C. 1995. Desenvolvimento e natureza: estudos para uma sociedade sustentável. Cortez, São Paulo, Brasil.

Clark, G. and D. Jacks. 2007. Coal and the industrial revolution, 1700–1869. European Review of Economic History. 1: 39-72.

Forstater, M. 2018. Illicit financial flows, trade misinvoicing, and multinational tax avoidance: the same or different? CGD Policy Paper. 123(29).

Fraser, J. 2019. Creating shared value as a business strategy for mining to advance the United Nations Sustainable Development Goals. The Extractive Industries and Society. 6(3): 788-791.

Garner, R.L. 1988. Long-term silver mining trends in Spanish America: A comparative analysis of Peru and Mexico. The American Historical Review. 9: 898-935.

Ghisellini, P., C. Cialani and S. Ulgiati. 2016. A review on circular economy: the expected transition to a balanced interplay of environmental and economic systems. Journal of Cleaner Production. 114: 11-32.

Gregory, C.E. 1984. A Concise History of Mining. Pergamon, Oxford.

Guilbert, J.M. and F. Park. 2007. The Geology of Ore Deposits. Waveland Press Inc., Illinois.

Guimarães, R.P. and Y.S.R. Fontoura. 2012. Rio+20 ou Rio-20?: Crônica de um fracasso anunciado. Ambiente & Sociedade. 15: 19-39.

Haugan, P.M., L. Levin, D. Amon, M. Hemer, H. Lily and F.G. Nielsen. 2019. What Role for Ocean-Based Renewable Energy and Deep Seabed Minerals in a Sustainable Future? World Resources Institute, Washington, DC.

Herrera, A.O., H.D. Scolnik, G. Chichilnisky, G. Gallopin and J.E. Hardoy. 1976. Catastrophe or New Society? A Latin American World Model. IDRC, Ottawa, Canada.

Hoi, C.K., Q. Wu and H. Zhang. 2013. Is corporate social responsibility (CSR) associated with tax avoidance? Evidence from irresponsible CSR activities. The Accounting Review. 88(6): 2025-2059.

Hopper, R.J. 1968. The Laurion mines: a reconsideration. The Annual of the British School at Athens. 63: 293-326.

IBRAM (Brazilian Institute of Mining). 2019. Annual Report 2018-2019.

IGF (Intergovernmental Forum on Mining, Minerals, Metals and Sustainable Development). 2017. IGF Guidance for Governments: Managing Artisanal and Small-scale Mining. Winnipeg: IISD.

John, A.V. 2013. By the Sweat of Their Brow: Women Workers at Victorian Coal Mines. Routledge, London and New York.

Lander, E. 2016. Com o tempo contado: crise civilizatória, limites do planeta, ataques à democracia e povos em resistência. *In*: G. Dilger, M. Lang and J. Pereira Filho. Descolonizar o imaginário. Editora Elefante, São Paulo.

Lebdioui, A. 2020. Uncovering the high value of neglected minerals: 'Development Minerals' as inputs for industrial development in North Africa. The Extractive Industries and Society. 7: 470-479.

Månberger, A. and B. Stenqvist. 2018. Global metal flows in the renewable energy transition exploring the effects of substitutes, technological mix and development. Energy Policy. 119: 226-241.

Manning, D.A.C. 1995. Introduction to Industrial Minerals. Springer-Science+Business Media. B.V. Chapman & Hall, London.

Martos, M.C. 2006. Bartolomé de Medina y el siglo XVI. Universidad de Cantabria, Calabria.

Matthews, R. and H. Fazeli. 2004. Copper and complexity: Iran and Mesopotamia in the fourth millennium BC. Iran. 42: 61-75.

Milanez, B. and R.S.P. dos Santos. 2020. Mineração e captura regulatória: a estratégia da Anglo American em Conceição do Mato Dentro (MG), Brasil. Revista Pós Ciências Sociais. 16(32): 69-91.

Morlin, G. 2017. Mensuração da fuga de capitais do setor mineral no Brasil. Red latinoamericana sobre Deuda, Desarrollo y Derechos.

Moon, C.J., M.K.G. Whateley and A.M. Evans. 2006. Introduction to Mineral Exploration. Blackwell Publishing, Oxford.

Mutunhu, T. 1981. Africa: The Birthplace of Iron Mining. Negro History Bulletin. 44: 5.

Nassar, N.T., T.E. Graedel and E.M. Harper. 2015. By-product metals are technologically essential but have problematic supply. Sciences Advances. 1: e:1400180.

Petts, G.E. 1990. Regulation of large rivers: problems and possibilities for environmentally-sound river development in South America. Interciencia. 15: 388-395.

Picanço, J. and M.J. Mesquita. 2018. Uma breve história do tempo geológico: a questão do Antropoceno. ClimaCom [online], 5: (2). Available from: http://climacom.mudancasclimaticas.net.br/?p=9378

Ross, M.L. 2015. What have we learned about the resource curse? Annual Review of Political Science. 18: 239-259.

Schaffartzik, A., A. Mayer, S. Gingrich, N. Eisenmenger, C. Loy and F. Krausmann. 2014. The global metabolic transition: regional patterns and trends of global material flows, 1950–2010. Global Environmental Change. 26: 87-97.

Seccatore, J., M. Veiga, C. Origliasso, T. Marin and G. De Tomi. 2014. An estimation of the artisanal small-scale production of gold in the world. Science of the Total Environment. 496: 662-667.

Shepherd, R. 1993. Ancient Mining. Elsevier Applied Science for the Institution of Mining & Metallurgy. London and New York.

Smil, V. 2010. Energy transitions: history, requirements, prospects. Praeger/ABC CLIO, Santa Barbara.

Sommer, M. 2007. Networks of commerce and knowledge in the Iron Age: the case of the Phoenicians. Mediterranean Historical Review. 22: 97-111.

Sonter, L.J., S.H. Ali and J.E. Watson. 2018. Mining and biodiversity: key issues and research needs in conservation science. Proceedings of the Royal Society B. 285(issue 1892).

Sovacool, B.B.K., S.H. Ali, M. Bazilian, B. Radley, B. Nemery, J. Okatz and D. Mulvaney. 2020. Sustainable minerals and metals for a low-carbon future. Science. 367: 30-33.

Steffen, W., R.A. Sanderson, P.D. Tyson, J. Jäger, P.A. Matson, B. Moore III, F. Oldfield, K. Richardson, H.J. Schellnhuber, B.L. Turner and R.J. Wasson. 2005. Global change and the earth system: a planet under pressure. The IGBP Series. Springer Science & Business Media.

Steffen, W., W. Broadgate, L. Deutsch, O. Gaffney and C. Ludwig. 2015. The trajectory of the Anthropocene: the great acceleration. The Anthropocene Review. 2: 81-98.

Svampa, M. 2019. As fronteiras do neoextrativismo na América Latina: conflitos ambientais, giro ecoterritorial e novas dependências. Editora Elefante, São Paulo.

Swain, E.B., P.M. Jakus, G. Rice, F. Lupi, P.A. Maxson, J.M. Pacyna, A. Penn, S.J. Spiegel and M.M. Veiga. 2007. Socioeconomic consequences of mercury use and pollution. Ambio 36: 46-61.

Temper, L., D. Del Bene and J. Martinez-Alier. 2015. Mapping the frontiers and front lines of global environmental justice: the E J Atlas. Journal of Political Ecology 22: 255-278.

Veiga M. M., R. Baker. 2004. Protocols for environmental and health assessment of mercury released by artisanal and small-scale gold miners. Global Mercury Project. United Nations Publications.

UN. 2015. Transforming our world: the 2030 Agenda for Sustainable Development, United Nations, New York. https://sustainabledevelopment.un.org/post2015/transformingourworld (accessed Aug 08, 2020).

UNEP (UN Environment Programme). 2020. Global Industry Standard on Tailing Management. Available from: https://globaltailingsreview.org/wp-content/uploads/2020/08/global-industry-standard_EN.pdf

World Bank. 2020. Minerals for Climate Action: The Mineral Intensity of the Clean Energy Transition. Climate Smart Mining (World Bank Group Report).

Worrall, R., D. Neil, D. Brereton and D. Mulligan. 2009. Towards a sustainability criteria and indicators framework for legacy mine land. Journal of Cleaner Production. 17: 1426-1434.

WEF (World Economic Forum). 2016. Atlas for Sustainable Development Goals (SDGs) to the mining sector. Available from: https://www.undp.org/content/dam/undp/library/Sustainable%20Development/Extractives/Mapping_Mining_SDGs_An_Atlas.pdf.

Wood, J.R., I. Montero-Ruiz and M. Martinón-Torres. 2019. From Iberia to the Southern Levant: the movement of silver across the Mediterranean in the Early Iron Age. Journal of World Prehistory. 32: 1-31.

Zhouri, A. 2018. Mineração, Violências e Resistências: um campo aberto à produção de conhecimento no Brasil. Editorial iGuana y Associação Brasileira de Antropologia (ABA), Marabá.

CRIAB Project – Conflicts, Risks and Impacts Associated with Dams: Looking for Sustainability and Human Rights

Jefferson de Lima Picanço[1]*, João Frederico da Costa Azevedo Meyer[2],
José Mario Martinez[2] and Claudia Regina Castellanos Pfeiffer[3]

[1] CRIAB, Geosciencies Institute, Unicamp, Brazil
[2] CRIAB, Institute of Mathematics, Unicamp, Brazil
[3] CRIAB, Laboratory of Urban Studies, Unicamp, Brazil

1. Introduction

The major accidents with tailings dams in Brazil, Samarco 2015 and Vale, in 2019, shocked and distressed the world with the number of deaths and the extent of environmental and social damages caused. This impact generated an intense debate with the inner causes behind these major accidents worldwide (Santamarina et al. 2019, Tanya and Allstadt 2018, Kossoff et al. 2014, Roche et al. 2017).

Unfortunately, disasters involving tailings dams are a relatively common global event and are relatively more frequent in mining sites and territories. Even though the total number of tailing dam failures are diminishing worldwide, great disasters are increasing in number (Bowker and Chambers 2015). Since 2014 we have had 28 major tailing dams' failures/spills all over the world (Wise Uranium Project 2020). According to this database, these disasters killed 548 and affected thousands of people, besides causing very serious and long lasting social and environmental impacts in several areas in many countries.

A disaster, whether caused by natural events or by anthropic actions (as well as inactions), involves not only the moment and the immediate aftermath of the catastrophic event but also those decisions and processes made over some time on both spatial and temporal scales that make this event possible (Valencio 2016, Zhouri et al. 2016). In the same way, after the event, the disaster's developments occur at many different levels, on quite different time scales. In addition to the simple

*Corresponding author: jeffepi@unicampr

failure of a soil and land dam structure, there are economic and political causes involved in the way these tailings dams are planned, operated, decommissioned, and eventually fail (Roche et al. 2017). Assessing the caused damage is also complex, involving knowledge of the local and down-river populations, territory, governance structures and the different social actors involved from the moment before the dam rupture until the post-disaster periods (Valencio 2016). These time/space scales can reach the dimension of thousands of kilometres and many years as in the disaster at Mount Poley (Peticrew et al. 2015), or the Samarco/2015 Disaster (Fernandes et al. 2016), or to cover different countries such as the disaster in Baia Mare/2003 (Macklin et al. 2003). The time involved in environmental recovery can be dozens of years, as in the case of Aznallcollar/1999 (Gallart et al. 1999). In the same way, they involve populations, both traditional and those neighbouring such sites as well as cities and towns downstream of those water bodies affected by contamination hundreds of kilometres away (Ricardo et al. 2018, Wilson et al. 2016). But it also involves political and institutional agents on local, regional and national levels, not to mention the different leaders of the mining industry and hedge funds operators around the world (Santos and Milanez 2017, Fonseca et al. 2017).

Rupture of tailings dams are phenomena that can be understood only from a transdisciplinary point of view. Only an overview can account for the different dynamics and processes that took or that are taking place in the mining territories and even so if this becomes possible in some ways.

These problems related to the construction, maintenance and eventually the rupture of these dams and the associated social, economic, and ecological consequences continue to be the object of intense study (Roche et al. 2017). Major disasters of this nature are singular facts and have the capacity to shock and generate empathy in large sectors of society. Sometimes, these feelings can generate activities and associations that act to mitigate social suffering and environmental and economic disastrous effects and seek solutions for the problems posed by these kinds of disasters (Zhouri et al. 2016). And this action can be a reflective one, directed at the historical and conceptual aspects that present themselves from the physical to the social environments. Or to move towards political actions, seeking conflict resolution in the struggle for social activism.

There are many research groups in Brazil dealing with mining-related problems. In fact, there are 158 officially registered groups in the Brazil Research Groups database CNPQ Research Group Directory that work with this theme (http://lattes.cnpq.br/web/dgp). Among these groups, reference should be made to GESTA/UFMG with long-term experience in the social study of conflicts related to mining activities (Zhouriet al. 2016) to the POEMAS/UFJF group with research focused on the relationship between the mineral economy and the performance of mining companies in the territory (Mansur et al. 2016) and the TERRA/UFJF group with an emphasis on the study of territories and related impacts (e.g. Mendes and Felippe 2019). Studying the impact of the Samarco/2015, disaster there is also the ORGANON group from Federal University of Espírito Santo (UFES). In northern Brazil, we must mention GETTAN/UFPA, which also works with these themes in the Amazon region (Almeida et al. 2019).

This text deals with one of these associations, CRIAB (the acronym in Portuguese for research and action group on conflicts, risks and environmental impacts associated with dams). Constituted immediately after the rupture of Vale's 01 dams in the Córrego do Feijão mine in Brumadinho, State of Minas Gerais, this group, still in the process of its formation, intends to make an effective contribution mainly in what regards the affected populations directly and immediately in mid-term and long-term consequences, involving both scientific knowledge and solidarity as well as relevant contributions in terms of policies and strategies for emergencies, contingencies, prevention actions and computational simulations of different scenarios.

CRIAB's purpose of helping the affected population, society and environment is also related to a responsible way of effectively operating in the sense of academic outreach. The workgroups that makeup CRIAB intend to act not as a bureaucratic NGO or a cold 'scientific' academic research team with no relation with the social actors present in these conflicts. To achieve these goals, the groups must be able to develop a careful and empathic relationship as to the needs of those involved in the human, social, environmental and political aspects and to care for their grief. And only after this is accomplished should CRIAB permit itself to undertake serious research in incisive and constructive models. To create these models the group is aligned with the objectives of sustainable development (SDO) of the UN (Griggset al. 2013). These 17 goals and 169 targets are today the most comprehensive set of policies and measures that could be easily understood and implemented. Although there are many relevant discussions about the assessment in obtaining these goals and targets (Hák et al. 2016)a preliminary set of 330 indicators was introduced in March 2015. Some SDGs build on preceding Millennium Development Goals while others incorporate new ideas. A critical review has revealed that indicators of varied quality (in terms of the fulfilment certain criteria, and even as so many aspects of sustainable development concepts in global southern hemisphere economic and social policies are refuted by some researchers, CRIAB understands that the working concepts are still very useful.

This text intends to contextualise the emergence of the CRIAB group in the scenario of the great ruptures of mining tailings dams in Brazil as well as to describe its process and carry out some reflections on its path. We will start with a contextualisation of mining tailings dams; after that, we will describe the disaster of the Córrego do Feijão mine and subsequently we will deal with the construction and consolidation of CRIAB and his challenges and further developments.

2. Tailings Dam Failures

Tailings dams are part of large-scale mining processes, as one of the last steps in the separation of ore from products that must be disposed of. The operations involve mining, transportation for processing, processing operations and the separation and transport of ore and the final disposal of tailings (Roche et al. 2017).

The beneficiation process involves the separation, by physical or chemical means, of ores from waste rocks. There are significant differences between the ore type and the bulk of tailings generated. These differences are also related to the ore grade, as well as the type of mineral processing in Mine facilities (Roche et al. 2017).

Gold ore tailings, for example, have much less volume than iron ore tailings. On the other hand, tailings from gold ores, as well as base metals (copper, zinc and nickel) have quite a large quantity of toxic substances, generating extremely acid tailings.

Tailings dams are structures that have the function of storing these different types of liquid or pulp residues derived from the processing of the ore (Bowles et al. 2007, USEPA 1994). However, since the beginning of the twentieth century, the size of these dams has grown significantly. A factor that influences the growth of these dams is the decrease in the ore grade, which has been consistently decreasing as the richer and higher-grade mines become exhausted (Davies and Martin 2014). At present, the most targeted prospects for mining are with low-grade large amounts of ore sites. On the other hand, these mines require increasingly larger and sophisticated methods of processing since the sterile/ore ratio has also been growing significantly (Davies and Martin 2014). As the tailings production is growing faster, most of these tailings are stored in higher and larger tailing dams (Bowkers and Chambers 2015). These increasing waste volumes are challenging the big mining companies as well as social and environmental agencies, concerning policymakers and threatening the neighbouring populations, economic activities and nature.

Tailings dams are very fragile structures, often built with the tailing's materials derived from the mining process. Its construction begins with a starting dike, made with soil and rock, in which the tailings are deposited. Many of these dams have a constructive history in multiple stages, depending on the mining or changes in the beneficiation process. According to the Global Tailings Portal (2020), 65% of the tailing dams worldwide are built by the upstream and downstream methods.

Tailings dams are frequently seen by the companies as a liability and considered merely as a cost issue (Davies and Martin 2014). In general, technologies used in the construction, operation and maintenance of these dams are less developed than the technologies used in the mining and material processing processes, which are the source of income for the project (Roche et al. 2017).

The most common causes of tailings dam failures are flooding, piping, overtopping liquefactions (Kossoff et al. 2014, Rico et al. 2008). Unusual weather conditions and poor management are also cited as the main causes in the last one hundred years (Azam and Li 2010). Failures are frequently a consequence of more than one factor. These disruptions can lead to a large spill of toxic substances into the environment with long-lasting impacts, difficult remediation and extremely costly in terms of financial, social, and ecological costs.

Often, at the global level, the regulation of these structures or their planning has little regulative principles and supervision. The stability of these structures requires continuous and meticulous monitoring and control during the design, construction, and operation of the dam. During the decommissioning processes of these structures, the high cost of remediation works is also relevant. As a result, in times of falling commodity prices, budget cuts in mining and processing projects often reach these structures before others (Davies and Martin 2015).

After these Brazilian mega-disasters, the International Council on Mining and Metals (ICMM) is assembling a prescriptive tailings dam standard to face these problems. This independent panel, composed by mining industries and UN experts, published a new standard, which involves respect to the rights of project-affected people and access to information, to develop a multidisciplinary knowledge to support and minimise risks at every phase of tailings dams lifecycles, from the planning to the closure (S&P Global Market Intelligence 2020a,b).

In this century, Brazil has already registered several disasters with tailing dams (Table 1). However, in recent years, a significant number of major disasters have caused much damage and impacted not only the environment and the economy but has also a seriously high number of casualties and dislodged population. The concern has led to a rise in the control and the inspection of these structures. However, the large number of dams existing in the face of the small number of inspection teams makes the work much more difficult.

The Brazilian National Dam Safety Policy is regulated by the law number 12,334/set/2010. The ore tailings dams in Brazil, according to this regulation, are in charge of the National Mining Agency (ANM). However, in 2017 only, two years after the Samarco disaster, the Integrated Mining Dam Safety Management System (SIGBM) was created. This database comprises 843 tailing dams registered National Mining Agency (ANM) database (SIBGM 2020). Most of this tailings dam are one-step, while dams built-in more steps are essentially built by the downstream method (20%), centreline method (9.1%) and upstream method (8%). Only 27 dams (3.2%) have volumes of more than 25 million m^3, 50 dams have heights up to 60 m (5.9%). From these, 259 tailing dams are considered of high potential risk.

3. Córrego do Feijão Tailings Dam Disaster

The Córrego do Feijão mine, where the tailings dam that caused the disaster was located, is in the municipality of Brumadinho in the metropolitan region of Belo Horizonte, capital of the state of Minas Gerais (Figure 1). The region is part of the Paraopeba River basin and integrates the Quadrilátero Ferrífero, one of the most important areas of iron ore production in the world. The Córrego do Feijão mine started in 1956, operated by Companhia de Mineração Ferro e Carvão, controlled since 1973 by Ferteco Mineração. In 2003, Ferteco became part of current Vale SA.

The Córrego do Feijão Mine's Dam 01 was built in 1974 and it was impounded in 1976. The dam has 11 rises until 2006, which took it to a height of 96 m and a volume of approximately 11 Mt tailings (Silva and Gomes 2013). Dam 01 was finally deactivated in 2015 (Robertson et al. 2019).

The rupture of dam 01 of the Córrego do Feijão mine occurred at 12:28 on January 25, 2019. The dam collapsed due to the liquefying of the contained material(Robertson et al. 2019). The collapse of the dam caused an avalanche that moved at a great speed to an ore beneficiation zone (Lima et al. 2020). Almost at the same time, the tailings began to move, forming a stream of debris with great destructive power besides its great speed too, which quickly hit the company's facilities and completely destroying them and killing hundreds of employees, mostly those in the cafeteria since it was lunchtime.

The flow continued downstream with lower albeit still destructive energy and speed, reaching several houses and rural areas until it became a deposit on the bed of the Paraopeba River, 8.4 km away (Lima et al. 2020). The Paraopeba river was partially blocked by the tailings. However, with the rupture of this dam, the finer

Table 1: Major Recent Tailings Dam Failure Disasters in Brazil (WISE 2020)

Date	Location	Parent company	Ore type	Type of incident	Release	Impacts
2019, Oct. 1	Nossa Senhora do Livramento, Mato Grosso,	VM Mineração e Construção, Cuiabá	Gold	Tailings dam failure	?	Tailings flowed 1-2 km, disrupting a power line
2019, Mar. 29	Machadinho d'Oeste, Oriente Novo, Rondônia,	Metalmig Mineração S/A	Tin	Failure of inactive tailings dam after heavy rain	?	Tailings spill damaged seven bridges, leaving 100 families isolated; no deaths or injuries reported
2019, Jan. 25	Córrego de Feijão Mine, Brumadinho, Minas Gerais,	Vale SA	Iron	Failure of tailings dam No. 1	12 million m^3	(In this chapter)
2018, Feb. 17	Barcarena, Pará,	Hydro Alunorte / Norsk Hydro ASA	Bauxite	Overflow of red mud basin after heavy rain	?	Highly alkaline and metal-laden liquids flooded the surrounding residential areas. On March 12, 2018, local environmental activist Paulo Nascimento was shot dead in front of his house.
2015, Nov. 5	Germano mine, Bento Rodrigues, distrito de Mariana, Região Central, Minas Gerais,	Samarco Mineração S.A.	Iron	Failure of the Fundão tailings dam due to insufficient drainage.	32 million m^3	Slurry wave flooded town of Bento Rodrigues, destroying 158 homes, at least 17 persons killed and 2 reported missing; slurry pollutes North Gualaxo River, Carmel River and Rio Doce over 663 km,
2014, Sep. 10	Herculano mine, Itabirito, Minas Gerais,	Herculano Mineração Ltda	Iron	Tailings dam failure	?	Two workers killed and one missing

(Contd.)

Table 1: *(Contd.)*

Date	Location	Parent company	Ore type	Type of incident	Release	Impacts
2009, April 27	Barcarena, Pará,	Hydro Alunorte/ Norsk Hydro ASA	Bauxite	Overflow of drainage channels after heavy rain		
2007, Jan. 10	Miraí, Minas Gerais,	Mineração Rio Pomba Cataguases Ltda	Bauxite	Tailings dam failure after heavy rain	2 million m³of mud, containing water and clay ("red mud")	The mudflow left about 4,000 residents of the cities of Miraí and Muriaé in the Zona da Mata homeless. Crops and pastures were destroyed, and the water supply was compromised
2001, Jun. 22	Sebastião das Águas Claras, Nova Lima district, Minas Gerais,	Mineração Rio Verde Ltda	Iron	Mine waste dam failure	?	Tailings wave travelled at least 6 km, killing at least two mineworkers, three more workers are missing

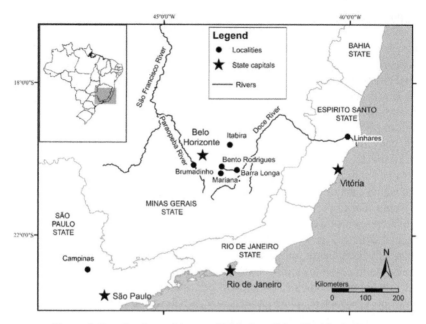

Figure 1: Localisations of Area and Main Localities Cited in the Text

material was taken by the river channel, increasing turbidity and causing the death of fauna and flora over a distance of hundreds of kilometres.

Since then, the population of the city has been trying to make life as it was before the disaster. But the circumstances are not favourable. Officially, there were 259 deaths with 11 persons still missing. Social damages include, in addition to deaths and disappearances, loss in agricultural lands and the water security. Another important aspect of the caused damages refers to the environment affected immediately as well as for long-term periods. Last but not least to be mentioned is the imposing presence of uncertainty about jobs and the income of a large part of the city's population. Grief and pain are present in every house of the area, almost visible, and strong and persistently manifested.

4. The CRIAB Group

The CRIAB group, the subject of this text, emerged as a consequence of the Córrego do Feijão mine disaster. Just after its occurrence, Professor Jefferson Picanço participated in the disaster response works in Brumadinho, as part of the CENACID group, of the Federal University of Paraná. This group, led by Prof. Renato Lima, aims to provide information to the different institutional and social actors in the post-disaster periods, to assist in rescue efforts and recovery work. The report of the mission in the Córrego do Feijão Mine can be obtained in (CENACID 2019). During the four-day period in which he was in this midst, Prof. Jefferson Picanço made daily reports of his participation, which were widely disseminated through UNICAMP's official site (Picanço 2019).

Upon returning from field activities in Brumadinho, Prof. Jefferson Picanço approached the dean of UNICAMP, Prof. Marcelo Knobel, with the proposal of creating a transdisciplinary group that could aggregate the different expertise existing in Unicamp to contribute to the phases of disaster mitigation and recovery (Dare 2019). For Professor Knobel, this is about "looking for skills to carry out an adequate research and being able to really collaborate not only with the victims but also in the prevention of future disasters" (Dare 2019). From this meeting, with the support of Unicamp's Dean of Graduation, Prof. Munir S. Skaf, a preparatory meeting was widely convened on February 12, 2019.

This meeting was attended to by more than a hundred people and was chaired by Prof. Munir Skaf. In spite of being an objective meeting, it was also very emotional, during which each of the participants presented and exposed their expertise and how they would like to participate in the group. The intention at this meeting was the effective creation of a group to act in the mitigation of environmental, economical and social damages as well as helping in collaboration with other local groups in the state of Minas Gerais, studies that could help in the recovery of the area. There was mention to possible actions in the most varied spectrum, from the areas of law, social psychology, medicine, social sciences, mathematics, statistics, physics and engineering, earth sciences and nature among others. Another characteristic was that several governmental and social organisations sent members of their professional teams proposing to be part of the group.

At this time, the group had an extensive multidisciplinary team of professors, researchers, technical-administrative staff and graduate and undergraduate students with representatives from almost all teaching and research units, nuclei and research centres at UNICAMP and other universities—USP and UNESP—and Research Institutes such as EMBRAPA (Brazilian Company of Research in Farming and Agriculture), the National Laboratory of Luz Synchrotron, the Boldrini Medical Center (a research and treatment organization for childhood cancers) and the CTI Renato Archer (research centre for Information Technology), among others.

5. CRIAB Activities

During this period, there were several meetings involving the entire CRIAB group that had an organisational character. These global meetings were held two to three times per semester, having gathered an average number of participants between 25 and 50 participants. Given their importance in the organisation of the group, some of these meetings will be briefly reported here.

After this first meeting, it became clear that in spite of the desired transdisciplinary outlook, the impressive variety of capabilities and aims would have to be separated into subgroups, a movement that happened in a natural and spontaneous manner.

Due to this attitude, in February 2019, a second organisational meeting was held at Unicamp's Institute of Geosciences, where the group's subsequent activities were defined. It was suggested that the group would be subdivided into five workgroups. The formed groups were: (a) WG of Education and Society, (b) WG of Mathematical Engineering; (c) WG of Physical and Biotic Environments, (d) Health WG and (e) Psycho-social and therapeutic WG. There was also the proposition of a group on economic and legal aspects, which did not prosper, unfortunately. These groups were informally conceived in order to encompass the main interests of the different groups of researchers and students. Today, in an active manner, only the Education and Society, Mathematical Engineering and Physical and Biotic Environment groups maintain active work and contacts.

A third general meeting, held on May 24, 2019, occurred in the Educorp/ Unicamp auditorium. In this meeting, the different WGs presented a diagnosis of the tailings dam situations with an emphasis on the Brumadinho region in view of the latest available data. Each group made a quick presentation of the latest news collected and the first research results presented, followed by discussions between all participants. Then there was a more general strategic discussion, involving the participants in order to think about identity for the group and its activities. It this meeting, the identity of the group was discussed for the first time with the choice of its name: Research and Action Group on Conflicts, Risks and Impacts related to Dams (the Portuguese acronym of which is CRIAB), which was chosen in order to identify intentions and areas of future actions. In a way, setting bounds on desires and will.

In this discussion, emphasising the academic design, the role of agent and participant in the processes was also highlighted. At the same time, as a group formed relatively far from the conflict area, CRIAB should have a collaborative role with other groups (both local and otherwise) that are already active in these areas. The group's scope was limited to the study of dams, although it was discussed whether it would also work with mining as a whole. Another issue is that of distances. Hundreds of kilometres away from the study areas with obvious problems of locomotion and logistics, it was discussed whether it could also participate in activities related to other types of dams around UNICAMP, that is also in the state of São Paulo. It should be said that after the dam failure in Brumadinho, at least three other dam failures have occurred in the States of Rondônia and Mato Grosso do Sul.

After these three first general meetings, the CRIAB group has been reorganised around Thematic Groups. In the two general meetings held in the second semester (September and December), the main discussions were about their organisation and activities. In the final meeting, on December 9, 2019, the objectives of each WG were presented in a more elaborated and precise format, and a proposed schedule of activities for the internal work of the WGs was presented as well. Other important and permanent issues discussed in these meetings was the search for funding. The December meeting was reserved for a general assessment of what was possible in each WG and a plan of what could be accomplished in the first half of 2020, both in the internal work of each WG and for CRIAB as a whole.

One of the first activities of the group was the organisation of an official UNICAMP Forum with the title of 'The Disaster of the Córrego do Feijão Mine: university and society in search of solutions'. With several participants from other universities and research groups as well as professionals from related areas, it was held at the Unicamp Convention Center on 8 May 2019 and, as expected, its main result was a reflective moment for discussion and convergence of studies and actions.

In this forum, specifically, the rupture of the dams in Brumadinho and the first results of the research groups that worked in the region were discussed. Prof. Renato

Lima (UFPR) spoke about the work of CENACID and in the construction of the Destructive Capacity map to assist in the understanding of the phenomenon and guide in the search for victims (and bodies). Professor Alfredo Costa (IFNM) showed multi scale data from the affected region immediately after the rupture of dam 01 along the Paraopeba river, reporting on the most mediated social and environmental impacts downstream.

The impact of the disaster in the region of the Paraopeba river, its similarities and differences with the Samarco 2015 disaster were presented and discussed. This debate, carried out by geologist Wilson Yomasa (IPT) and lawyer Fabrício Soler, sought to reflect on the disasters associated with other tailings dams involving legal and technical milestones, short, medium and long-term impacts in social, environmental and economic aspects and consequences. And, besides, the risk situations for communities and regions in the presence of other tailings dams.

During the forum, problems were pointed out, such as the lack of governance in disaster risk management, with disaster security public and private policies that are insufficient, poorly maintained and otherwise drastically underdeveloped.

As Professor Juliano Gonçalves (UFSCAR) stated there is an urgent need for institutional strengthening in terms of environmental policies, as well as a major investment in monitoring and organizational structure in order to make possible the broadening and the deepening of these issues and the creation of a much better dialogue with the vulnerable communities. The role of mining companies in the post-disaster was analysed as were acts undermining public policies and the population organisation. Were there more effective public policies, according to Prof. Leila Ferreira from the Center for Environmental Studies and Research (NEPAM) at Unicamp, it would be difficult for these disasters to become commonplace and to have such overwhelming magnitudes.

Participation in this permanent forum, from the speakers to the audience that attended the event, was very important, as it was a starting point for CRIAB's actions and official affirmation. It guided the discussions and galvanised people around a project that seemed to them extremely relevant urgent.

6. The Workgroups

The *Mathematical Engineering WG* started its activities at about the same time as CRIAB. Initially formed by professors from the Institute of Mathematics, Statistics and Scientific Computing, including physicists and statisticians, it was soon joined by engineers from several Unicamp colleges, besides professional actors from other universities as well as from other organisations. The group currently has approximately 15 stable members, including professors and undergraduate and graduate students. This last set is clearly expanding.

The objective of this WG is two-fold: first, to apply modern mathematical, statistical and computational developments to the prevention, prediction, and mitigation of extreme events of a social and environmental type, especially as in the rupture of dams; and secondly, appraise possible technological actions in the sense of the internet of things.

In general, fluid mechanics phenomena are well represented by systems of differential equations in partial derivatives (PDE). Unlike simple equations, which can sometimes be solved analytically with paper and pencil, most PDE systems can only aspire to the construction of approximate solutions that are obtained on the computer through sophisticated programming and numerical techniques. See, for example, Leveque. These are often large problems that require the use of advanced computational architectures. However, the problem of modelling is not limited to solving equations whose correspondence to real-life phenomena should be well consolidated. The most frequent is that the PDE system is a problem the data of which we do not know or if we do know something about it, it is that such information is very incomplete. Furthermore, the mathematical form of the systems of equations can be questioned, constituting an additional unknown. Situations of this type characterise what, in applied mathematics, is called 'inverse problems'. Their resolution needs the interaction between 'PDE solvers', usually called 'numerical analysts' and optimisation specialists. In the article on Numerical Analysis of the Encyclopaedia Britannica, the famous mathematician, now deceased, Gene Golub, defines a direct problem as one in which the unknown is the solution while an inverse problem would be one in which the unknown is the problem itself or part of it.

On the other hand, mathematical models also differ in their degree of empiricism. In different applied areas there is talk, under different names, of 'theoretical models', when their suitability for well-established physical laws is very large and 'empirical models' when physical laws are largely ignored (or even unknown), models in which it is only feasible to reproduce the available observations in the best possible way. Most models based on neural networks are included in this category. There is a range of possibilities that leads from one type of model to another, characterising what are sometimes called semi-empirical models. In practice, strictly theoretical models are very rare as are empirical models. In the general classification of dam models, semi-empirical models are called semi-physical. (See Froehlich D. 1995, 2016).

At the same time, mathematical models can be descriptive (for understanding the modelled problem and learning more about it) or predictive (with which scenarios and hypotheses can be tried out or experimented with). Obviously, the desired objective is the development of predictive models because these are the ones that have practical utility and allow the forecast of the real phenomena with some acceptable approximation. However, the descriptive models also fulfil a very important function that of helping to comprehend the phenomena in the way in which they were effectively observed and often constitute previous steps for the elaboration and evaluation of predictive models. For example, in modelling dam breaks it is common to find models that reproduce a specific break that has already occurred with some accuracy. Obviously, these descriptive models have been adjusted to observations of the phenomenon itself and are obtained after its occurrence. See, for example, (Pirulli et al. 2017)

Statistics seeks to establish relationships and correlations between variables and calculate or estimate their validity in probabilistic terms. See, for example Fai et. al. (2020). Finally, technological advances in terms of remote sensing and network theory can mean, if well implemented, considerable advances in terms of predicting and even preventing future disasters. Pure and applied mathematics,

Statistics, Physics, Fluid Mechanics, Hydraulics, Computing and Data Sciences are disciplines that the Mathematical Engineering WG seeks to cover with a view to a deep understanding of the effective causes, of the development, of possible effects and of options for repairing dam failures are themes related to scientific methodology and academic responsibility and commitment. The WG, at the same time, seeks, together with the rest of CRIAB, to influence community and state authorities with the powers to make the right decisions in terms of preserving human lives and the environment. An additional objective is the training of human resources at all levels and in various areas of knowledge, aiming at preparing professionals for the transdisciplinary approach to natural and social phenomena.

The main activity of the Mathematical Engineering WG has been the holding of weekly seminars, in general, taught by the stable members but which occasionally includes visitors from other Brazilian universities, research institutions and social agents.

The seminars deal mainly with issues related to mathematical modelling of dam failure and related phenomena as well as an introduction to specific software for computer simulations. At the same time, the statistical treatment and possible prevention and mitigation technologies that are emerging for the near future are being studied. In the first half of 2020, even under the limitation of the pandemic, the seminar was held regularly with lectures on modelling the dependence between leaked volume and distance reached in the rupture of dams, initial conditions for one-dimensional models of fluid flow, environmental contamination by toxic elements in the Paraopeba River basin, remote sensing for tailings dams, physical-mathematical models of fluids with application to dam accidents, classification and evaluation of models for dam rupture, the effect of the presence of upstream sediments in the analysis of hypothetical dams, mathematical tools for discontinuous and chaotic phenomena, types of failures that occur in dams and, finally, turbulence problems, all of which arise in dam failures.

Due to the pandemic, some of the planned activities could not be carried out. Most notably, several Science Days on the Street, considered fundamental for the awakening of popular awareness of ecological issues and environmental disasters, could not be held, unfortunately, and on the other hand, fieldwork that had been planned to corroborate or refute mathematical-computational developments developed in the WG also had to be postponed. These activities will certainly take place in the coming months when the restrictions arising from the end of covid19 pandemic isolation policies.

The *Physical and Biotic Environment WG* was formed by researchers in the areas of Geosciences, Engineering and Biological Sciences, involving researchers from the Institute of Geosciences and the Faculty of Technology, both at UNICAMP, and the above mentioned EMBRAPA of which there are three Labs in the region of Campinas. Among its main objectives are the characterisation of the original physical environment (soils, rocks, water and biota) and the socio-environmental impacts resulting from the rupture of tailings dams, the consequences on a large spectrum and the implications for human and environmental health followed by the assessment of the many impacts caused by loss of topsoil, water and biodiversity quality besides the appearance of toxic wastes in natural spaces. These problems are very difficult for society to overcome, be it in terms of space, in human development and with respect to serious health issues. Among the main research topics are: (a) identification of bioavailable substances in the water column, in the soil and the mud along the path reached by the tailings in the natural environment; (b) determination of the flow and the transport modelling at the soil-water-atmosphere interfaces; (c) identification of the main topics related to soil quality, water resources, and atmosphere. Studies to mitigate these impacts are planned, in addition to proposals for decontamination, recovery, and remediation of degraded areas.

The group held several meetings, in which it took stock of the main data available so far on the state of the environment in the Brumadinho region. A meeting was also held at the EMBRAPA Lab in Jaguariúna (a Campinas neighbouring town), where CRIAB researcher and collaborator Mariana Guerra presented the main ecotoxicity studies being developed in EMBRAPAS laboratories. Despite its potential, this WG was initially formed by countless graduate and undergraduate students, who returned in the medium term to their research activities since the group is not yet a viable way of attracting researchers. In other words, the dispersion of researchers by different agencies and institutes, and even in different cities, hindered the holding of larger face-to-face meetings. Today, the group's activities are related to Maíra Silva's Ph.D. work, entitled "Assessment of Environmental Contamination by Potentially Toxic Elements Present in the Paraopeba-Mg River Basin". This project, in its initial phases and expected to end in 2024, aims to evaluate the physical and chemical environment of the benthic environment of the Paraopeba River directly downstream from the area affected by the tailings. Quantitative and qualitative analyses of the action of heavy metals in the aquatic environment are to be carried out with the study of several trustworthy indicators. Another important contribution was a detailed qualitative description of the tailings flow during the Brumadinho event in 2019 (Lima et al. 2020). In this work, the impacts were characterised and categorised in terms of the flow, the energy involved as well as the impacts on the physical/biotic environment besides the man-made environment. This study, due to its preliminary character, serves as an important starting point for numerous activities to be generated by all CRIAB Work Groups.

The general objective of the Education and Society Work Group is to collaborate in the construction of ways to live, survive and resist in territories that are affected by the presence of dams, whether these dams have or not failed or are or not in danger of doing so. The thematic group's specific goals are: (a) to contribute to the reconstruction and strengthening of social and identity ties within populations affected by tailing dams, especially in the region of Brumadinho and other places in Minas Gerais State, such as Barão de Cocais, Barra Longa, Bento Rodrigues and Mariana counties (see Figure 1); (b) to give visibility of the asymmetric power relations in territories with the presence of tailing dams (whatever situation these might be in); (c) social perception about the presence of dams in living spaces; construction, organisation and guarantee of ample digital access to and from spaces of testimony, memories, registered and oral, archives that contribute to the social articulation of the affected communities; d) association between major initiatives in progress, especially between universities and public institutions that are consistent with the purposes of this WG.

In a group marked by a diverse range of interdisciplinary as well as highly transdisciplinary areas and theoretical-analytical approaches, it is common to place education—understood in a very comprehensive and inclusive way—as a practice that can affect the material conditions of existing social groups, their environments and their dynamics.

To achieve these goals, the Education and Society WG defined three intertwined main axes, which are: (a) Memory; (b) Archive and Communication; (c) Education. The WorkGroup prepared the first initiatives in liaison between these three axes. In that way, the Group started its activities, in terms of the deployment and a specific outlining of the Work Groups further developments.

In a synthetic way, our focus on 'Memory' is the memory of trauma, the memory of risks, memory of indifferences as a space for reconstruction and strengthening of social bonds. In the 'Archive and Communication' axis, our focus has been, from the Group's interpretative point of view, the organisation of a digital space that gives visibility to a set of projects, actions, movements, laws, rules, regulations that concern 'Conflicts, Risks, and Impacts Associated with Dams', as well as being a space for disseminating everything that the CRIAB Research and Action Group has been undertaking. In the 'Education' axis, our focus is planned as both a relationship with the formal education system, establishing a collective work to build a local curriculum that discusses proposed themes, as well as a relationship with non-formal education to also address specific issues involved in 'Conflicts, Risks, Impacts Associated with Dams'.

Finally, an important place of work in this WG is how to deal with different positions, not only disciplinary, interdisciplinary besides transdisciplinary but also theoretical positions and convictions. What gives us points in common are our goals and our social positions in relation to the political power of education; put in another way, we take into account that there are different ways to achieve collectively constructed objectives. Each of the various axes overlaps some of these modes, theoretically, disciplinarily and transdisciplinary identifying the analysed problem. This is done with no formal decision to conceptualise a problem or situation in terms of a precise and exhaustive formulation. That is, we assume at all times that, this being a collective, multidisciplinary and transdisciplinary enterprise, composed of different theoretical approaches, our reflections and elaborations will be continuously affected and influenced by both this important diversity as well as this extremely potent multiplicity when treated with respect, considering and taking into account whatever is presented and must be integrally considered. This is, therefore, the ethical-political position that guides us in our collective action: the elaboration of these central axes (which we identified as intertwined), which make some excerpts from certain disciplinary and transdisciplinary and theoretical positions, and which presupposes that will be affected by the sensitive listening to all its participants, e.g., researchers from this group and the populations affected by the existence by the risk and by the disastrous failure of these tailings dams.

7. Discussions

Our university's decision to commit itself and to involve professors, researchers, students and staff and the mobilisation caused by the terrible catastrophe in

Brumadinho led to the formation of the CRIAB Group, and this has already been mentioned in previous texts (Coll 2019). The occurrence of these great disasters does not fail in affecting socially responsible groups and sectors, bringing up empathy, emotion and the desire to 'do something about this'. On the other hand, media contributes to a diminishing feeling about disasters like the one we have been studying, by starting with an intense news cover that gradually reduces in intensity and time, petering out finally – as a consequence, public interest wanes in the disaster's medium and long term effects, equally serious and heavily affecting local society and nature as well as all the downriver menaces. This is when the disaster becomes 'slow' and effectively needs more attention, be it economical, social, in terms of public policies, as well as in scientific efforts and historical registers. These moments are when negotiations and agreements among the several engaged agents and subjects become much harder, have a tendency to be stretched out and this is when economical enterprises make efforts to reduce both their costs and their several and multiple responsibilities.

CRIAB appeared during the 'hot' moments of the disaster, that is when the several aspects of social shock were still occurring, moments which mobilise many social agents as well as academic researchers and groups in universities, research centres, scientific societies and agencies.

As it becomes more and more difficult to organise, put in action and maintain programmed agendas, many team members return to their many daily tasks, gradually reducing work, interest and energy in continuing the related work and eventually stopping their contributions altogether. On the other hand, a good part of the researchers, being as they were so seriously involved with the creation of CRIAB and the need to face the challenges that had appeared (and still are appearing) ended up by creating a commitment that in spite of the having to make time for the extra work, alongside all the rest, eventually formed a threshold of dedication for the group to continue to work and to work hard. The necessity of future actions, even when effective conditions for the work were still inexistent made many participants of the ample research group to stay at it and to resist, continuing to participate regularly in all activities.

Interactive cooperation between researchers of the different workgroups and their different areas of interest of their multiple capacities and expertise allied to the working group's regular meetings and diverse interchanges, whether face-to-face, by e-mail or editing documents did very much to make the larger group blend into a group the diversity of which, even though not yet characterised by its transdisciplinarity, is rapidly becoming so.

Initially, this professional interchange between researchers of all the different areas of interest and work is simple and, in general, easy. A statistician, for example, can discuss and solve problems related to the gathering of biochemical data. Researchers that work with mathematical modelling can provide necessary instruments for colleagues at work with empirical information obtained from dam failures. Many other examples and possibilities exist. Another essential factor was the inclusion of graduate students (and some undergraduates) whose energy and outlooks made the interdisciplinary relations between all participants more explicit, organised, and vital. The need to formally discuss many subjects with students from many different origins and interests has interfered in a fully positive manner, generating and promoting the presence of an excellent, albeit challenging, working ambience for the work of the whole group.

Creating common scientific languages—or even scientific dialects—is a very ambitious endeavour, a work effort that must continue to be carried out carefully and systematically. And different ways, as well as different professional stances, practices and values in working even within the different groups, raise important questions, situations, necessities, interests. The semantical meaning of terms, reasoning, actions and expressions (and even words)ends up having an extra value in the understanding of the different situations, consequences, motives and, of course, communication between researchers. An appropriate definition of disaster, as proposed by several authors in the sociology of disasters (e.g., Valencio 1990, Zhouri et al. 2016), is placed under questioning and critical evaluations vis-*à*-vis a definition of a criminal disaster as some of these mentioned authors have pointed out. Another important aspect of this joint effort developed in the midst of a significant diversity is the tension between universal research and local and regional actions, a tension present in CRIAB's name as well as in its work. Besides this tension as well as other ones, which are construed and mediated by discussions and achieved through consensus are fundamental in guiding the whole group forward in its research proposals and action in order to alleviate mainly the local populations' suffering, hardships, lack of hope and, many times, their despair in the environment brought about by disasters caused by dams and their risks and failures.

In order to maintain this kind of contact with the affected individuals, families, communities or groups, the group has a firm agreement that all relationships with the affected population great care must be always present and practised. During the short, medium and long-term periods when severe effects are still being experienced, any action not previously and adequately planned and not carefully executed may cause disruptions due to memories of terror, fear, despair, helplessness and feelings of suffering and loss that certainly interfere with the purposes of the group's actions and purposes. Having the trauma lived by the individuals and their groups exposed to each new researcher's attitudes, present another cruel aspect of the situations lived and, therefore, relieved to be avoided.

So all this careful preparation, planning and actual implementation receive paramount importance in a contact with local and affected communities according to the UN's sustainable development objectives. In other words, these are paths and decisions to be tread with extreme human respect and social care. With the complete human and communal sustainability in perspective.

8. Conclusions

In lieu of conclusions, besides the acceptance of a very serious relationship between science, politics and survivals (human, social, economic, environmental and cultural), to point out main aspects of this joint survival, there is no discussion that the work undertaken until now is but a starting point, from which derive many future challenges, and in which there must be a human and scientific commitment to life.

The possibility of multidisciplinary projects in the near future, from the group itself or/and in partnership with other national and international researchers, will be a good alternative and a bet for the future. Only in joint projects, which require unitary responses, can the group embrace a more interdisciplinary practice.

On the other hand, the interaction with research groups that are in direct contact not only with Brumadinho and Mariana populations but also with other mining areas that could potentially be subject to disasters related with dams failures (Itabira, Mariana, Antônio Pereira, Barão de Cocais, among others) is also the next possibility of action. Interaction with these research groups and those affected populations will certainly affect the dynamics of the CRIAB group, change it, and perhaps change its members or partners significantly. However, it is a path of no return, if the group wants to fulfill the objectives that are decided upon.

On the other hand, it is worth noticing that the trial of coordinating efforts aimed at creating and sharing knowledge between universities and populations affected by dams with disasters is a major challenge. In addition to the territory to be studied, the new syntaxes that must derive from the conversations between those involved, it is crucial to say, are an attempt to develop and involve bonds and affections. The same original affection that animated and sustained CRIAB since its foundation should be directed in the near future to actions that aim at a more in-depth discussion about mining practices in the region and change the practices of Mining companies in their relationship with the surrounding populations. Several problems and issues involving sustainability issues must necessarily arise from this change. The intention is that the deepening of these issues will strengthen the human rights of the populations living in mining areas with respect for the rights for the well-being of these populations, respect for the environment and its future perspectives and for an economic arrangement that does not sacrifice the necessary resources for future generations.

Acknowledgements

The authors acknowledged the institutional support provided by Unicamp to the set up of CRIAB group, headed by the Dean, Professor Marcelo Knobel, and Unicamp's Dean of Graduation, Prof. Munir S. Skaf. We also warmly acknowledge our colleagues from CRIAB group, who with kindness and strength are building this idea.

References

Almeida, A.F.B., E.A. Melo, I.T.R. Nepomuceno and V.C. Benvegnu (orgs). 2019. Mineração e Garimpo em Terras tradicionalmente ocupadas: conflitos sociais e mobilizações étnicas. 1. Ed. - Manaus: UEA Edições/ PNCSA, 2019.

Aires, U.R.V., B.S.M. Santos, C.D. Coelho, D.D. da Silva and M.L. Calijuri. 2018. Changes in land use and land cover as a result of the failure of a mining tailings dam in Mariana, MG, Brazil. Land Use Policy. 70: 63-70.

Azam, S. and Q. Li. 2010. Tailings dam failures: a review of the last one hundred years. Geotechnical News. 28(4): 50-54.

Bowker, L.N. and D.M. Chambers. 2015. The risk, public liability, and economics of tailings storage facility failures. Earthwork Act. 1-56.

Bowles, D.S., F.L. Giuliani, D.N. Hartford, J.P.F.M. Janssen, S. McGrath, M. Poupart, D. Stewart and P.A. Zielinski. 2007. ICOLD bulletin on dam safety management. IPENZ Proceedings of Technical Groups. 33(2).

BRASIL 2010. Lei 12334 de 20 de setembro de 2010. In: http://www.planalto.gov.br/ccivil_03/_ato2007-2010/2010/lei/l12334.htm ; Accessed 15 October 2020.

Carmo, F.F., L.H.Y. Kamino, R.T. Junior, I.C. de Campos, F.F. do Carmo, G. Silvino, M.L. Mauro, N.U.A. Rodrigues, M.P. de Souza Miranda and C.E.F. Pinto. 2017. Fundão tailings dam failures: the environment tragedy of the largest technological disaster of Brazilian mining in global context. Perspectives in Ecology and Conservation. 15(3): 145-151.

CENACID (CENTRO DE APOIO CIENTÍFICO A DESASTRES-UFPR) 2019. Emergency CENACID mission to support the disaster response related to the tailings dam disruption of the Córrego do Feijão Mine in the municipality of Brumadinho-MG (in Portuguese). In: http://www.cenacid.ufpr.br. Accessed 12 September 2019.

Coll, L. 2019. Pesquisadores dedicam-se ao tema de Brumadinho e de barragens de rejeitos. In: https://bit.ly/2IDf2Ml. Accessed 8 October 2020.

Dare, E.F., 2019. Grupo multidisciplinar se reúne para estudar Brumadinho. In: https://www.unicamp.br/unicamp/noticias/2019/02/12/grupo-multidisciplinar-se-reune-para-estudar-brumadinho. Accessed 15 October 2020.

Davies, M. and T. Martin. 2009. Mining market cycles and tailings dam incidents. *In*: 13th International Conference on Tailings and Mine Waste, Banff, AB. http://www. infomine. com/publications/docs/Davies2009. pdf. Accessed 27 July 2019.

Fais, L.M.C.F. and V.A. González-López, D.S. Rodrigues and R.R. Moraes. 2020. A copula based representation for tailings dam failures. 4 open 3 12. DOI: 10.1051/fopen/2020011

Fernandes, G.W., F.F. Goulart, B.D. Ranieri, M.S. Coelho, K. Dales, N. Boesche, M. Bustamante, F.A. Carvalho, D.C. Carvalho, R. Dirzo and S. Fernandes. 2016. Deep into the mud: ecological and socio-economic impacts of the dam breach in Mariana, Brazil. Natureza & Conservação. 14(2): 35-45.

Froehlich, D.C. 2016. Predicting peak discharge from gradually breached embankment dam. Journal of Hydrological Engineering. 21(11).

Gallart, F., G. Benito, J.P. Martin-Vide, A. Benito, J.M. Prió and D. Regüés. 1999. Fluvial geomorphology and hydrology in the dispersal and fate of pyrite mud particles released by the Aznalcóllar mine tailings spill. Science of the Total Environment Dec 6. 242(1-3): 13-26.

Global Tailings Dam Portal. 2020. https://www.grida.no/publications/472; Accessed 15 August 2020.

Griggs, D., M. Stafford-Smith, O. Gaffney, J. Rockström, M.C. Öhman, P. Shyamsundar, W. Steffen, G. Glaser, N. Kanie and I. Noble. 2013. Sustainable development goals for people and planet. Nature. 495(7441): 305-307. https://doi.org/10.1038/495305a

Hák, T., S. Janoušková and B. Moldan. 2016. Sustainable Development Goals: A need for relevant indicators. Ecological Indicators. 60: 565-573. https://doi.org/10.1016/j.ecolind.2015.08.003

Kossoff, D., W.E. Dubbin, M. Alfredsson, S.J. Edwards, M.G. Macklin and K.A. Hudson-Edwards. 2014. Mine tailings dams: characteristics, failure, environmental impacts, and remediation. Applied Geochemistry. 51: 229-245.

Lima, R.E., J.L. Picanço, A.F Silva and F.A. Acordes. 2020. An anthropogenic flow type gravitational mass movement: the Córrego do Feijão tailings dam disaster, Brumadinho, Brazil. Landslides. 1-12.

Macklin, M.G., P.A. Brewe, D. Balteanu, T.J. Coulthard, B. Driga, A.J Howard and S. Zaharia. 2003. The long term fate and environmental significance of contaminant metals released by the January and March 2000 mining tailings dam failures in Maramureş County, upper Tisa Basin, Romania. Applied Geochemistry. 18(2): 241-257.

Mansur, M.S., L.J. Wanderley, B. Milanez, R.S.P.D. Santos, R.G. Pinto, R.J.A.F. Gonçalves and T.P. Coelho. 2016. Antes fosse mais leve a carga: introdução aos argumentos e recomendações referentes ao desastre da Samarco/Vale/BHP Billiton. 17-49.

Mendes, L.C. and M.F. Felippe. 2019. Alterações geomorfológicas de fundo de vale na bacia do Rio do Carmo (MG) decorrentes do rompimento da barragem de Fundão. Caminhos de Geografia. 20(69): 237-252.

Petticrew, E.L., S.J. Albers, S.A. Baldwin, E.C. Carmack, S.J. Déry, N. Gantner, K.E. Graves, B. Laval, J. Morrison, P.N. Owens and D.T. Selbie. 2015. The impact of a catastrophic mine tailings impoundment spill into one of North America's largest fjord lakes: Quesnel Lake, British Columbia, Canada. Geophysical Research Letters. 42: 3347-3355. doi:10.1002/2015GL063345.

Picanço, J. 2019. Diario de Brumadinho. In: https://www.unicamp.br/unicamp/noticias/diario-de-brumadinho. Accessed 15 October 2020.

Pirulli, M., M. Barbero, M. Marchelli and C. Scavia. 2017. The failure of the Stava Valley tailing dams (Northern Italy): numerical analysis of the flow dynamics and rheological properties. Geoenvironmental Disasters. 4: 3.

Rico M., G. Benito and A. Diez-Herrero. 2008a. Floods from tailings dam failures. Journal of Hazardous Materials. 154(1-3): 79-87.

Robertson, P.K., L. Melo, D.J. Williams and G.W. Wilson. 2019. Report of the Expert Panel on the Technical Causes of the Failure of Feijão Dam I. December 12 2019. http://www.b1technicalinvestigation.com/pt/. Accessed 29 dec 2019.

Roche, C., K. Thygesen and E. Baker, 2017. Mine tailings storage: safety is no accident. A UNEP Rapid Response Assessment. United Nations Environment Programme and GRID-Arendal, Nairobi and Arendal. www.grida.no. ISBN: 978-82-7701-170-7

Santamarina, J.C., L.A. Torres-Cruz and R.C. Bachus. 2019. Why coal ash and tailings dam disasters occur. Science. 364(6440): 526-528.

Santos, R.S.P.D. and B. Milanez. 2017. The construction of the disaster and the "privatization" of mining regulation: reflections on the tragedy of the Rio Doce Basin, Brazil. Vibrant: Virtual Brazilian Anthropology. 14(2).

Schoenberger, E. 2016. Environmentally sustainable mining: the case of tailings storage facilities. Resources Policy. 49: 119-128.

S&P Global Market Intelligence. 2020a. After Brazil tailings disaster, standards need to be more 'prescriptive'. In: https://bit.ly/3lRpqhD; Accessed 15 October 2020.

S&P Global Market Intelligence. 2020b. Global standard on waste dams released in response to multiple mining disasters. In: https://bit.ly/3dwLlb0; Accessed 15 October 2020.

Silva, W.P.D. and R.C. Gomes. 2013. Tailings liquefaction analysis using strength ratios and SPT/CPT results. Soils and Rocks. 36(1): 37-53.

SNISB (Sistema Nacional de Informações sobre Segurança de Barragens). 2020. In: http://www.snisb.gov.br/portal/snisb, Accessed 15 October 2020.

Tanya, H. and K.E. Allstadt. 2018. An updated method for estimating landslide-event magnitude 1847. 1836–1847. https://doi.org/10.1002/esp.4359.

USEPA (US Environmental Protection Agency). 1994. Design and evaluation of tailings dams. Technical Report. 40-49.

Valencio, N. 2016. Elementos constitutivos de um desastre catastrófico: os problemas científicos por detrás dos contextos críticos. Ciência e Cultura. 68(3): 41-45.

WISE (WISE Uranium Project-Tailings Dam Safety). 2020. Chronology of major tailings dam failures. http://www.wise-uranium.org/mdaf.html. Accessed 14 February 2020.

Zhouri A., N. Valencio, R. Oliveira, M. Zucarelli, K. Laschefski and A.F. Santos. 2016. O desastre da Samarco e a política das afetações: classificações e ações que produzem o sofrimento social. Ciência e Cultura. 68(3): 36-40.

Habitability and Health: A Relationship with Energy Efficiency

Andrea Lobato-Cordero[1]* and Sônia Regina da Cal Seixas[2]

[1] Universidade Estadual de Campinas (Unicamp), Campinas, Brazil
[2] UNICAMP, Brazil

1. Introduction

Addressing sustainability from a multi-interactive vision is an opportunity to integrate the inhabitants in different areas of action. In the case of a house, created to improve living conditions, this idea of protection is lost as a consequence of improper design or construction. Actions to provide new housing, renovate existing housing or save energy in the built environment are popular and widespread. However, situations such as climate change or a pandemic remind us that sustainability starts with human beings and their ability to maintain their lives in healthy conditions.

From this perspective, it is essential to analyze what the role of a house is, the household characteristics and the inhabitant's health. Initially, the house and the actions to improve its services were contextualized with international documents, such as the Sustainable Development Goals (SDGs).It is worth to mention that these goals are part of current legal documents in several countries, which in turn are part of official commitments to achieve global objectives.

Furthermore, there is empirical evidence that supports the idea that the architecture, the design strategies, the materials used in the envelope, the inhabitants, and so forth modify the indoor conditions of a house. The indoor environment in a house is directly related to energy consumption and outdoor conditions. For this reason, improving the indoor environment is the main objective of the energy efficiency programs. On the contrary, an improper indoor environment could endanger the health of the occupants by exposing them to different risks.

Finally, a discussion about the interrelation between energy consumption, the air quality and the indoor temperature, and their impact on the well-being of the inhabitants of a house is proposed in this chapter. Research related to COVID-19 in the indoor environment of a house was considered. Although there is still uncertainty

*Corresponding author: andreaplc@hotmail.com

about various aspects of this virus, it is important to integrate it due to its alignment with this study.

2. Sustainable Approach to the Role of a House

Through a review of topics that address environmental conditions, climate change, energy consumption, housing and the health of its occupants, this section will examine the role of a house. Although a large number of topics can be connected in this chapter, mainly those related to Climate Action (SDG 13), Enhanced sustainability of cities (SDG 11), Energy (SDG 7) and Good health and Well-Being (SDG 3) will be addressed. However, it is important not to lose sight of human beings as the driving force and receiver of actions to improve life in different contexts.

The Sustainable Development Goals adopted by the United Nations General Assembly were consolidated as a normative framework with the capacity to be applied in a transversal manner. The 2030 Agenda for sustainable development, proposed by the United Nations (UN), is an action plan represented by interlinked goals, targets, and indicators. According to Breuer et al. (2019) with Kanie and Biermann (2017), the 17 SDGs and the 167 associated targets "combine the human development process of the Millennium Development Goals (MDGs) and the sustainable development Rio+ process. They also expanded the range and depth of topics covered substantially and signaled the necessity for a shift in governance strategies" (Breuer et al. 2019).

Theoretical aspects of SDGs are part of the practical processes, integrating international, national and local actions. Measuring and understanding development, as well as environmental interventions designed to have a positive impact on the well-being of inhabitants, requires local points of view to lead to concrete actions (Schleicher et al. 2018). According to Knox (2015), actors involved in promoting human rights and protecting the environment are supported by SDGs. Vulnerable people to environmental harm are the principal objective to define indicators and compare information, which allows measuring the impact of different interventions to avoid the document to be solely theoretical. Therefore, it is crucial to identify needs, limitations and existing knowledge of each local environment in order to understand how to use them for policy-making and support research on SDG interdependencies (Breuer et al. 2019).

Analyzing scientific literature related to the interaction of SDGs is a way to evaluate the implementation of the 2030 Agenda and orient future actions. A mapping of SDGs interaction shows that policy innovation and issues of integrated monitoring and evaluation were not considered. Few studies consider geographic scales and spill-over effects or interactions across SDG indicators. Some studies employ participatory methods or adopt a systemic approach to the 2030 Agenda, but "none of the studies in the reviewed sample consider interactions between the actors responsible for implementing the SDGs" (Bennich et al. 2020).

In this sense, Schapper and Lederer (2014) emphasize the need to establish the micro-macro link as a mechanism of interaction between actors and institutions. Likewise, the interrelation of SDGs and their unidirectional or bi-directional impact is a confirmed argument. An analysis of the positive and negative interactions between the seventh SDG Energy, and the other 16, based on a review of scientific literature, found more positive than negative interactions, both in number and magnitude. Nevertheless, the access to energy and energy efficiency contribute to the challenges of "achieving poverty alleviation (1SDG), better human health (3SDG), greater water availability and quality (6SDG), enhanced sustainability of cities (11SDG), natural resources protection (12SDG), reduced climate change (13 SDG), and strong and fair institutions (16SDG)" (McCollum et al. 2018).

Promoting and protecting human rights are the articulating axes of the SDGs. The proposal on human rights due to climate change, presented in 2015 by the Swedish International Development Cooperation Agency, brings together various aspects and actors at different levels of governance, as an example guide to improve the results of initiatives against climate change. In particular, development programs executed in various environments would have the greatest benefit if people were included as active participants and not as passive recipients (Sida 2015).

The Intergovernmental Panel on Climate Change (IPCC), as the international body assessing of science related to climate change, provides a scientific basis to governments at all levels to develop climate-related policies. It postulated that the temperature increase from 1.5°C to 2°C andthe effects of climate change will affect people's well-being in different aspects: economic, social, physical, psychological (UNHR 2015). But the effects of humans on the environment have escalated. About 30 to 50% of the planet's land surface is exploited by humans producing the so-called anthropogenic emissions. Carbon dioxide emissions have changed the global climate significantly. Toxic substances are released into the environment; even some of them that are not toxic have damaging effects (Crutzen 2002). According to Mapp and Gabel (2019), with Philip Alston report (2019), 50% of carbon emissions were generated by 10% of the richest people in the world. Thus, the application of strategies to promote, respect, protect and fulfill human rights, including the mitigation of the Greenhouse effect is, directly and indirectly, related to the development of life and its surroundings.

The global warming presents risks to natural and human systems because any increase in global temperature is projected to negatively affects human health (Hoegh-Guldberg et al. 2018). An executive summary for The Lancet Series mentions that climate change will harm human health and successful strategies to mitigate its effect will restrict that harm. Health benefits of tackling climate change as policies intended to reduce the output of greenhouse gases will offset some of the costs of climate change mitigation measures and will benefit both the environment and public health (The Lancet 2009).

Cases analyzed by Kumar (2018) in North America, refer to the inequality of health conditions in social groups. This reinforces the direct relationship between climate change and the impact on health. Moreover, it impacts the physical infrastructure, human settlements, livelihoods, security, among others, causing the conceptualization and landing through local governments and private actors. International guidelines, build normative documents that will govern in developed and developing countries which would allow the assessment of local vulnerability on the guidelines of exposure, sensitivity and adaptability, defined by the IPCC. Seixas and Nunes (2017) confirm how the subjectivity of the inhabitants has a complementary role in studying the effects of climate change in different social contexts, and how it can compromise the projected results from a uniquely epidemiological perspective. They refer a study of fishing communities in the densely populated municipality of Caraguatatuba at the North Coast region of the State of São Paulo-Brazil. In this place, external agents reduced fish stocks and caused a greater effort by fishermen to maintain the diets and livelihood of their families. Data collected by health care professionals in the same period showed that 60% of patients were diagnosed with depression (Seixas and Nunes 2017).

The increase of temperature in specific areas, known as Urban Heat Island (UHI), is more evident in urban areas than in rural areas. Among the urban features related to housing are buildings and traffic (whose urban effect is the direct addition of heat) and construction materials (whose urban effect is increased thermal admittance and water-proofing of surfaces) (Howard 1988). In addition, approximately an "80% of emissions from material production were associated with material use in construction and manufactured goods" (UNEP 2020). Therefore, if the stages of manufacture and use of construction materials are included in the effects of the increase in temperature in urban areas, "can also affect other stages of the life-cycle of residential buildings, leading to synergistic reductions of energy use" (UNEP 2020). An analysis about residential buildings performance simulation that included the urban heat island effect in South American cities (Guayaquil in Ecuador, Lima in Peru and Valparaiso in Chile) reported that diurnal and nocturnal temperatures greatly vary across the analyzed cities, hence the UHI intensity. The energy demand increases between 15% to 200% when UHI is incorporated in building performance simulation, which is commonly used to calculate heating and cooling demand. In fact, "building operation is doubly related to the urban environment: on the one hand, buildings generate heat which warms up the environment, and on the other hand, the urban environment alters building performance by the influence of UHI" (Palme et al. 2017).

Other researchers studied the vulnerability to extreme heat in the state of New York. The study developed a heat vulnerability index and identified four components: (1) social/language vulnerability, (2) socioeconomic vulnerability, (3) environmental/urban vulnerability, and (4) elderly/social isolation (Nayak et al. 2017). Differences in regional sociodemographic and land cover characteristics can convert most vulnerable "urban areas with high housing density, less open space, and high proportions of elderly, minority populations, and lower in-come households" (Nayak et al. 2017).

Housing is a key component of the conformation of urban areas. In Latin America "the quantitative housing deficit represents only 6% of the total deficit in urban areas, while the largest deficit is related to the coverage and quality of services, house ownership and other qualitative aspects. The qualitative deficit represents 94% of the total deficit, 90% of the new housing solutions implemented, with policies that have not to improve the quality of the existing stock and its environment" (BID 2018).

The Economic Commission for Latin America and the Caribbean (ECLAC), CEPAL by its Spanish acronym, published a review of 14 non-monetary indicators of poverty to monitor the progress of the first SDG: end poverty in all its forms everywhere. This document considers three dimensions: education, work and social protection, and health care. It also shows how these indicators are approached in developed and underdeveloped countries in relation to "housing materials, overcrowding, housing tenure, durable goods, access to safe water and improved sanitation, access to clean sources of energy, garbage collection and nearby sources of pollution, (public) transportation, child attendance to school and adult schooling, employment, social security and access to health care" (Santos 2019). However, the applicability of the indicators based on the unit of identification of these, at the individual or household level, is put into context and emphasize the benefit of anthropogenic data in health aspects.

Global Burden of Disease analysis 2017 mentions that "total burden of disability increased by 52% between 1990 and 2017" (IHME 2018). Disability and development data, shows respiratory infections and tuberculosis in second place of the global burden of disease, by years of life lived with any short-term or long-term health loss. According to development levels, lower respiratory infections in second place of the global burden of disease by years of healthy life lost to premature death and disability in low socio-demographic Index (SDI) countries. Latin America and the Caribbean (LAC) is one of the GBD super-regions with a greater difference between observed and expected healthy life expectancy in 2017. Future health trends of this study 2016-2040, present lower respiratory infections as leading causes of early death in third place (IHME 2018). Financing Global Health 2019, press in July 2020, examines spending related to the SDGs with a focus on SDG 3 linked SDG 11 and SDG 16 to promote healthy societies. It provides estimates of spending on health, "development assistance for health, spending for HIV/AIDS, tuberculosis, and malaria, as well as projections of future health spending" and the related to pandemic preparedness, ongoing COVID-19 (IHME 2020). While countries experience economic growth, government health spending remained the leading source of total global health spending. Latin America and the Caribbean saw a decrease of 18% in Development Assistance for Health (DAH) between 1995 and 2017 (IHME 2020). It concluded that "COVID-19 – because of its burden and effects on daily life and in-person social interaction –" evidences the need to strengthen health financing as a public good to face the crisis (IHME 2020). In addition, "the importance of building robust health systems that can absorb shocks, maintain extensive disease surveillance, and provide care for large populations in emergency settings" (IHME 2020).

Besides, Article 25 of the Universal Declaration of Human Rights, refers to the house as part for living adequate for the health and well-being, for everyone (UN 1948). World Health Organization (WHO) guidelines of housing and health affirm that "Improving housing conditions can save lives, prevent disease, increase quality of life, reduce poverty, help mitigate climate change and contribute to the achievement of the Sustainable Development Goals, including those addressing health (SDG 3) and sustainable cities (SDG 11)" (WHO 2018).

From this perspective, extreme situations such as the new COVID-19, a disease caused by the SARS-CoV-2 coronavirus with an impact on public health worldwide, has happened. It mainly causes respiratory illnesses that range from a mild illness to severe illness and death; besides, some people infected with the virus never develop symptoms. WHO, on March 11, 2020, communicated in a press conference, in charge of the Director-General Dr. Tedros Adhanom Ghebreyesus, "concern for the alarming levels of propagation and gravity, as well as the alarming levels of inaccuracy of COVID-19, evaluating that could be characterized as a pandemic". In addition, he reiterated the call "for countries to adopt a comprehensive society approach, based on a comprehensive strategy to prevent infections, save lives and minimize the impact" (WHO 2020a).

According to WHO, "People can catch COVID-19 from others who have the virus. The disease is spread mainly from person to person through small droplets from the nose or mouth, which are expelled when a person with COVID-19 coughs, sneezes, or speaks. These droplets are relatively heavy, do not travel far and quickly sink to the ground. People can catch COVID-19 if they breathe in these droplets from a person infected with the virus. This is why it is important to stay at least one meter away from others. These droplets can land on objects and surfaces around the person such as tables, doorknobs and handrails. People can become infected by touching these objects or surfaces, then touching their eyes, nose or mouth. For this reason, it is important to wash hands regularly with soap and water or clean with alcohol-based hand rub" (WHO 2020b). On 9 July 2020, WHO published an updated Scientific Report on the modes of transmission of COVID-19, which includes: contact, droplet, airborne, fomite, fecal-oral, bloodborne, mother-to-child, and animal-to-human transmission (WHO 2020c).

The Update COVID-19 Strategy, published by WHO, on April 14, 2020, collects implemented measures to control the transmission of the virus and the stage in which each country. Among the protection measures applied are hands washing, avoiding touching the face, practicing good respiratory etiquette (take precautions when breathing, sneezing, coughing among others), individual-level distancing and cooperating with physical distancing measures and movement restrictions when called on to do so. Nevertheless, physical distancing measures and movement restrictions often referred to as "shutdowns" and "lockdowns" have had a socioeconomic impact (WHO 2020d).

This change in the dynamics of the work activities, education, health, and services, in general, makes the house a suitable or safe place to work or continue with online studies. Evidently, people perform their work activities from their houses because it is assumed that the safest place is being inside the house.

According to estimates of the home-based work potential as a result of the COVID-19 pandemic by International Labour Organization (ILO), approximately 23% of workers than live in LAC countries would have "the infrastructure that would allow them to effectively perform their work from home" (ILO 2020). This study makes reference to homeworking and teleworking, where the second is a narrower concept "than home-based work, in that it is understood as applying to employees who carry out their work remotely from home" (ILO 2020). Thus, in residential inhabited spaces, energy consumption can vary significantly due to the number, age and permanence of family members, number of household appliances, size of housing, construction materials, climate among others. This is evident in the COVID-19 Pandemic Impact Analysis published by OLADE, in the Energy Sector of Latin America and the Caribbean, where it is mentioned that "the energy consumption of the LAC residential sector increases around 20% compared to 2019 and 19% compared to the reference scenario in 2020. The other sectors decrease their consumption percentages, which range from 9% compared to 2019 and 10% with respect to the referential scenario in 2020" (OLADE 2020).

In consequence, the intensity in the use of a house could contribute to or weaken actions in search of sustainability. However, the importance of its role as a safe and protective space for its inhabitants is occasionally reduced probably due to its historical permanence, assumptions of physical transformation capacity to meet unplanned needs. An analysis of sanitary, economic and social crisis by impacts of COVID-19, mentions the need for scientific data and information to prepare us for upcoming crises and toward sustainable development. SDG 3 reports that COVID-19 has saturated health services due to the rapid increase in cases, thus compromising the health progress achieved up to 2019. One of the effects in SDG 7 is the limited access to the energy of households to carry out basic activities. SDG 11 shows the "vulnerability of slum dwellers and those living in informal settlements. Many of these urban residents already suffer from inadequate housing with limited or no access to basic infrastructure and services, including water, sanitation and waste management" (UNDESA 2020).

The exposed approaches are articulated with the house as the space that allows life inside it. Factors such as the climate and its increase in temperature have promoted actions to mitigate or adapt life to these conditions. This is where the house becomes the receiving entity for actions to reduce energy consumption, seeking the efficient use of resources that allow maintaining or improving the conditions inside the houses. Also, the current necessity of working from house, both during and potentially after the pandemic, housing could alleviate overcrowding, making it more feasible, or boosting productivity while working from house. These actions would contribute to companies and workers if "social dialogue is used to identify and address specific challenges with respect to work-life balance and productivity, so that the needs of both parties are best met" (ILO 2020).

However, the return of these efforts depends on the house user on how he accepts, integrates and acts in front of the changes. This shows the level of vulnerability to which they may be exposed because of the conditions of their house, health being a delicate resource that could be affected if we give it its place. It is evident that house and user cannot be separated if positive results are sought as a product of sustainable development.

3. Factors That Influence the Household Environment

Influencing factors such as climate conditions, limited energy resources and disease, produce effects on household actions. Sustainable development, defined in the 1987 Report "Our Common Future" by the World Commission on Environment and Development, as "meeting the needs of the present generation, without compromising the ability of future generations to meet their own needs" (AGNU 1987). Sustainable development has been the main guideline in the conception of a house over the years. An example is a vernacular architecture, considered that is defined as the type of architecture that characterizes a place where it is built. Architecture generates habitable spaces that protect its occupants from the exterior environment. However, it should be clear that "architecture is not an autonomous entity but appears integrated into the system and must be at the service of people" (Jebens-Zirkel 2020).

Climatic and socio-cultural conditions and the availability of materials are parameters that define the particular construction systems in each geographical area (Neila 2015). The climatic differences between the countries have led to the generation of regulations to divide their geographic area according to the weather conditions in these areas and establish recommendations and constructive strategies for habitability in a house (ABNT NBR 15220-3 2005, Palme et al. 2017, MIDUVI 2018).

This capacity for adaptation has been related to current bioclimatic architecture that has been defined by Neila Gonzalez F.J. as "the architecture that represents the use of materials and substances with sustainability criteria, without putting their use at risk by future generations; it represents the concept of optimal energy management of high-tech buildings, through the capture, accumulation and distribution of passive or active renewable energies and, the landscape integration and use of indigenous and healthy materials, of ecological criteria and eco-construction" (Neila 2004).

The interaction between a house and its environment was already postulated in the Middle Ages by Marco Vitruvii in the Treatise of Architecture. It is a philosophical study on nature in which the architect is held responsible to develop and apply knowledge about the climate of the localities where the houses are going to be built since they can be healthy or harmful for human beings (Rodríguez-Ruiz 1997). Also, the indoor environmental quality (IEQ) and the reduction of negative effects on the environment are the results of good construction practices. Adequate temperature and humidity, air quality and air circulation are part of the indoor environmental qualities in a building. However, generate and maintain these IEQs affect the energy consumption and pollution of the environment. Energy conservation strategies such as passive or active techniques are alternatives that allow capture, accumulation and use of natural energies (solar, wind, geothermal, biomass, and so forth). The indoor quality ambient is correlated with the exterior environment, so they should have the same importance considering the possible damage to the environment as the damage to the occupants inside (Neila 2000).

Housing and its inhabitant's data are also collected from indicators about secure tenure, access to services, habitability, location and culture, grouped housing and basic services dimensions, but these have differences in developing and developed countries (Santos 2019). In developing countries, "considered indicators are typically related to availability of services and habitability: housing materials (most commonly floor, and sometimes walls and roof), access to sufficient space (overcrowding), access to clean water and improved sanitation, and, occasionally, energy sources" (Santos 2019). In developed countries, "indicators of housing materials and energy are ignored (presumably virtually all dwellings satisfy the minimum standards). Indicators can be: indicators related to the availability of different services and infrastructure (having bath or shower in the household, having an indoor flushing toilet, hot running water), indicators related to habitability (presence of leaks on the roof, damp in walls, floors or foundation, rot in window frames or on the floor, possibility of keeping the house warm, house is too dark/not enough light, overcrowding, and the presence of some items in the household such as dishwasher, telephone, color TV, microwave, washing machine), indicators related to location (pollution and other environmental problems (self-reported), noise from neighbors or from the street, crime, violence or vandalism in the area were also frequently included)" (Santos 2019).

The data collected is different in developed and developing countries. An investigation examined the effect of hypothetical strategies to improve energy efficiency in the UK and India housings and evidenced diverse needs. In settings developed in the UK, greenhouse-pollutant abatement strategies would be related to changes in the indoor environment, such as low winter temperatures, outdoor air quality, and so forth. In a country like India "the cookstove intervention shows the very great potential for improvement of public health by interventions that also have appreciable bearing on climate change mitigation" (Wilkinson et al. 2009). But according to Santos (2019) "the indicators related to location (pollution and other environmental problems (self-reported)" are reported only in developed countries. However, in India, diseases like an acute lower respiratory infection in children, chronic obstructive pulmonary disease, and ischemic heart disease are associated with indoor household fossil fuel combustion used inside the house to cooking (Wilkinson et al. 2009).

It is important to refocus the indicators towards the benefit of the health of the inhabitants, based on experiences from different environments. In Latin America and the Caribbean cities, there are overcrowded houses with dirt floors, without sanitation and waste collection services. Urban infrastructure and equipment are recognized at the neighborhood and housing level, which has a direct and indirect effect on the quality of life as well as on the value of the house. But not "owning a house can create concern and lower levels of life satisfaction, which has a negative impact on psychological or physical health" (Bouillon et al. 2012). Therefore, it is important to promote urbanization and innovative construction technologies for social housing to benefit the poorest households.

Currently, the architectural challenge is the climatic comfort without the assistance of devices that consume electricity but considering that energy savings are related to the severity of the climate in which the building is located (Manzano-Agugliaro et al. 2015). Both passive and active strategies are used to reduce operational energy consumption in a building, but the results depend on the type and mix of active and passive measures used, climatic conditions of the place, and materials used in the construction (Ramesh et al. 2010). Lisa White from the Passive House Institute US (PHIUS) defines a passive building as a "design methodology defined by a set of principles that prioritize energy conservation and best practices. Passive building is associated with lower energy use and, specifically, lower space-conditioning loads. This methodology produces other benefits: comfort, improved indoor air quality, durability, and resilience. Passive building principles can be applied to all building typologies—from single-family homes to multi-family apartment buildings, offices, and skyscrapers" (White 2020). Active strategies need energy and are applied later as a complement to passive strategies in a building. These strategies can be: "mechanical system components such as air-conditioning, heat pumps, radiant heating, heat recovery ventilators, and electric lighting; they include systems that generate energy such as solar electric and solar thermal panels, wind turbines, and geothermal energy exchangers" (Sustainable 2020).

In addition to the passive and actives strategies, the shape characteristics in a house respond to building strategies to achieve protection from solar radiation, incorporation of thermal mass or cooling of the air through materials in hot climates. In temperate climates, the building strategies will focus on flexibility against solar radiation. This means that strategies to both capture and protect from solar radiation can be applied. Similarly, the design and materials of the envelope will vary to achieve thermal mass or insulation from the walls; they will also provide cooling and ventilation. For cold climates, the building strategies are thermal insulation mainly in the envelope, energy conservation, use of slow-heating interior finishing materials, regulated ventilation to eliminate excess moisture and solar collection (Neila 2015).

The indoor temperature in controlled environment studies have associated data about people and built environment in order to obtain results that are closer to reality. However, interdisciplinary research about perceived thermal comfort including "cultural and behavioral aspects, age, gender, space layout, possibility of control over the environment, user's thermal history and individual preferences" (Rupp et al. 2015) are limited. Currently, it is necessary aboard climate change, its thermal effects in indoor comfort and energy-efficient buildings (Rupp et al. 2015). A report on the health benefits of combating climate change places energy consumption in houses as a problem that promotes actions in developed and developing countries. For the first ones, the concern is focused on reducing energy consumption associated with air conditioning systems among others; but in non-developed countries, the problem is related with the quality of the air inside the houses, mainly by low-income families cooking stoves. It is evident that "the domestic use of energy is responsible for a significant amount of greenhouse gases" (Watts et al. 2017).

From this perspective, houses design "determines how much material they use, the energy used in their manufacturing and operations, their durability, and their ease of reuse and recycling. Building codes and standards connect building design to policy. They can encourage or constrain material efficiency" (UNEP 2020). "Increasing user intensity shifts the policy focus from choice and use of materials to how people live. Policy instruments such as taxation, zoning and land use regulation play a role, but so do consumer preferences and behavior" (UNEP 2020).

The evaluation of energy performance of a building should consider a reference of the type of building, its occupancy and users' behavior with the purpose that the expected and actual consumption do not have significant differences. Therefore, buildings planned as nearly zero energy consumption and based solely on the technical parameters associated with electric energy consumed by equipment and lights household, they could have results that are far from reality (Carpino et al. 2017). The occupants of buildings are of vital importance to guide their impact on energy consumption, occupant-related building energy codes and standards requirements "—for prescriptive and performance paths alike—may mislead designers towards suboptimal building designs" (O'Brien et al. 2020). O'Brien et al. (2020) reviewed 23 codes or standards from different regions, 19 from countries located in the northern hemisphere and four starting from the equatorial line toward the southern hemisphere with Brazil being the only country in South America. In the first phase, quantitative requirements like schedules, densities, set points and general code objective are considered. Despite the fact that the buildings are occupied for offices, the variables analyzed are similar to the residential ones, so the results are relevant and "showed considerable variations across the codes with regard to the occupancy, lighting and equipment power density values. While these can likely be partly assigned to cultural and contextual differences…" (O'Brien et al. 2020). The second phase compared code requirements for identifying similarities and differences, "code requirements and underlying philosophies about occupants are diverse, they are generally quite simplistic … requirements do not adequately acknowledge design as a way to positively influence occupant behavior because they assume that behavior is the same in reference and design buildings" (O'Brien et al. 2020).

An overview of different certification schemes applied in office, commercial and social houses buildings, located in countries with latitudes distant to the equator both north and south, collects information about post-certification performance. Case studies indicated that "there are major discrepancies between design and measured performance in certified building" (Afroz et al. 2020). They recommend "implement data-driven building operation strategies" collected through "detailed meter and submeter infrastructure and occupant feedback collection practices" for certification programs (Afroz et al. 2020). In addition, building performance improvement strategies considering COVID-19 pandemic should include "requirements to minimize the risk of exposure to infectious aerosols" (Afroz et al. 2020).

Quantifying the household-level relationship between performance occupation and well-being would generate information to improve public policies scope. A literature analysis emphasizing thermal comfort shows that "occupant-centric thermal comfort control has great potential to maximize building occupants' thermal comfort while saving energy", a 22% energy saving and 29.1% comfort improvement approximately (Xie et al. 2020). However, the temperature inside the buildings interacts with other elements of the indoor environment. In consequence, in addition to generating comfort, it is important to analyze the health benefits for its inhabitants.

The actions or strategies applied to achieve sustainability must be related to the benefits of the user and mainly its health. Solely focusing on energy efficiency activities without considering the needs and comfort of the users can be considered an incomplete approach. In the case of housing, the users become a consumer sector where actions implemented and actions not implemented do not have standardized results. Moreover, these results become less interesting if we compare them with the commercial sector where, the efficient use of energy will have a direct economic impact, as a result of reduction of time, personnel among others.

A static approach between energy use and housing user could do not reflect the interaction of the influencers of the changes that are already taking place. Climate change has a transverse effect and extreme events such as the COVID-19 pandemic mark the priorities that need to be addressed to maintain people's health and lives. In addition, they show levels of social inequality that put at risk the effectiveness of plans or programs to maintain the lives of the inhabitants of developing countries and the quality of life of the inhabitants of developed countries.

4. Indoor Environment Conditions and Health Effects

Assuming that human beings as the driving force and receiver of actions to improve life, we could ask what are the effects of indoor environmental conditions on a house? Giving an answer to this question requires a detailed analysis supported by evidence obtained using the integrative sustainability approach. This analysis focuses on risk factors "any trait, characteristic, or exposure of an individual", that increases the likelihood of illness or injury (WHO 2020e) in the occupants of a house.

Information on indoor conditions in a house can be related to different aspects or needs at the country level. However, despite the existence of relevant data of great use for various areas of research, the protocols that allow their interrelation are weak or non-existent in some cases. As mentioned above, through indicators, positive results or actions can be identified to achieve representative goals according to the level of development of each region.

The European Union through the Statistical Office (Eurostat) "offers a whole range of important and interesting data that governments, businesses, the education sector, journalists and the public can use for their work and daily life" (Eurostat 2020). Data from Eurostat (European Community Household Panel – ECHP) was used in a simplified analysis of the impact of inequality on relative health. This analysis showed that the interaction coefficients between income inequality and household income quintiles are significant. Also, it found "consistent evidence that income inequality is negatively related to self-rated health status in the European Union for both men and women, particularly when measured at national level. However, despite its statistical significance, the magnitude of the impact of inequality on health is very small" (Hildebrand and Van Kerm 2009). European Union Statistics on Income and Living Conditions (EU-SILC) was the information source for the construction of the Healthy Homes Barometer. The aim was evaluating the relationship between health and housing conditions across EU28 and its Member States. The data were grouped into three clusters: socio-economy, state of building and health. For each of the clusters, different variables were associated to personal level: household level, house type (single-family homes and multi-family homes), and degree of urbanization (urban, suburban and rural). The results indicated that the probability that adults report bad health is significantly higher in housings with leaking roof, reported dampness and lack of daylight. The bad economic situation increases the probabilities of having at least one of the risk factors because they relate with the inability to maintain the house warm in multi-family buildings (Hermelink and John 2017).

Available information will depend on the country's forms of government, interests or priorities. However, data presented in the scientific literature have a broad contribution. Below are detailed case studies, that show risk factors indoor social housing and COVID-19 inhabitants' cases in a general house.

Social housing responds to the need to reduce the quantitative housing deficit, a policy applied in developed and developing countries. "The main contents of a social interest housing policy are to provide equitable access to lower-income households to housing with adequate standards in terms of land lots, basic services, built surfaces, materials and finishes" (CEPAL 2000), but it will depend on the budget allocated to this activity so the quality of the materials will have an effect on the conditions inside these houses. The following are results regarding the internal temperature analyzed and regulated through the construction materials. Also, data that relate the thermal effects with the health of the inhabitants of interior spaces will be presented.

An architectural element in social interest housing is the roof, where the characteristics of the material used can be associated with reduced energy consumption and thermal comfort. Results of the use of high solar reflectance paints applied to the roofs of low-income housings in Guayaquil-Ecuador influenced indoor thermal behavior. The case study applied high solar reflectance paint in metal roofs that are not thermally insulated and that they are built with a single material that separates the interior from the exterior. The estimated reduction in indoor temperature was from 2.3°C, up to 3.7°C at noon (Lee et al. 2017).

Indoor thermal comfort deficit in the social housing built in Ecuador is estimated, considering the same design and construction materials; it does not have heating, ventilation or air conditioning (HVAC) systems and there is no thermal insulation regardless the climatic conditions. The results showed areas with a significant thermal comfort deficit in the Andean Highlands (Gallardo et al. 2017). Lobato-Cordero et al. (2019) mentions that the conditions of temperature and relative humidity inside the houses of social interest would make possible the increase or decrease in the transmission of respiratory diseases. Based on the correlation between the environmental conditions of the interior with the exterior, reference is made to the materials of the envelope, infiltrations, the percentage of glazing and height of the ceiling. In addition, overcrowding, the socioeconomic status and the absence of laws which regulate hygiene in houses will negatively impact the health and economic situation of the occupants, who are mostly the elderly, children and women.

Temperature and relative humidity influenced by the design and material of a house in rural Africa could affect the density of mosquitoes in the interior. The results show that the highest temperature and CO_2 concentration, generated by the house occupants, combined with the permeability of the villa and the number of mosquitoes in the interior is higher. Indoor conditions were modified through the maintenance of the traditional pajama box by metal plates, closing eaves and the placement of the permeable door. If the indoor temperature and the CO_2 concentration are reduced, the signals used by malaria mosquitoes to locate human blood are reduced. Also, reducing the indoor temperature by 1.5°C was considered comfortable by the inhabitants of the house. In this case, 33.3°C-35.1°C of indoor temperature was considered uncomfortable and 32.5°C-33°C temperature comfortable (Jatta et al. 2018).

In this chapter, COVID-19 inhabitants' cases in a house were considered, even though the virus transmission is still being investigated. Studies refer to indoor environment as alert outbreak points to the exchange of air between the people who inhabit it. In this regard, exposure times, contact and frequencies, among other parameters that could contribute to the evaluation of strategies to prevent infections are analyzed. Public gathering spaces are the ones that have received the most of the attention. However, housing plays a leading role. Factors such as the number of inhabitants, health status, access to medical care, economic availability, among others, can influence in the quality of the interior environment of a house and consequently the health household. Below is research on viral respiratory infections, and mainly SARS-CoV-2 viruses in indoor environments.

An overview of the role of relative humidity (RH) in airborne transmission of SARS-CoV-2 in indoor environments, based on literature considered H1N1, Influenza, SARS-CoV-1, MERS, SARS-CoV-2 viruses and proposes three scenarios in which RH affects virus transmission (Ahlawat et al. 2020).

- Scenario (a) fate of microorganisms inside the viral droplets, the percentage of humidity and the survival capacity of some microorganisms inside the viral droplets and suspended aerosols (Ahlawat et al. 2020).
- Scenario (b) (Ahlawat et al. 2020, with Chan et al. 2011) mentions "lower temperatures and low humidity support prolonged survival of virus on contaminated surfaces" and that "the virus transmission has often occurred in well air-conditioned environments such as hospitals or hotels in some countries which have an intensive use of air-conditioning" (Ahlawat et al. 2020).
- Scenario (c) "There is a significant contribution of dry indoor air in both disease transmission and poor resident health. During cold winters, outdoor air is drawn indoors and then heated to a comfortable temperature level. This process will significantly lower the indoor RH, which creates an extremely dangerous situation for indoor residents, particularly during the COVID-19 pandemic. When the indoor RH is less than 40%, humans become more vulnerable to viral respiratory infections making SARS-CoV-2 virus more infectious in the inhaled air" (Ahlawat et al. 2020). They also mentioned precautions such as: "increasing natural ventilations like opening of windows during indoor stay, using proper face masks (face shields along with face mask could provide better results), avoid staying in direct periphery of the infected or other people, and maintaining social distancing" (Ahlawat et al. 2020), for resident's health.

A study on the transmission of Covid-19 to the interior of habitable spaces carried out in 320 cities of China between January 4 and February 11, 2020, identified six categories of outbreaks: houses (apartments and villas), transportation (train, private car, high-speed train, bus, passenger plane, taxi, cruise ship, and so forth), restaurants, entertainment centers (gymnasiums, toys, tea houses among others), shopping places (commercial center and supermarkets) and miscellaneous (hospitals, hotel rooms, unspecified community, thermal power plants, and so on). The house is the dominant outbreak point with 79.9% (254 of 318 outbreaks), containing between three and five cases, after this transport with 34% (108 outbreaks). In addition, categories of infected individuals were considered in the function of their relationship: family members, socially connected and socially unconnected. In this sense, the discussion on SARS-CoV-2 aerosol transmission in indoor environments is suggested, recommended by the National Health Commission of the People's Republic of China. Nowadays, these interior environments are where life and work take place with direct contact between its occupants, high-contact surfaces and different level of ventilation causing an interior phenomenon that rises the transmission of the virus (Qian et al. 2020).

A study on COVID-19 showed the dynamics of contagion among the 7 members of a family of whom six travelled through the cities of China (Shenzhen-Wuhan) between March 2019 and 2020. Of the six members who were present in Wuhan, four presented symptoms, two came to the hospital diagnosed with coronavirus among other respiratory illnesses, two presented symptoms, but there was no need for hospitalization. The seventh member of the family who did not travel also had symptoms that got worse. The study proposed that there are several possible transmission scenarios, it mentions "that the transmission from person to person in family homes or hospitals, and the long-distance propagation of this new coronavirus are possible; therefore, vigilant control measures are justified, in this early stage of the epidemic", considering that "the coronavirus is a virus of single-stranded RNA with positive direction and surroundings, with rapid mutation and recombination capacity" (Chan et al. 2020).

The health of people is a need that cannot be prioritized based on their economic level. Therefore, analyzing the environmental conditions inside a house and the impact on health contributes to the redefinition of actions in search of sustainability.

This analysis proposes that the internal temperature of a house plays a main role both in terms of energy and health. The approach towards thermal comfort would be limiting if it is not contrasted with the effects of environmental conditions inside the houses, such as temperature and humidity. Regulating or controlling internal conditions for energy-saving purposes are strategies that have been validated and implemented in indoor environments.

International regulations and reports to improve Indoor Environmental Quality (IEQ), (IPMVP 2002) recommend temperature ranges (ASHRAE 55 2017), ventilation (ASHRAE 62.1 2019), pollution levels, among others, that allow the development of activities in buildings according to their use. However, data on the health effects of the occupants are limited. Consequently, it is necessary to verify and validate recommended values.

5. Final Considerations

House becomes a receptor of natural or provoked events, where the user is responsible for managing them in search of sustainability in different aspects. Although the user's behavior improves their quality of life, this could not be achieved if their health is at risk. If the exposure to health risk in a house that does not provide protection is high, its users may have limitations to perform their daily activities. Besides, people with health problems will seek to improve or solve them by keeping rest, seeking medical or hospital care, depending on the severity so regular attendance at work is interrupted.

Processes to address health problems must be accompanied by preventive measures that improve the health of the inhabitants of a home. Considering sustainable development, the transversality with which these preventive measures are projected can have positive impacts on different economic sectors of a country. That is why the energy efficiency strategies should consider the impacts on the health of the inhabitants of a house, since the analysis focused only on thermal comfort is limited, considering the transverse impact postulated by the SDGs.

Habitability and health could be the result of energy efficiency in a house, but the analysis has demonstrated the need for a clear methodology to identify the factors that contribute to achieving sustainability. Promoting the efficient use of resources could generate real savings that allow reinvesting in areas with greater weakness. In addition, it would prevent that sustainability be assumed as a change in the direction of resources, where the inefficient sector benefits from the savings of the efficient sector. In this way, the interaction proposed in the SDGs is visible, leading to improvements supported on needs of each country.

Acknowledgments

To the Program of Alliances for Education and Training (Programa de Alianzas para la Educación y la Capacitación – PAEC) between the Organization of American States (OAS) and the Coimbra Group of Brazilian Universities (Grupo Coimbra de Universidades Brasileiras – GCUB). And to the Post-Graduation Program in Planning of Energy Systems of the State University of Campinas (Universidade Estadual de Campinas – UNICAMP).

References

ABNT – Associação Brasileira de Normas Técnicas. 2005. NBR 15220-3 Thermal performance in buildings, Part 3: Brazilian bioclimatic zones and building guidelines for low-cost houses. Code: ABNT NBR 15220-3:2005.

Afroz, Z., H.B. Gunay and W. O'Brien. 2020. A review of data collection and analysis requirements for certified green buildings. Energy and Buildings. 226: 110367. doi: https://doi.org/10.1016/j.enbuild.2020.110367

AGNU – Asamblea General de las Naciones Unidas. 1987. Presidente del 65° período de sesiones. Desarrollo Sostenible. Available at: https://www.un.org/es/ga/president/65/issues/sustdev.shtml. Accessed: 11 August 2020.

Ahlawat, A., A. Wiedensohler and S.K. Mishra. 2020. An Overview on the role of relative humidity in airborne transmission of SARS-CoV-2 in indoor environments. Aerosol Air Qual. Res. 20: 1856-1861. doi: https://doi.org/10.4209/aaqr.2020.06.0302

ASHRAE – American Society of Heating, Refrigerating and Air-Conditioning Engineers. 2017. Standard 55 – Thermal Environmental Conditions for Human Occupancy.

ASHRAE – American Society of Heating, Refrigerating and Air-Conditioning Engineers. 2019. Standard 62.1 – Ventilation for Acceptable Indoor Air Quality.

Bennich, T., N. Weitz and H. Carlsen. 2020. Deciphering the scientific literature on SDG interactions: a review and reading guide. Sci. Total. Environ. 728: 138405. doi: https://doi.org/10.1016/j.scitotenv.2020.138405

BID – Banco Interamericano de Desarrollo. 2018. Vivienda ¿Qué viene?: de pensar la unidad a construir la ciudad. IDB-MG-659. Available at: http://dx.doi.org/10.18235/0001594

Bouillon, C., A. Blanco, V. Fretes, C. Boruchowicz, K. Herrera, N. Medellín, et al. 2012. Un espacio para el desarrollo: Los mercados de vivienda en América Latina y el Caribe. Development in the Americas (DIA) – Banco Interamericano de Desarrollo. Available at: https://publications.iadb.org/es/publicacion/un-espacio-para-el-desarrollo-los-mercados-de-vivienda-en-america-latina-y-el-caribe

Breuer, A., H. Janetschek and D. Malerba. 2019.Translating sustainable development goal (SDG) interdependencies into policy advice. Sustainability. 11: 2092. doi: https://doi.org/10.3390/su11072092

Carpino, C., D. Mora, N. Arcuri and M. De Simone. 2017. Behavioral variables and occupancy patterns in the design and modeling of nearly zero energy buildings. Build. Simul. 10: 875-888. doi: https://doi.org/10.1007/s12273-017-0371-2

CEPAL – Comisión Economica para America Latina y el Caribe. 2000. Políticas de viviendas de interés social orientadas al mercado: experiencias recientes con subsidios a la demanda en Chile, Costa Rica y Colombia. Series de la CEPAL, Financiamiento para el Desarrollo. Available at: https://www.cepal.org/es/publicaciones/5304-politicas-viviendas-interes-social-orientadas-al-mercado-experiencias-recientes

Chan, J.F.W., S. Yuan, K.H. Kok, K.K.W. To, H. Chu, J. Yang, et al. 2020. A familial cluster of pneumonia associated with the 2019 novel coronavirus indicating person-to-person transmission: a study of a family cluster. The Lancet. 395: 514-523. doi: https://doi.org/10.1016/S0140-6736(20)30154-9

Crutzen, P.J. 2002. Geology of mankind. Nature 415: 23. Available at: https://www.nature.com/articles/415023a

Eurostat – Statistical office of the EU. 2020. European Commission, Knowledge for Policy, Organisation. Available at: https://ec.europa.eu/knowledge4policy/organisation/eurostat-statistical-office-eu_en. Access: 20 August 2020

Gallardo, A., G. Villacreses, M. Almaguer, A. Lobato-Cordero and M. Cordovez. 2017. Estimating the indoor thermal comfort deficit in the social housing built in Ecuador by integrating Building Information Modelling and Geographical Information Systems. Proc. 15th IBPSA Conference, San Francisco, USA, 1397-1404. Available at: http://www.ibpsa.org/proceedings/BS2017/BS2017_354.pdf

Hermelink, A. and A. John. 2017. The relation between quality of dwelling, socio-economic status and health in EU28 and its Member States – Scientific report as input for Healthy Homes Barometer. Ecofys.

Hildebrand, V. and P. Van Kerm. 2009. Income inequality and self-rated health status: evidence from the european community household panel. Demography. 46: 805-825. doi: https://doi.org/10.1353/dem.0.0071

Hoegh-Guldberg, O.D., J.M. Taylor, M. Bindi, S. Brown, I. Camilloni, A. Diedhiou, et al. 2018. Impacts of 1.5°C global warming on natural and human systems. *In*: Global Warming of 1.5°C. An IPCC Special Report on the impacts of global warming of 1.5°C above pre-industrial levels and related global greenhouse gas emission pathways, in the context of strengthening the global response to the threat of climate change, sustainable development, and efforts to eradicate poverty. Available at: https://www.ipcc.ch/sr15/chapter/chapter-3/

Howard, L. 1988. The Climate of London. International Association for Urban Climate, 6-7. Available at: https://www.urban-climate.org/documents/LukeHoward_Climate-of-London-V1.pdf

IHME – Institute for Health Metrics and Evaluation. 2018. Findings from the Global Burden of Disease Study. Seattle, WA. Available at: http://www.healthdata.org/policy-report/findings-global-burden-disease-study-2017

IHME – Institute for Health Metrics and Evaluation. 2020. Financing Global Health 2019: Tracking Health Spending in a Time of Crisis. Seattle, WA. ISBN: 978-0-9976462-8-3. Available at: http://www.healthdata.org/policy-report/financing-global-health-2019-tracking-health-spending-time-crisis

ILO – International Labour Organization. 2020. Working from Home: Estimating the worldwide potential. Policy Brief. Available at: https://www.ilo.org/global/topics/non-standard-employment/publications/WCMS_743447/lang--en/index.htm

IPMVP – International Performance Measurement & Verification Protocol Committee. 2002. International Performance Measurement & Verification Protocol, Concepts and Practices for Improved Indoor Environmental Quality, Volume II. DOE/GO-102002-1517. Available at: https://www.nrel.gov/docs/fy02osti/31505.pdf

Jatta, E., M. Jawara, J. Bradley, D. Jeffries, B. Kandeh, J.B. Knudsen, et al. 2018. How house design affects malaria mosquito density, temperature, and relative humidity: an experimental study in rural Gambia. Lancet Planet Health. 2: E498-E508. doi: https://doi.org/10.1016/S2542-5196(18)30234-1

Jebens-Zirkel, P. 2020. Pautas principales de arquitectura y urbanismo sostenible - Las personas en el centro – solidaridad y alma. Available at: https://jebens-architecture.eu/knowledgebase/pautas-principales-de-arquitectura-y-urbanismo-sostenible/ Accessed: 20 August 2020

Knox, J.H. 2015. Human rights, environmental protection, and the sustainable development goals. 24 Washington International Law Journal. 517-536. Wake Forest Univ. Legal Studies Paper. Available at: https://ssrn.com/abstract=2660392

Kumar, N. 2018. Cities, Climate Change, & Health Equity - Wellesley Institute, Toronto, ON, Canada Available at: https://www.wellesleyinstitute.com/wp-content/uploads/2018/06/Cities-Climate-Change-Health-Equity-WIJune-2018-fv.pdf

Lee, S., J. Macías, G. Soriano and A. Lobato-Cordero. 2017. Aplicación de techos de alta reflectividad solar en residencias de inter és social en Guayaquil: Caso de estudio "Socio Vivienda II". Proc. Congreso de Investigación Desarrollo e Innovación en Sostenibilidad Energética, 65-69. Available at: https://www.researchgate.net/profile/Massimo_Palme/publication/327068227_Evaluacion_de_la_intensidad_de_la_Isla_Urbana_de_Calor_en_la_ciudad_de_Guayaquil/links/5b763404a6fdcc87df817c5b/Evaluacion-de-la-intensidad-de-la-Isla-Urbana-de-Calor-en-la-ciudad-de-Guayaquil.pdf

Lobato-Cordero, A., E. Quentin and G. Lobato-Cordero. 2019. Spatiotemporal analysis of influenza morbidity and its association with climatic and housing conditions in Ecuador. Journal of Environmental and Public Health. 2019: 6741202. doi:https://doi.org/10.1155/2019/6741202

Manzano-Agugliaro, F., F.G. Montoya, A. Sabio-Ortega and A. García-Cruz. 2015. Review of bioclimatic architecture strategies for achieving thermal comfort. Renew. Sus. Energ. Rev. 49: 736-755. doi: http://dx.doi.org/10.1016/j.rser.2015.04.095

Mapp, S. and S. Gatenio Gabel. 2019. The climate crisis is a human rights emergency. J. Hum. Rights Soc. Work. 4: 227-228. doi: https://doi.org/10.1007/s41134-019-00113-0

McCollum, D.L., L.G. Echeverri, S. Busch, S. Pachauri, S. Parkinson, J. Rogelj, et al. 2018. Connecting the sustainable development goals by their energy inter-linkages. Environ. Res. Lett. 13: 033006. doi: https://doi.org/10.1088/1748-9326/aaafe3

MIDUVI – Ministerio de Desarrollo Urbano y Vivienda. 2018. Eficiencia Energetica en Edificaciones Residenciales (EE). Norma Ecuatoriana de la Construcción – NEC, Habitabilidad y Salud – HS.Code: NEC-HS-EE. Available at: https://www.habitatyvivienda.gob.ec/wp-content/uploads/downloads/2019/03/NEC-HS-EE-Final.pdf

Nayak, S.G., S. Shrestha, P.L. Kinney, Z. Ross, S.C. Sheridan, C.I. Pantea, et al. 2017. Development of a heat vulnerability index for New York State. Public Health. 161: 127-137. doi: https://doi.org/10.1016/j.puhe.2017.09.006

Neila Gonzalez, F.J. 2000. Arquitectura bioclimática en un entorno sostenible: buenas prácticas edificatorias. Textos sobre Sostenibilidad. 89-99.

Neila Gonzalez, F.J. 2004. Arquitectua Bioclimatica en un Entorno Sostenible. Editorial Munilla-Lería. ISBN: 84-89150-64-8

Neila Gonzalez, F.J. 2015. Miradas Bioclimaticas a la Arquitectura Popular del Mundo. Editorial: Garcia Maroto Editores. ISBN: 9788415793687

O'Brien, W., F. Tahmasebi, R.K. Andersen, E. Azar, V. Barthelmes, Z.D. Belafi, et al. 2020. An international review of occupant-related aspects of building energy codes and standards. Build. Environ. 179: 106906. doi: https://doi.org/10.1016/j.buildenv.2020.106906

OLADE – Organización Latinoamericana de Energía. 2020. Análisis de los impactosde la pandemia delCOVID-19 sobre el Sector Energético de América Latina y el Caribe. Available at: http://biblioteca.olade.org/opac-tmpl/Documentos/old0452.pdf

Palme, M., A. Lobato-Cordero, A. Gallardo, J.P. Kastillo, R.D. Beltrán, G. Villacreses, et al. 2017. Estrategias para mejorar las condiciones de habitabilidad y el consumo de energía en viviendas. Instituto Nacional de Eficiencia Energetica y Energias Renovables – INER, Ecuador. Available at: https://www.researchgate.net/publication/314092871_Estrategias_ para_mejorar_las_condiciones_de_habitabilidad_y_el_consumo_de_energia_en_ viviendas

Palme, M., L. Inostroza, G. Villacreses, A. Lobato-Cordero and C. Carrasco. 2017. From urban climate to energy consumption. Enhancing building performance simulation by including the urban heat island effect. Energ. Buildings 145: 107-120. doi: http://dx.doi. org/10.1016/j.enbuild.2017.03.069

Qian, H., T. Miao, L. Liu, X. Zheng, D. Luo and Y. Li. 2020. Indoor transmission of SARS-CoV-2, medRxiv 2020.04.04.20053058. doi: https://doi.org/10.1101/2020.04.04.200530 58. (This article is a preprint and has not been peer-reviewed [what does this mean?]. It reports new medical research that has yet to be evaluated and so should not be used to guide clinical practice. Access: 13 Jun 2020.

Ramesh, T., R. Prakash and K.K. Shukla. 2010. Life cycle energy analysis of buildings: an overview. Energ. Buildings. 42: 1592-1600. doi: https://doi.org/10.1016/j. enbuild.2010.05.007

Rodríguez-Ruiz. D. 1997. M. Vitruvii Pollionis De Architectura. Editorial Alianza Editorial S.A. ISBN: 84-206-7133-9

Rupp, R.F., N.G. Vasquez and R. Lamberts. 2015. A review of human thermal comfort in the built environment. Energ. Buildings 105: 178-205. doi: http://dx.doi.org/10.1016/j. enbuild.2015.07.047

Santos, M.E. 2019. Indicadores no monetarios para el seguimiento de las metas 1.2 y 1.4 de los Objetivos de Desarrollo Sostenible: estándares, disponibilidad, comparabilidad y calida. Serie Estudios Estadísticos, No 99 (LC/TS.2019/4), Santiago, Comisión Económica para América Latina y el Caribe (CEPAL). ISBN: 1680-8789.

Schapper, A. and M. Lederer. 2014. Introduction: human rights and climate change: mapping institutional inter-linkages. Camb. Rev. Int. Aff. 27: 666-679. doi: http://dx.doi.org/10.10 80/09557571.2014.961806

Schleicher, J., M. Schaafsma and B. Vira. 2018. Will the sustainable development goals address the links between poverty and the natural environment? Curr. Opin. Env. Sust. 34: 43-47. doi: https://doi.org/10.1016/j.cosust.2018.09.004

Seixas, S.R.C. and R.J. Nunes. 2017. Subjectivity in a context of environmental change: opening new dialogues in mental health research. Subjectivity. 10: 294-312. doi: https:// dx.doi.org/10.1057/s41286-017-0032-z

Sida – Swedish International Development Cooperation Agency. 2015. A human rights based approach to environment and climate change. 1-11. Available at: https://www.sida.se/ globalassets/sida/eng/partners/human-rights-based-approach/thematic-briefs/human-rights-based-approach-environment-climate-change.pdf

Sustainable. 2020. Passive Design and Active Building Strategies. Available at: https://www. sustainable.to/strategies. Accessed: 22 August 2020.

The Lancet. 2009. The health benefits of tackling climate change. An Executive Summary for The Lancet Series 1-8. The Lancet. Available at: https://www.thelancet.com/pb/assets/ raw/Lancet/stories/series/health-and-climate-change.pdf

UN – United Nations. 1948. United Nations Universal Declaration of Human Rights.

UNDESA – United Nations Department of Economic and Social Affair. 2020. The Sustainable Development Goals Report. ISBN: 978-92-1-101425-9

UNEP – United Nations Environment Programme. 2020. Resource efficiency and climate change, material efficiency strategies for a low-carbon future. A report of the International Resource Panel – IRP. Nairobi, Kenya.

UNHR – United Nations Human Rights. 2015. Understanding Human Rights and Climate Change. Submission of the Office of the High Commissioner for Human Rights to the 21st Conference of the Parties to the United Nations Framework Convention on Climate Change. Available at: <https://www.ohchr.org/Documents/Issues/ClimateChange/COP21.pdf>

Watts, N., M. Amann, S. Ayeb-Karlsson, K. Belesova, T. Bouley, M. Boykoff, et al. 2017. The Lancet Countdown on health and climate change: from 25 years of inaction to a global transformation for public health. The Lancet. 39: 581-630. doi: https://doi.org/10.1016/S0140-6736(17)32464-9

White, L. 2020. Passive building on the rise. High Performing Buildings Magazine. 16-25. Available at: https://www.hpbmagazine.org/passive-building-on-the-rise/

WHO – World Health Organization. 2018. Housing and health guidelines, Geneva. License: CC BY-NC-SA 3.0 IGO. ISBN 978-92-4-155037-6

WHO – World Health Organization. 2020a. WHO Director-General's opening remarks at the media briefing on COVID-19 – 11 March 2020. Available at: https://www.who.int/dg/speeches/detail/who-director-general-s-opening-remarks-at-the-media-briefing-on-covid-19---11-march-2020. Accessed: 15 August 2020

WHO – World Health Organization. 2020b. Q&A on coronaviruses (COVID-19). How does COVID-19 spread? – 17 April 2020. Available at: https://www.who.int/emergencies/diseases/novel-coronavirus-2019/question-and-answers-hub/q-a-detail/q-a-coronaviruses. Accessed: 15 August 2020

WHO – World Health Organization. 2020c. Transmission of SARS-CoV-2: implications for infection prevention precautions – 9 July 2020. Available at: https://www.who.int/news-room/commentaries/detail/transmission-of-sars-cov-2-implications-for-infection-prevention-precautions. Accessed: 15 August 2020

WHO – World Health Organization. 2020d. COVID-19 Strategy Update – 14 April 2020. Available at: https://www.who.int/docs/default-source/coronaviruse/covid-strategy-update-14april2020.pdf?sfvrsn=29da3ba0_19. Accessed: 15 August 2020

WHO – World Health Organization. 2020e. Temas de salud, Factores de riesgo. Available at: https://www.who.int/topics/risk_factors/es/ Accessed: 26 August 2020

Wilkinson, P., K.R. Smith, M. Davies, H. Adair, B.G. Armstrong, M. Barrett, et al. 2009. Public health benefits of strategies to reduce greenhouse-gas emissions: household energy. The Lancet, Health and Climate Change. 374: 1917-1929. doi: DOI:10.1016/S0140-6736(09)61713-X

Xie, J., H. Li, C. Li, J. Zhang and M. Luo. 2020. Review on occupant-centric thermal comfort sensing, predicting, and controlling. Energ. Buildings. 226: 110392. doi: https://doi.org/10.1016/j.enbuild.2020.110392

Data or Misconceptions? Understanding the Role of Economic Expertise in the Development of Sustainable Marine Aquaculture in Santa Catarina, Brazil

Thomas G. Safford* and Marcus Polette

[1] Associate Professor & Director, UNH International Affairs Program,
  Department of Sociology, 345C McConnell Hall, University of New Hampshire,
  Durham, NH 03824 - USA

[2] Escola do Mar, Ciência e Tecnologia, Universidade do Vale do Itajaí,
  Rua Uruguai 458 - Fazenda - Itajaí - CEP 88302 - Santa Catarina - Brasil

1. Introduction

From climate change and pollution to commodity prices and consumer preferences, inter-related environmental and economic challenges shape sustainable development around the globe. In no arena are these interconnected concerns more apparent than in the use and conservation of marine resources. The world's oceans absorb 30% of global CO_2 emissions and contain some of the Earth's most diverse ecosystems (GAUN 2015). Similarly, oceans provide more than three billion people with their livelihoods and primary source of protein, and the value of marine resources and industries is estimated at US\$3 trillion (GAUN 2015). However, environmental degradation and precipitous declines in fisheries threaten global food security and the economic well-being of communities reliant on the oceans. In response, planners have forwarded marine aquaculture, or mariculture, as a sustainable development alternative (Bostock et al. 2010, FAO 2016, Subasinghe et al. 2009).

Mariculture closely aligns with the United Nation's 2030 Agenda for Sustainable Development. Sustainable Development Goal (SDG) 14 strives to "Conserve and sustainably use the oceans, seas, and marine resources for sustainable development" (GAUN 2015). Similarly, mariculture connects with the Agenda's social and economic objectives which prioritize promoting responsible consumption and production

*Corresponding author: Tom.Safford@unh.edu

(SDG 12) and sustainable economic growth and employment (SDG 8) (GAUN 2015). Goals 14, 12, and 8 highlight the importance of advancing alternative uses of marine resources that preserve the ocean environment and promote employment and sustained economic growth. Across the UNSDGs, sub-targets emphasize data-driven decision making and increasing scientific knowledge and research capacity to support sustainable development planning (GAUN 2015). Extensive scientific research shows that marine aquaculture can be an environmentally sustainable ocean use (FAO 2016, Frankic and Hershner 2003, Subasinghe et al. 2009). However, investigation of the economic aspects of mariculture or examination of how access to business-related data and expertise may affect its sustainability is limited (Burbridge et al. 2001, Cleaver 2006, Costa-Pierce 2008, Pereira and Rocha 2015). To fill this gap, we investigate the economic dimensions of ocean aquaculture planning and development in Santa Catarina, Brazil.

2. Mariculture as a Sustainable Development Alternative for Santa Catarina

Like many coastal areas around the world, environmental degradation and collapsing fisheries led Brazilian planners to consider new marine development options. They identified mariculture as a strategic alternative largely due to its perceived social and environmental sustainability (Diegues 2006, EPAGRI 2015, MPA 2015, Mungioli 2012, Valenti et al. 2000). Perhaps nowhere were these trends more apparent than in the state of Santa Catarina in southern Brazil (See Fig. 1). Commercial fishing long-supported the state's coastal economy, but as fisheries declined there were few employment options (Rodrigues et al. 2010, Rossi-Wongtschowski 2007). Planners believed mariculture could be an optimal development alternative.

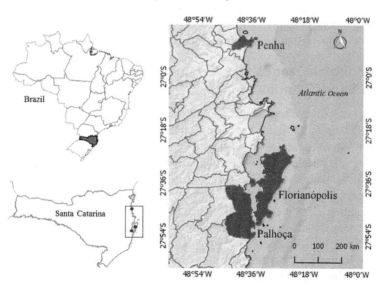

Figure 1: Map of the Study Region in Santa Catarina, Southern Brazil (Safford et al. 2019)

With limited experience growing seafood, governmental officials looked to the scientific community for guidance. Using a collaborative science-based approach, agronomic scientists and planners identified shellfish as ideal for the region and this spurred the expansion of mariculture (de Andrade 2016, Rodrigues et al. 2010, Safford et al. 2019, Sidonio et al. 2012, Suplicy and Novaes 2015, Suplicy et al. 2017, Vianna et al. 2012). The ocean environment in Santa Catarina was well-suited for farming oysters and mussels, but the social context also seemed propitious to planners. Tourism thrived in the state and eating seafood at coastal restaurants was a quintessential *Catarinense* experience (Corrêa and Müller 2016, Lins 2006, 2007, Santos 2011). As the industry grew, shellfish became staples at these establishments, spurring demand and helping Santa Catarina emerge as a national leader in marine aquaculture (EPAGRI 2015, MPA 2014, 2015, Rodrigues et al. 2010, Suplicy et al. 2017). Policymakers and supporters championed mariculture for its economic contributions as well as its sustainability (EPAGRI 2015, Pereira and Rocha 2015, Suplicy et al. 2017).

Shellfish farms expanded in Santa Catarina from the 1990s forward (Ferreira and Oliveira Neto 2007, Vinatea and Vieira 2005, Jacomel and Campos 2014, Rodrigues et al. 2010, Sidonio et al. 2012, Suplicy et al. 2017, Vianna et al. 2012). As mariculture spread, both environmental and economic threats to its sustainability emerged. In response, governmental authorities and the scientific community partnered with farmers to apply scientific innovations to address threats to productivity and mitigate environmental risks. However, there was not a parallel response in the economic arena and commercial uncertainty for growers persists. With the economic viability of the industry seeming less certain, it is important to ask how professionals engaged with mariculture development viewed its practicability, accessed business-related data and expertise to inform decision-making, and assessed its economic sustainability. Answering this question will not only provide insights into the sustainability of mariculture in Santa Catarina but also illustrate how both social and environmental science can support achieving the UNSDGs.

2.1 Economic Trends and Mariculture Development in Santa Catarina

Promoting sustainable economic development underpinned the establishment of shellfish farming in Santa Catarina. Nonetheless, access to economic data and analyses to inform development was limited. There is no evidence that economists or business scientists from universities completed market research or applied their expertise in similar ways as the agronomists who were key collaborators in the science-based development of the mariculture industry.

Information about the economic trajectory of mariculture that does exist comes from the Santa Catarina state agricultural extension agency(EPAGRI). These data show that the state harvested only a few thousand tons of shellfish in the mid-1990s (Alves dos Santos et al. 2010). Production then spiked and reached a peak of 23, 495 tons in 2012, but then fell to 13, 699 tons in 2017, a 41% decline from its high point (Alves dos Santos et al. 2013, 2018).

Shifts in the number of growers and the value of harvests parallel these changes in harvests. The number of shellfish growers peaked at 844 in 2000 and then steadily declined, even as tons of shellfish harvested increased through the mid-2000s (Alves dos Santos et al. 2013, Alves dos Santos and Winckler da Costa 2016). During the years with the largest harvests in the mid-2000s, the number of growers remained in the 600s, declined to the mid-500s, and was 552 in 2017, a decline of 35% from the high point in 2000 (Alves dos Santos et al. 2013, 2018, Alves dos Santos and Winckler da Costa 2016).

Similarly, mariculture employment rose steadily through the 2000s, but then dropped precipitously from 3,388 to 1,915 individuals employed from 2014 to 2017, a 43% decline (Alves dos Santos et al. 2018, Alves dos Santos and Winckler da Costa 2015). Finally, estimates of the commercial value of shellfish also fluctuated over time. The value of shellfish harvests in Santa Catarina rose continuously through the 2000s from R$21,606,609 (US$11,137,427) in 2009 to a high of R$78,895,697 (US$25,450,224) in 2015 and had a value of R$66,152,000 (US$20,106,990) in 2017 (Alves dos Santos et al. 2010, 2018, Alves dos Santos and Winckler da Costa 2016)[1]. Together, trends in harvests, growers, employment and revenues show considerable volatility and parallel declines into the late 2010s, suggesting significant risks and uncertain economic returns from mariculture.

These production data offer important insights into the socioeconomic features of mariculture. However, many shellfish farmers in Santa Catarina are *clandestinos*, or unlicensed growers who informally harvest and sell their mussels and oysters. Thus, the actual number of growers, employment, tonnage, and value of harvests may differ from the official statistics given the covert nature of clandestine shellfish farming. Also, with limited metadata available to assess how EPAGRI collected and analyzed its data, these statistics should be used cautiously when drawing conclusions about the economic features of mariculture.

Professionals from EPAGRI also executed a small study comparing growers, mariculture researchers, and extensionist agents' views regarding issues affecting the success of mariculture (Ventura et al. 2011). However, the structure of the authors' survey instrument focused primarily on production-related and regulatory issues rather than business management concerns. Findings from this study show concern about the commercial aspects of mariculture differed across the three sets of actors and were of greatest among growers (Ventura et al. 2011). Results from this research point to the importance of understanding how and why views about the economic aspects of mariculture vary across these key stakeholder groups. Finally, the lack of longitudinal farm-level business data, such as the costs of inputs, wholesale versus retail prices, or transportation and labor costs make it difficult to definitively assess the broader economic benefits, potential risks, and overall sustainability of shellfish farming in Santa Catarina.

---

[1]   To account for inflation, harvest value data were converted to US Dollars using exchange rates for $1 US Dollar to Brazilian Reals on 1 July 2009, 2015, and 2017 (https://fxtop. com/en/historical-currency-converter. php).

3. Research Design

In this study, we first assess how socialized beliefs among stakeholders about the likely economic contributions from mariculture affected the development of the industry. Next, we examine to what extent government officials, scientists and growers prioritized and valued accessing economic expertise and business-related data to support science-based mariculture planning. Finally, we apply insights from the sociological study of institutions to inform our analysis of the social forces shaping consideration of threats to mariculture commerce and economic sustainability and assess its effectiveness in helping achieve UNSDGs 14, 12, and 8.

Sociologists define institutions as socially constructed frameworks that reflect beliefs about appropriate actions and behaviors and offer stability and meaning to social life (Powell and DiMaggio 1991, Safford 2010, Scott 2008). As institutional influences significantly shape group processes, we utilize a conceptual framework from institutional sociology to guide our research. Sociologists often delineate three types of institutions to structure their analyses – regulative, normative, and cultural-cognitive (Scott 2008). Regulative institutions appear as codified rules that both enable and limit individual and group behaviors. Generally accepted social norms that guide social interactions and practices form normative institutions. Finally, cultural-cognitive institutions reflect culturally derived understandings and myths about appropriate behaviors, actions, and social roles (Scott 2008). These three types of institutional influences function both independently and in combination. Individuals consciously conform to regulative and normative institutions, while cultural-cognitive influences are less overt and subconscious (Scott 2008). These institutional effects are shaped by social interactions and are 'bounded' by the information available to individuals and organizations, often making seemingly rational decisions and behaviors result in suboptimal outcomes (Safford and Norman 2011, Simon 1997).

Given the limited research on how economic expertise and knowledge are applied in mariculture planning, we conducted an in-depth case study using a grounded inductive approach (Yin 1994). Data collection consisted primarily of qualitative interviews with individuals involved with mariculture development in Santa Catarina in 2014-2015. Secondary sources (e.g., government documents, scientific articles, and websites) provided contextual information related to planning activities, economic trends, and risks to economic sustainability.

A multi-level typology with three over arching categories of mariculture-related professionals—governmental actors, scientific researchers, and shellfish producers—guided our selection of interview respondents. Governmental actors were subdivided by level (federal, state, and municipal) and based on their distinct planning, management, and regulatory responsibilities. Scientific researchers were split into two sub-categories, production scientists involved with agronomic inquiry and impact scientists focused on environmental and human impacts on and from mariculture. We also interviewed a small number of scientists from the social, economic, and business sciences whose research linked to seafood commerce.

Finally, we interviewed both mussel and oyster growers with operations of all sizes. As farmers in Santa Catarina sell their shellfish directly to consumers as well as through distributors, we included several wholesalers in our producer category. We confined our interviews with shellfish producers to three municipalities, Florianópolis, Palhoça, and Penha (See Figure 1). These municipalities have the largest number of farms and include areas where oysters (Florianópolis) and mussels (Palhoça and Penha) are the primary species cultivated (EPAGRI 2015). We completed 65 semi-structured interviews with the number of respondents within each category and subgroup varying based on their overall population.

Our respondent typology guided the organization of data, which were coded for references to economic and business-related themes. A concept-mapping technique was used to identify patterns in the way regulative, normative and cultural-cognitive institutional influences shaped consideration of the economic aspects of mariculture development (Trochim 2000, Yin 1994). Concept-mapping can be an effective analytical approach for investigating similarities and differences in the responses of sub-groups of individuals involved in shared endeavors such as mariculture planning (Safford and Norman 2011, Trochim 2000).

Our analyses show two key factors most affecting stakeholders' consideration of the economic dimensions of mariculture development in Santa Catarina, (1) a resolute belief that mariculture will generate economic benefits for coastal communities and (2) attitudes regarding the value of economic expertise and business-related data within development planning. These two factors provide a framework for organizing our findings and highlighting key patterns in respondents' consideration of risks and the economic sustainability of shellfish farming.

4. An Unwavering Belief in the Economic Benefits of Mariculture Development

There was a near consensus among respondents from all typological categories that enthusiasm about the potential economic benefits of mariculture drove development. This largely unquestioned belief that mariculture was profitable influenced these stakeholders' ability to assess market trends and threats to the industry's economic sustainability. While stakeholders' focus on the anticipated economic benefits of mariculture is consistent with the employment and commerce-related objectives of SDGs 8 and 12, data from our study suggest a disconnect between this belief and the actions necessary to meet these socioeconomic goals as well as SDG 14's broader objective to promote the sustainable marine resource development.

4.1 Beliefs about the Economic Benefitsof Mariculture: Governmental Actors

EPAGRI was pivotal in stressing mariculture's potential socioeconomic contributions and this focus is embedded in the agency's mission statement, "Knowledge, technology, and extension for the sustainable development of rural areas, for the benefit of society (EPAGRI 2019)." All our EPAGRI respondents had internalized this edict, stating that mariculture's societal benefits were the primary impetus for development. Many also noted their direct interactions with coastal residents facing economic struggles as a factor driving their work and interest in promoting sustainable economic alternatives. The pervasiveness of this value orientation created solidarity among extensionists and shellfish growers which they stated helped them persevere when encountering challenges during the development process.

Interestingly, a pattern emerged in our data connecting beliefs about technological innovations, extension, and social wellbeing. Most EPAGRI respondents suggested that the rudimentary nature of existing growing techniques was limiting the profitability of mariculture and that technological innovations could resolve this problem. They shared a near ideological belief that if mariculture could be modernized it would be lucrative and generate community-wide economic benefits. However, while they detailed how scientific research showed mechanization could increase productivity, EPAGRI professionals did not identify business-related studies to support their beliefs about the monetary returns from shellfish commerce.

Marine resource management officials voiced similar taken-for-granted beliefs about mariculture's economic contributions. There was a near consensus among these professionals that technical constraints, not market dynamics, limited the success of the industry. They also emphasized gathering agronomic and biological data and applying production science insights to ensure profitability. These beliefs appear to have been socially constructed. Respondents from federal, state and local marine resource management organizations consistently cited EPAGRI staff and university scientists when asserting that technical innovations would spur the commercial success of shellfish farming.

When probed about evidence of mariculture's commercial potential, many government respondents noted the success of mariculture in countries like Chile, although none identified studies of the economic sustainability of Chilean mariculture. Similarly, they pointed to burgeoning gastronomic tourism in Santa Catarina. However, when asked about analyses evaluating economic returns from gastro-tourism or efforts to partner with restaurants or commerce organizations to investigate market trends, none could point to such endeavors.

While governmental actors shared 'take it for granted beliefs' about the expected returns from mariculture, some identified price volatility, competition, and lack of credit as threats to the industry. Nonetheless, these commerce-related concerns did not dissuade governmental actors from their core belief in mariculture as an optimal economic development alternative. When probed about threats to the sustainability of the industry, most governmental respondents focused on structural issues related to licensing and permitting, and few noted concerns about the market instability, the sustained employment, or the limited community-wide economic contributions.

The social force of the assumed economic benefits of mariculture also appears in data from governmental respondents from the seafood safety and public health arenas. While their associated agencies had broader responsibilities, these officials shared concern about the economic wellbeing of coastal communities with their marine management counterparts and these beliefs shaped risk assessments and decision making. Professionals involved with regulating water quality and seafood safety voiced concern about health threats from marine biotoxins (red tides), vibrio bacteria, and anthropogenic contaminants. They collectively recognized that uncertainty about these risks endangered the economic foundation of the mariculture industry, but they also understood that extended closures could deter consumers, devastate growers and hurt already marginalized communities. Several public health respondents suggested they were more careful analyzing data, reporting findings, and suggesting closures, given that alarmist press coverage and regulatory responses could severely impact communities economically dependent on shellfish.

These actors also advocated for expanding science to support mariculture commerce. They noted that increased precision in water quality analyses would reduce uncertainty related health risks and minimize economic shocks. However, in comparison with their marine resource counterparts, these actors were more cognizant that the economic future of the industry depended not only on productivity but also on understanding risks and threats to consumption. Public health officials also more openly acknowledged that data documenting market impacts from closures or consumer behavior in response to health risks were scarce. Finally, seafood safety officials consistently noted the need for more research on these economic threats to ensure the sustainability of the mariculture industry. This recognition of gaps in socioeconomic data and the need for economic expertise to inform planning was a key difference from their marine management colleagues.

4.2 Beliefs about the Economic Benefits of Mariculture: Scientific Researchers

Scientific researchers widely accepted that mariculture would bolster economic development, but the influence of this belief on their behaviors and assessments of risks varied. Production scientists considered mariculture science pivotal for promoting sustainable development. Most of these scientists felt the risks associated with rapidly applying novel scientific insights were outweighed by the economic benefits to growers and communities. The depth of this shared belief in the economic contributions from science-based mariculture also motivated them to work closely with growers and planners to speed the development process.

When queried about data confirming the economic returns from mariculture, none of these respondents cited economic analyses to support their beliefs. Most scientists based their views on anecdotal evidence or knowledge derived from on-the-ground partnering with growers. There was consensus among production scientists that market-related data could be beneficial, but the lack of economic analyses did not impede them from promoting mariculture. Finally, when asked about risks related to shifts in consumer preferences or shocks from marine biotoxins or pollution events, production scientists acknowledged these were threats. However, they could not identify studies of the economic impact of past events and none considered these risks grave enough to undermine the economic sustainability of mariculture.

Impact scientists also collectively believed that mariculture offered an ideal economic development alternative, and they were similarly unaware of market studies or economic risk analyses. Nonetheless, the internalization of this belief affected these scientists in distinct ways. Researchers focused on topics relevant to seafood safety were apprehensive that studies showing water quality problems or health dangers could lead to industry-wide economic shocks. This created a dilemma for many of these respondents, who feared being responsible for creating economic hardship for communities dependent on mariculture. Several impact scientists stated they were more cautious in their research practices and sought to ensure the highest degree of certainty in their findings given that they might lead to regulatory actions and potentially impact livelihoods. Theses scientists also voiced concern about gaps in understanding of the economic implications of seafood safety concerns; however, none noted collaborations with social scientists to understand the socioeconomic implications of their findings or recommendations.

Finally, the economists and business scientists interviewed for this study logically understood the value of social science expertise and data for supporting planning. They saw the burgeoning mariculture industry as an interesting development alternative, but none were aware of studies assessing its contributions to the economies of coastal communities, the financial returns to growers, or economic risks. Some of these professionals identified undergraduate and graduate-level theses focused on mariculture, but not peer-reviewed publications investigating risks or the economic contributions from the industry. Sustainable economic development was a research emphasis for a number of these respondents, but they concentrated on land-based agriculture or macro-level economic trends and none noted engaged research examining the economic sustainability of development initiatives as their emphasis.

4.3   Beliefs about the Economic Benefits of Mariculture: Shellfish Producers

Irrespective of farm size or type of shellfish grown, growers uniformly indicated that they entered mariculture because they believed it would be lucrative for them and their communities. Analysis of the origins of these beliefs shows that social interactions with scientists and governmental actors were pivotal. Most growers learned to farm shellfish from EPAGRI agents or agronomic scientists, and those professionals also informed them about markets and expected revenues from mariculture. Relatedly, growers indicated that extension agents and university scientists were actively involved in the creation of grower cooperatives focused on improving economic returns to communities. This engagement was viewed as an indicator of scientists' expertise in the business of mariculture. Many growers suggested that these actors were trusted experts and they would not have promoted mariculture if it were not profitable, reinforcing their belief in the commercial potential of shellfish farming.

While the appeal of mariculture as a sustainable development opportunity was logical, growers and distributors direct experiences generated heightened awareness about economic risks. Growers believed mariculture was a lucrative endeavor, but often contextualized this belief with concerns about the increasing costs of production, declining prices, inconsistent demand, and regulatory uncertainty. Many had in-depth knowledge of farm-level impacts from economic volatility and several farmers lamented the surprisingly low returns from their farms. Nonetheless, many stated that other growers must be more successful, otherwise government officials and scientists would not be championing the economic benefits of mariculture.

Growers awareness of economic threats parallels other respondents. Most concentrated on risks in the ocean environment such as red tides or pollution-related mortality, and they focused scant attention on seafood safety concerns and their implications for market shocks. Again, the expert status of scientists and government officials influenced these beliefs. Many farmers suggested that if public health concerns were grave enough to impact demand, governmental officials would alert them. Also, since mariculture management focused on conditions in the ocean, and not seafood safety post-harvest, growers believed threats to the industry were at the production stage and not at the point-of-sale. Since neither scientists nor governmental officials emphasized concerns about market volatility or threats to the economic sustainability of mariculture, growers assumed these risks were minimal.

While most growers steadfastly believed that mariculture was a profitable sustainable development alternative, it is important to note that on-the-ground trends were leading some to question this belief. These growers suggested that, initially, economic returns had been robust. However, more recently revenues declined and a number wondered whether those championing the activity understood the mariculture business. Interestingly, some had considered leaving the industry, but professionals from EPAGRI and production scientists suggested that their economic difficulties likely resulted from regulatory problems or the lack of appropriate production technology. This discourse appears to have dissuaded growers from exiting the industry and diverted their attention away from economic risks to mariculture's sustainability.

Finally, when asked about threats to the success of the mariculture, respondents from across our three categories also identified structural issues, such as clandestine farming, unregularized growing areas, and bureaucratic concerns with licensing. Some also pointed to inefficient production techniques as barriers to success and focused on the need for production science innovations to ensure sustainability. Mitigating economic risks was not central nor was gathering and analyzing the shellfish-related market or commercial data considered a priority. Our interview data show that mariculture stakeholders shared an unwavering belief in the economic development potential of shellfish farming. Nonetheless, deeply embedded misconceptions about mariculture commerce and economic data constrained their behaviors and actions toward achieving the socioeconomic objectives inherent to SDGs 8 and 12.

5. Attitudes Concerning the Importance of Economic Expertise and Data

Mariculture is a commercial activity; however, our research suggests neither market data nor economic expertise was not readily available to inform development decision-making, and business scientists were not engaged in planning like their production science colleagues. When respondents were asked to identify economists or business scientists specializing in mariculture in the region, only one was suggested. Similarly, when queried about access to mariculture-related economic data, the only source noted were the EPAGRI annual production reports. Nonetheless, the fact that economists were not involved with development planning, nor market data available, did not mean that economic knowledge did not influence the industry's development or consideration of risks.

Attitudes about what constituted relevant information related to seafood commerce and who had mariculture-related business expertise became key factors shaping the consideration of the industry's development potential and economic risks. SDGs 14 and 12 target increasing ocean science research, expanding science-based decision making to support sustainable marine development, and strengthening scientific capacity to move toward more sustainable production and consumption (GAUN 2015). In the subsequent sections, we examine how attitudes about the need for marine economic expertise and business data further influenced mariculture planning in Santa Catarina, providing broader insights into the social forces shaping the development of sustainable marine aquaculture.

5.1   Attitudes about Economic Expertise and Data: Governmental Actors

Governmental actors from EPAGRI and marine resource agencies consistently stated that part of their organizations' missions was to promote sustainable economic development. However, there was no institutionalized governmental effort to gather precise economic or market data to assess mariculture's economic contributions nor assess potential market-related risks. In addition, social scientists were rare at these agencies illustrating gaps in both expertise and data.

These respondents had well-defined attitudes about which experts should participate in development planning. While these actors considered agronomic expertise critical for increasing productivity and promoting development, they did not view social scientists' engagement as similarly essential to support planning. When asked about research related to the profitability of mariculture, planners pointed to studies highlighting the efficiencies of agronomic innovations and stated they were unaware of parallel studies of the longitudinal costs of mechanization, how profitability might vary by farm size or mechanization's implications for employment that could be central to achieving broader sustainable development objectives.

The origins of these attitudes about expertise and information appear to be social in nature. Extension and marine resource management programs were extensive in the region and were predicated on actionable scientific data and collaboration between scientists and planners. Programs focused on commercialization and business practices were less common and thus similar relationships with social scientists did not exist. While production scientists' expert status was internalized among governmental officials, economists and business scientists were not readily identified in parallel supporting roles. Governmental respondents also believed investigating consumer demand for shellfish or evaluating marketing strategies was external to mariculture planning and that growers, distributors, or restaurants should collect and analyze these types of information themselves rather than governmental entities.

As noted earlier there was near unanimity among governmental respondents that mariculture was a lucrative industry that had brought and would continue to bring socially and environmentally sustainable development to needy communities. Many noted the proliferation of farms along the coast and the growth of gastro-tourism as evidence of these economic benefits. Governmental actors also cited farmers' new cars or houses as indicators of the development returns from mariculture. None identified independent analyses of the economic benefits of shellfish farming at the grower or community levels to support these contentions.

While general attitudes about social science expertise were that it was less essential for planning, governmental actors did voice concern about economic shifts such as declining shellfish prices and the exodus of poorer farmers from the industry. Many of these respondents also acknowledged that larger farms appeared to be more profitable, raising questions about mariculture's ability to deliver socioeconomic improvements among marginalized communities. When discussing these troubling economic trends, some governmental actors voiced interest in business-related studies and acknowledged their potential value for planners and growers alike. Nonetheless, none indicated they had solicited or funded such studies nor were they aware of other governmental agencies initiating such research collaborations.

Governmental actors highly valued agronomic expertise and actively engaged scientific researchers to gather actionable data regarding environmental threats (e.g., red tides); however, they did not prioritize partnering with economists to analyze how these concerns might impact mariculture-related businesses. Rather, many of these actors focused on the burdensome and costly Brazilian regulatory structure and high taxes as the principal threat to the industry's economic success and this may have diverted attention from the need for broader economic risk analyses. The lower price of imported mussels from Chile was often cited as an indicator of these regulatory costs, but respondents could not point to studies comparing taxation or production costs in Chile versus Brazil. Similarly, governmental actors in the public health area pointed to the cost of complying with food safety regulations as encouraging clandestine production, but social scientific studies of these costs do not appear to have been available to inform policymaking and decision making.

Socialized notions about the limited need for engaging economists and accessing business-related data constrained governmental actors' ability to assess mariculture commerce and identify risks to its economic sustainability. However, some recognized this gap needed to be filled. EPAGRI staff created guides to assist growers with business practices and extension agents disseminated how-to business management materials where they were available. When asked about collaboration with business scientists to further develop these materials, respondents indicated they were unaware of individuals with these capabilities. Many also noted that there was not a clear mandate or institutional emphasis to focus on researching the economic aspects of mariculture within EPAGRI. Two respondents stated that the agency's hiring rules required employees have degrees in the agronomic or veterinary sciences. This inhibited capacity building in business administration and economics within the agency and likely made collaboration with university social scientists more difficult. While structural forces de-emphasized the economic dimensions of mariculture, some professionals from EPAGRI did attempt to assist with marketing-related materials and growers readily accepted and embraced this assistance.

5.2 Attitudes about Economic Expertise and Data: Scientific Researchers

Across the scientific community, respondents supported data-driven decision-making and science-based approaches to development planning. Production scientists also stressed the benefits of engaged scientific inquiry and they prioritized collaborating with both governmental actors and growers. Nonetheless, partnerships with social scientists were rare. This lack of inter-disciplinary collaboration was not necessarily deliberate nor signal a devaluing of social science expertise. Many production and impact scientists recognized that partnering with economists could be beneficial, but most did not have connections with such individuals and norms inherent to science-based planning did not foster engaging social scientists to inform consideration of the social or economic aspects of development initiatives.

Respondents from across our scientist category identified the need for rigorous mariculture-related scientific inquiry. However, none of these respondents identified data-driven analysis of the commercial or economic aspects of mariculture as similarly critical. When probed about economic expertise that might be available to assist growers, many indicated they felt comfortable providing business advice, noting that their history working in mariculture gave them industry know-how. When scientists were asked about data utilized to support their business advice, most redirected the question and emphasized knowledge came from their time working with the industry. It is interesting, however, that while many scientists worried that farmers were making ill-informed decisions about growing practices based on anecdotal knowledge, they did not have similar reservations about their assessments of business approaches without supporting commercial data or analysis. In some cases, scientists did counsel growers to seek assistance with business management practices from entities like Senac, a government agency that provides business workshops and training. Nonetheless, there is no evidence of any formal effort to integrate business experts from Senac or universities into science-based mariculture planning.

One of the most telling examples of the disconnect between production science assistance and economic expertise relates to the production of juvenile oyster seed at the Federal University of Santa Catarina (UFSC). Seeds were distributed to growers at little or no cost and scientists consistently cited this program as an example of the pivotal role of applied agronomic science in supporting mariculture development. However, when asked about the cost of the program and its implications for the economic sustainability of mariculture, scientists were unaware of such studies. Some recognized the program likely created a reliance on the university, but they believed this was outside of their purview as researchers. Paradoxically, a number of these same production scientists mentioned the necessity of continuing the program as they believed many oyster growers would go out of business if they had to produce their seed.

These types of incongruities in scientists' conceptions of the required expertise for collaborative science-based planning were also evident in the evaluation of risks to the mariculture industry. Scientists were concerned about pollution, red tides, shellfish diseases, and ocean acidification from climate change. There was consensus that expanded scientific inquiry was needed and integrating scientists with expertise on these issues into future planning would be critical. Nonetheless, while all these concerns have implications for the economic sustainability of mariculture, none in our scientist category noted the importance of engaging social science experts or gathering economic data to inform these risk analyses and foster resilience.

Finally, while a few small studies of market factors shaping mariculture existed, these were primarily theses executed by students in schools of business administration and food science, and the results were not widely distributed. Interview data from professors of business administration suggest that opportunities for further engagement of both students and faculty existed. However, the lack of social and institutional connections between these areas of universities and those related to the agronomic sciences appears to have inhibited broader collaboration. As an example, social scientists mentioned that while the dean of the agricultural college at UFSC was actively championing sustainable mariculture development, there was no parallel engagement by leaders from the business school.

Social science respondents also noted, that in the case of economics, applied studies were not prioritized by university administrators. They indicated that internal institutional pressures pushed faculty toward more macro-economic studies with theoretical contributions. Internal norms also led economists to prioritize scholarly products rather than engaged research. Similarly, other social scientists suggested that private sector actors in the more dominant business sectors in Santa Catarina—e.g., land-based agriculture and manufacturing—also pressured faculty and administrators to focus their research on topics of importance to those industries. Social scientists were aware of the emergence of mariculture, but they had not interacted, nor regularly communicated with, scientists from the agronomic sciences, extension agencies, or governmental authorities coordinating planning to inquire about opportunities to provide expertise.

5.3   Attitudes about Economic Expertise and Data: Shellfish Producers

One of the first questions all respondents in the producer category were asked was how they learned to run their business. Professionals from EPAGRI were one of the most often cited sources of business know-how. Additionally, many growers noted that production scientists from universities assisted with insights regarding how and where to sell their products. None cited guidance from business scientists or similarly trained government officials. There was a socialized notion that practical experience was how one learned about the mariculture business.

When queried about the overall business know-how within the mariculture community, most growers conceded that business expertise was limited. However, our data show that large-scale licensed growers had considerable business savvy, maintained balance sheets, and analyzed market trends. These producers were acutely aware of the value of market data and careful business administration. There was also consensus among large-scale growers that managing expenses and labor costs and analyzing commercial trends was essential for the sustainability of mariculture. Interestingly, growers with a heightened awareness of the importance of business management focused less on lamenting price fluctuations and pointed to controlling production costs and expanding markets as pivotal for profitability.

While larger licensed growers complained about the costs of complying with seafood safety regulations, they also understood the value of scientific information. Some of these growers had invested in their own health risk monitoring and production science. The ability to tailor the science to their farm, and to have exclusive access to data, was considered a strategic advantage. These growers also benefited from government and university science, like mechanical innovations and oyster seed production, but they understood farm-specific data, be it agronomic or economic, was valuable.

Small-scale growers had more limited access to data and business expertise and few had received business administration training. Most of these farmers were *clandestinos*, and this made government-sponsored assistance complicated. Many professionals from EPAGRI and government agencies worried about supporting unlicensed production and improving unauthorized businesses at the expense of regularized farmers. Similarly, formalized growers were acutely aware of the *clandestino* issue and criticized government officials for subsidizing 'illegal' competitors with extension assistance and low-cost oyster seed. They also lamented that *clandestinos* selling at low prices without sanitary controls posed health risks that could impact the entire industry while depressing markets and decreasing already thin margins.

Near-term revenues were prioritized by small farmers, and they appeared hesitant to invest in technologies or approaches that might require longer time horizons to see economic returns. Similarly, few of these farmers accounted for the cost of their labor or considered labor productivity in assessing the profitability and sustainability of shellfish farming. Most *clandestinos* focused on avoiding taxes and the costs of certification as strategies to increase revenue. Nonetheless, none could cite precise data on rates of taxation, licensing fees, or differences in prices from certified versus clandestine shellfish. Small farmers lamented diminishing returns from mariculture and worried about the rising cost of fuel and the low prices received from middlemen, but they did not have access to data to strategically adapt.

In addition, many *clandestinos* consciously chose to not keep accounting information for fear it could be used to impose fines or force payment of taxes. These growers also indicated that larger operators were increasingly serving as middlemen, purchasing much of their harvest and consolidating growing areas where small farmers had left the industry. Larger growers confirmed these patterns, suggesting that data on market trends indicated that for mariculture to be profitable it needed to be scaled up. These trends raise questions about the industry's future contributions to sustainable economic development and employment.

When discussing the business aspects of shellfish farming, one of the most prevalent data points cited was the wholesale prices received for oysters and mussels. While there was considerable knowledge of wholesale prices and those received from direct-to-consumer sales, there was little consideration of how price versus other factors affected consumer demand. Given the bounds of their business knowledge, growers collectively focused on price fluctuations rather than other aspects of commerce. Marketing expertise and information regarding consumer behavior were scarce and sales appeared to depend on relationships with restaurants or the reputation of growers rather than business data.

Finally, the deeply held belief that mariculture was an optimal development alternative played a key role in driving attitudes about the necessity of economic expertise and business data. Assumptions about the inevitable profitability of mariculture diverted planners, scientists, and growers' attention away from the underlying business conditions toward structural or production-related explanations for market volatility and downturns. Clearly the sustainability of mariculture depends on regulatory certainty and the application of innovative production science. However, mobilizing economic expertise and gathering and analyzing business-related data may be equally important. These experiences in Santa Catarina are particular to this locale, but the patterns and social forces shaping access to economic data and expertise illustrate the importance of social scientific data and expertise in supporting development initiatives striving to achieve the sustainability objectives of the UNSDGs.

6. Discussion

If aquaculture is to help achieve the marine environmental objectives of SDG 14, it must also forward economically sustainable production and consumption and generate productive employment as envisioned in Goals 12 and 8. In addition, enhancing research capacity and the availability of scientific knowledge to inform decision-making is central to all the UNSDGs (GAUN 2015). While myriad studies have shown that environmental science is pivotal in achieving SDGs, our investigation of mariculture in Santa Catarina demonstrates that social scientific study of its economic dimensions and applying business expertise to inform decision making may be equally critical for promoting sustainable aquaculture development.

Conceptual tools from institutional sociology provide a mechanism for analyzing how social forces shape stakeholders' beliefs about the business of mariculture and attitudes about the utilization of economic expertise and data to support planning (Scott 2008). Some of the most readily apparent institutional influences are regulative in form. Seafood safety rules and taxation significantly shaped the business of mariculture. Licensed growers complained of the burdensome costs of sanitary controls and *clandestinos* believed avoiding these regulatory costs, as well as taxes, were effective business management strategies. These emphases diverted attention away from the underlying economics of mariculture and identifying commercial practices that would forward sustainable production and consumption.

Similarly, planners and scientists singularly focused on regulatory barriers to explain market declines and this constrained their ability to analyze broader economic concerns. Beliefs about the economic implications of regulative institutional forces across our respondent groups exemplify bounded rationality, where limited collaborative interactions with actors with economic or business expertise led to isomorphic responses and strategies (Powell and DiMaggio 1991, Safford 2010, Scott 2008, Simon 1997). Principally, structural divisions at universities and government agencies constrained interdisciplinary interactions and the ability for stakeholders to understand and value of social science expertise and its application to inform analysis of mariculture commerce and business-related outcomes.

Our findings also suggest that un-regulated clandestine growing does affect price fluctuations and hinder efforts to collect reliable business-related data. Without accurate data, preconceptions about employment opportunities, revenues from shellfish, as well as market volatility and risks, drive decision making and raise questions about whether mariculture is consistent with the social improvements and the sustainable production and consumption targeted in SDGs8 and 12. Were there clear policies, regulations, and reporting procedures for harvest data, revenues, and employment, misconceptions would likely be less prevalent and economic risks more easily mitigated. Mariculture in Santa Catarina clearly generates revenues and social benefits, but without access to reliable data and rigorous analyses, uncertainty will remain whether it can deliver inclusive and sustainable economic development.

Normative and cultural-cognitive institutional influences are also pivotal in shaping the economic dimensions of mariculture. Norms inherent to science-based sustainable development paradigms prioritize engaging production scientists and data-driven decision-making (Safford et al. 2019). On the surface, this approach appears consistent with SDG 14's emphasis on increasing ocean-related scientific knowledge and its use to inform planning. However, in the Santa Catarina case, marine-related social scientific knowledge and expertise were notably absent from mariculture planning. This gap in part reflects a socialized belief among stakeholders that science-based planning does not involve economic science and that collecting and applying commerce data was the purview of individual businesses and not development planning.

While shared norms among scientists and planners stressed objective scientific practice, these actors did not apply this norm to business inquiry nor convey its importance to their grower partners. Governmental actors, scientists, and farmers all believed on-the-ground business experience naturally led to an understanding of mariculture markets and commerce, thus giving this belief normative power. Socialization also led shellfish farmers to view extension agents and production scientists as experts in all aspects of mariculture (Safford et al. 2019). Thus, rather than seeking business expertise external to the development community, they looked to EPAGRI agents and production scientists for business advice. Relatedly, interactions with agronomists reinforced growers focus on productivity and technical improvements when faced with lower returns, rather than assessing market trends or input costs. These well-intentioned and seemingly logical emphases on technical innovations to improve mariculture commerce lacked the key input of business science and once again illustrate bounded rationality (Safford 2010, Scott 2008, Simon 1997).

Growers readily recognized the limits to their business knowledge; however, norms de-emphasizing the importance of business expertise reduced their resilience to shifting economic conditions. Similarly, growers looked to scientists and planners for ques on how to assess risks. Norms within the planning community-focused efforts on assessing the origins, spread, and health risks associated with environmental concerns such as pollution and red tides. While stakeholders recognized the economic implications of health-related threats, internalized norms within this community focused on measures such as closure protocols rather than shifts in consumer demand. Normative institutional influences regarding expertise drove these asymmetries in utilizing social versus other scientific knowledge among planners and they appear to inhibit mariculture in Santa Catarina from achieving the commercial and employment objectives of SDGs 12 and 8. These patterns in our data suggest a need to reframe customary practices within science-based planning to ensure that both practical, as well as scientific insights, are applied to inform consideration of the economic dimensions of sustainable development initiatives.

Perhaps the most important institutional influence on the economic dimensions of mariculture development, the deeply embedded belief that mariculture is lucrative and economically sustainable, is cultural-cognitive in form. A key feature of cultural-cognitive institutions is that they are unquestioned and taken-for-granted (Safford 2010, Scott 2008). Despite a lack of business-related data, declining harvests, and the collapse of many mariculture businesses, planners, scientists, and growers were resolute in their belief that mariculture would deliver economically sustainable development. Bounded rationality led stakeholders to suboptimal isomorphic responses to commercial challenges and constrained their ability to identify strategic interventions to address economic difficulties (Safford 2010, Scott 2008).

The value of collaboration is inherent to all the UNSDGs, so it is important to emphasize that solidarity among planners, scientists and growers was key to the successful creation of a mariculture industry in Santa Catarina. While to some degree these social bonds diffused beliefs, attitudes, and practices that constrained engaging business expertise and assessment of the economic dimensions of mariculture, they illustrate the power of social forces in shaping sustainable development. This finding suggests that should the range of actors engaged in mariculture expand, norms shift, and the internal culture of the planning community change opportunities exist to refocus mariculture in Santa Catarina. This could help institutionalize resilience and forward a more economically sustainable development approach, helping mariculture achieve both the social and environmental objectives found in SDGs 14, 12, and 8.

7. Conclusion

The global oceans sustain billions of people as well many of the Earth's most important ecosystems. Yet because of their vastness, much about the marine world remains unknown. Farming the ocean is the new frontier for development and this makes science-based mariculture planning critical. However, because of its newness, there is greater uncertainty and more risks. Adaptation, resilience, and a detailed understanding of how both environmental and economic threats affect ocean ecosystems and the communities reliant on this new activity will be critical for the sustainability of mariculture in the future. The UN 2030 Agenda for Sustainable Development provides ambitious social, economic, and environmental objectives and success in reaching its goals and targets depends on mobilizing wide-ranging scientific expertise from across the natural, physical, and social sciences. Our study of mariculture in Santa Catarina shows that ensuring economic as well as environmental expertise and data are available to support planning is essential for its future success.

The creation of a shellfish industry in Santa Catarina was possible because of collaborative science-based planning that prioritized scientific engagement in development efforts (Safford et al. 2019). Failure to similarly mobilize economic and business expertise obscures risks to the industry and threatens its sustainability.

Planners provided seafood farmers with detailed knowledge of the ocean environment and worked to apply the best agronomic innovations to forward development. However, without access to business expertise and understanding of market trends, growers are vulnerable to economic risks and commercial failure. Social and cultural practices inherent to planning processes are, in part, responsible for the limited engagement of economists and business scientists. However, norms within the social science community also appear to dissuade these scientists from engaged research and interdisciplinary collaboration. Recognizing these barriers transcend disciplines and exist at the individual, group, and organizational level may be the first step toward the transdisciplinary science-based approaches needed to ensure sustainable development.

Perhaps the most elusive aspect of sustainable development is finding mechanisms to simultaneously advance human and environmental wellbeing. While the scientific community has tirelessly amassed data to establish parameters and targets for the environmental dimensions of development, similarly detailed data and analysis of the social and economic aspects are scarce. Investigating environmental threats has been paramount, but to ensure resilience the social and economic dimensions of these issues need to be equally well understood. Expertise and data are inextricably linked, and why and how they inform planning are social in origin. The development of mariculture in Santa Catarina shows social interactions are crucial for all types of scientific engagement. Socialization among planning stakeholders has institutionalized ideas about what data is required for decision making, the types of expertise that are internal or external to the development community, and the social statuses and roles of agronomists and economists have been viewed as distinct. This study shows they are in fact interdependent and mutually indispensable.

Mariculture development in Santa Catarina illustrates both the opportunities challenges inherent to efforts to meet the ambitious objectives and targets within the UN 2030 Agenda for Sustainable Development. The Agenda's final goal, SDG 17 highlights the importance of partnerships and cooperation in achieving the collective UNSDGs (GAUN 2015). While the specifics of this case are particular to shellfish mariculture in Santa Catarina, there are broader lessons. Environmentally, economically, and socially sustainable development is difficult, but not unattainable; interdisciplinary collaboration is challenging but not impossible. Social scientific investigations like our own and the others in this volume, that diagnose and explain why difficulties persist and how they may be overcome are an important step toward achieving those overarching objectives.

Acknowledgements

Financial support for this project came from the USIA Fulbright Program and the University of New Hampshire College of Liberal Arts. The authors recognize Dr. Paulo Vieira for his invaluable assistance with research design and Jacqueline Prudêncio for her aid with interview planning and map development. Finally, we are grateful to our interviewees in Santa Catarina who generously gave their time and insights to support this project.

References

Alves dos Santos, A., A.L.T. Novaes, F.M. Silva, G.S. Rupp, R. Ventura, G.L. Mello and S. Winkler. 2010. Síntese Informativa da Maricultura. EPAGRI/CEDAP. Florianópolis.

Alves dos Santos, A., A.L.T. Novaes, F.M. Silva, R. Ventura de Souza, S. Winkler da Costa and J. Guzenski. 2013. Síntese Informativa da Maricultura 2012. EPAGRI/CEDAP. Florianópolis.

Alves dos Santos, A. and S. Winkler da Costa. 2015. Síntese Informativa da Maricultura 2014. EPAGRI/CEDAP. Florianópolis.

Alves dos Santos, A. and S. Winkler da Costa. 2016. Síntese Informativa da Maricultura 2015. EPAGRI/CEDAP. Florianópolis.

Alves dos Santos, A., N.D.C. Marchiori and E.G.D. Giustina. 2018. Maricultura. pp. 169-170. *In*: Empresa de Pesquisa Agropecuária e Extensão Rural de Santa Catarina (EPAGRI). Síntese Anual da Agricultura de Santa Catarina 2017-2018. EPAGRI. Florianópolis.

Bostock, J., B. McAndrew, R. Richards, K. Jauncey, T. Telfer, K. Lorenzen, D. Little, L. Ross, N. Handisyde, I. Gatward and R. Corner. 2010. Aquaculture: global status and trends. Philosophical Transactions of the Royal Society B: Biological Sciences. 365(1554): 2897-2912.

Burbridge, P., V. Hendrick, E. Roth and H. Rosenthal. 2001. Social and economic policy issues relevant to marine aquaculture. Journal of Applied Ichthyology. 17(4): 194-206.

Cleaver, K.M. 2006. Aquaculture: Changing the Face of the Waters Meeting the Promise and Challenge of Sustainable Aquaculture. World Bank. Washington.

Corrêa, A.J. and S.G. Müller. 2016. A influência da ostra na origem, formação e manutenção da via gastronômica do Ribeirão da Ilha–Rota das Ostras–Florianópolis-SC. **Ágora**. 18(1): 119-130.

Costa-Pierce, B.A. (Ed.). 2008. Ecological Aquaculture: The Evolution of the Blue Revolution. Wiley. New York.

de Andrade, G.J.P.O., 2016. Maricultura em Santa Catarina: A cadeia produtiva gerada pelo esforço coordenado de pesquisa, extensão e desenvolvimento tecnólogico. Extensio: Revista Eletrônica de Extensão. 13(24): 204-217.

Diegues, A.C. 2006. Para uma aquicultura sustentável do Brasil. Banco Mundial. São Paulo.

Empresa de Pesquisa Agropecuária e Extensão Rural de Santa Catarina (EPAGRI). 2015. Plano estratégico para o Desenvolvimento Sustentável de Mariculture Catarinense (2015-2020). EPAGRI, Florianópolis.

Empresa de Pesquisa Agropecuária e Extensão Rural de Santa Catarina (EPAGRI). 2019. EPAGRI – Quem Somos. https://www. epagri. sc. gov. br/index. php/a-epagri/quem-somos/ Retrieved May 31, 2019.

Ferreira, J.F. and F.M. de Oliveira Neto. 2007. Cultivo de Moluscos em Santa Catarina. *In*: Barroso, G.F., L.H. da Silva Poersch and R.O. Cavalli. (Orgs.) Sistemas de cultivos aqüícolas na zona costeira do Brasil: recursos, tecnologias, aspectos ambientais e socioeconômicos (Série Livros 26, 87). UFRJ. Rio de Janeiro.

Food and Agriculture Organization of the United Nations (FAO). 2016. State of World Fisheries and Aquaculture 2016. FAO. Rome.

Frankic, A. and C. Hershner. 2003. Sustainable aquaculture: developing the promise of aquaculture. Aquaculture International. 11(6): 517-530.

General Assembly United Nations (GAUN). 2015. Transforming Our World: The 2030 Agenda for Sustainable Development. Division for Sustainable Development Goals. GAUN. New York.

Jacomel, B. and L.M.D.S. Campos. 2014. Produção sustentável e controlada de ostras: ações em Santa Catarina (Brasil) rumo aos padrões internacionais de comercialização. Revista de Gestão Costeira Integrada. 14(3): 501-515.

Lins, H.N. 2006. Sistemas agroalimentares localizados: possível "chave de leitura" sobre a maricultura em Santa Catarina. Revista de Economia e Sociologia Rural. 44(2): 313-330.

Lins, H.N. 2007. Interactions, learning and development: an essay on tourism in Florianópolis. Turismo-Visão e Ação. 9(1): 107.

Ministério de Pesca e Aquicultura – Governo do Brasil (MPA) 2015. Plano de Desenvolvimento da Aquicultura Brasileira – 2015/2020. MPA. Brasília.

Mungioli, R.P. 2012. Panorama da aquicultura no Brasil: desafios e oportunidades. BNDES Setorial. 35: 421-463.

Pereira, L.A. and R.M.D. Rocha. 2015. A maricultura e as bases econômicas, social e ambiental que determinam seu desenvolvimento e sustentabilidade. Ambiente & Sociedade. 18(3): 41-54.

Powell, W.W. and P.J. DiMaggio (Eds.). 1991. The New Institutionalism in Organizational Analysis. Chicago Press. Chicago.

Rodrigues, A.M.T., W.G. Matias, M. Polette, D.S. Occhialini, E.L. Micheletti and R. Dalbosco. 2010. A evolução da ocupação do espaço marinho do litoral catarinense pela malacocultura (1995-2005). Revista CEPSUL-Biodiversidade e Conservação Marinha. 1(1): 18-28.

Rossi-Wongtschowski, C.L.D.B., R.Á. Bernardes and M.C. Cergole (Eds.). 2007. Dinâmica das frotas pesqueiras comerciais da região Sudeste-Sul do Brasil. USP. São Paulo.

Safford, T.G. 2010. The Political-Technical Divide and Collaborative Management in Brazil's Taquari Basin. The Journal of Environment & Development. 19(1): 68-90.

Safford, T.G. and K.C. Norman. 2011. Planning salmon recovery: applying sociological concepts to spawn new organizational insights. Society & Natural Resources. 24(7): 751-766.

Safford, T.G., P.F. Vieira and M. Polette. 2019. Scientific engagement and the development of marine aquaculture in Santa Catarina, southern Brazil. Ocean & Coastal Management. 178: 104840.

Santos, R.R. 2011. Desenvolvimento do Sistema de Planejamento e Controle da Produção Comercial de ostras na fazenda marinha Paraíso das Ostras. TCC (graduação) - Curso de Engenharia de Aquicultura. UFSC. Florianópolis.

Scott, W.R. 2008. Institutions and Organizations: Ideas and Interests. Sage. Thousand Oaks.

Simon, H.A. 1997. Administrative Behavior: A Study of Decision-making Processes in Administrative Organizations. Free Press. New York.

Sidonio, L., I. Cavalcanti, L. Capanema, R. Morch, G. Magalhães, J. Lima and R. Mungioli. 2012. Panorama da aquicultura no Brasil: desafios e oportunidades. BNDES. 35: 421-463.

Subasinghe, R., D. Soto and J. Jia. 2009. Global aquaculture and its role in sustainable development. Reviews in Aquaculture. 1(1): 2-9.

Suplicy, F. and A.L.T. Novaes. 2015. Caracterização socioeconômica da maricultura catarinense e perspectivas para o futuro deste setor. Panorama da Aquicultura. 25: 38-43.

Suplicy, F.M., L.F.D.N. Vianna, G.S. Rupp, A.L. Novaes, L.H. Garbossa, R.V. de Souza, . . . and A.A. dos Santos. 2017. Planning and management for sustainable coastal aquaculture development in Santa Catarina State, south Brazil. Reviews in Aquaculture. 9(2): 107-124.

Trochim, W.M. (2000). The Research Methods Knowledge Base. Atomic Dog: Cincinnati.

Valenti, C.W., C.R. Poli, J.A. Pereira and J.R. Borghetti. 2000. Aquicultura no Brasil: bases para um desenvolvimento sustentável. CNPQ/MCT. Brasília.

Vianna, L.F.D.N., J. Bonetti and M. Polette. 2012. Gestão costeira integrada: análise da compatibilidade entre os instrumentos de uma política pública para o desenvolvimento da maricultura e um plano de gerenciamento costeiro no Brasil. Revista de Gestão Costeira Integrada. 12(3): 357-372.

Ventura, R., A.L. Vicente, A.A. Santo, A.L.T. Novaes, F. Muller and A. Ostrensky. 2011. Malacocultura em Santa Catarina: Maricultores, extensionistas e pesquisadores apontam problemas e demandas. Panorama da Aqüicultura. 21: 36-41.

Vinatea, L. and P.H.F. Vieira. 2005. Modos de apropriação e gestão patrimonial de recursos costeiros: o caso do cultivo de moluscos na Baía. Boletim do Instituto de Pesca, São Paulo. 31(2): 147-154.

Yin, R.K. 1994. Case Study Research: Design and Methods. Sage. Thousand Oaks.

Part IV
Mobility and Sustainability

Mobility and Sustainability: Individual and Collective Rights

Ennio Peres da Silva
NIPE/UNICAMP, Brazil

1. Introduction

Mobility and sustainability are two concepts linked by numerous parameters, which are technical, economic, social and cultural. This multidimensionality makes approaching this relationship something quite complex, being it usual to look at them as segments, taking limited analyzes among some of these parameters. Here, in the restricted space of this article, the energy aspect of mobility (fuel expenditure) and the respective emissions of pollutants and greenhouse gases, which impact its sustainability, will be considered.

The consumption of fuels used on mobility implies the use of natural resources, most of which are now non-renewable (fossil fuels as products of oil and natural gas). Thus, the indiscriminate use of such limited natural resources impose restrictions on their use by future generations, which is contrary to one of the foundations of sustainability. The emission of pollutants and greenhouse gases by the use of these fuels produce contamination of the atmosphere, notably in large urban centers, accelerating the processes of climate change, which equally reduces the quality of life of current and future populations (Lelieveld et al. 2019).

Thus, seeking conditions for sustainable mobility imposes the need for regulations and the introduction of legal limits in each society for the maintenance of acceptable living conditions today and in the future. These limits imply, in certain situations, in the restriction of individual freedoms and rights for the benefit of a greater, collective good. However, the establishment of laws, norms, etc., at the national, regional or municipal level often lead to questions about their scope, exaggerations or the improper intrusion of the public power into people's lives, intentionally or not, seeking only the solution of the problem or sometimes using these mechanisms to overstep rights or introduce certain ideologies.

The social conflicts arising in the context of mobility and its sustainability issues finds similarities in several other conflicts due to the use of natural resources, such

Email: lh2ennio@ifi.unicamp.br

as water, land, etc., as well as numerous environmental problems, ranging from the prohibition of smoking in certain areas to contamination of rivers, lakes and oceans by wastes from various natures (chemical, mining, oil industries, etc.). Therefore, some parallels can be drawn, facilitating the approach and understanding of the problems of mobility and sustainability.

According to the Oxford English and Spanish Dictionary Lexico, mobility is defined as "The ability to move or be moved freely and easily" (Oxford 2020). Therefore, mobility is directly associated with the concept of movement. The science that deals with this theme is called 'Mechanics', which is, specifically, the branch that studies the movement of bodies, one of the areas of physics. In this field, there are four fundamental ideas used: space (distance or displacement), time, velocity (movement per unit of time) and acceleration (variation of speed in time). Unfortunately, when it comes to human mobility, understood as the movement of people and goods in a given space, the complexity of the theme requires the consideration of a much larger number of variables.

In fact, individual or collective mobility is only effective from the occurrence of a large number of factors, including the availability of physical means (means of transport), financial means (travel costs), safety, comfort, among others. For each of these factors, there are a number of constraints, without the satisfaction of which mobility cannot be accomplished. Therefore, one should understand the complexity of a thorough analysis of this situation, which has resulted in studies, evaluations and analyzes of certain aspects of this problem. The same will be done in this space, where only some aspects related to certain environmental impacts of some means of transport will be addressed, seeking to identify relative degrees of sustainability among them.

However, the adaptation of physical quantities on the subject of mobility can provide useful quantitative information for the analyzes and understanding of the problem involved. For this, it is considered that the physical space refers to sidewalks, streets, avenues, roads, highways, railways and other structures of vehicle locomotion, where sea and air routes can also be included. The speed normally used in mobility studies is the average speed, calculated by the ratio between the distance traveled (route or path) and the time elapsed in that displacement. In these cases, two means are used: the average velocity itself obtained in each of the routes performed, and the average of the average velocities since it is common for certain displacements to be repeated periodically, suffering variations in values due to traffic conditions, which is they compose statistical data that needs to be analyzed in their median values. On the other hand, the concept of acceleration is not normally used in mobility studies, and rare are the cases in which this magnitude can be useful.

Another useful physical magnitude that can be used on the analysis of issues involving the displacement of people is the kinetic energy. Defined as half of the product between the mass of the body that moves and its velocity squared, this magnitude is part of the Law of Conservation of Mechanical Energy, which means that if an object has speed in relation to the reference (in the case of earth's mobility the reference is always the soil), then its kinetic energy differs from zero and as such, there was an expenditure of energy, either through an applied force or the expenditure of fuel in the case of motor vehicles.

The importance of the kinetic energy is to indicate that the greater the mass of the bodies, the greater the energy expenditure for them to move. Similarly, the higher the speed of the bodies, the greater the energy expended since the relationship between these quantities is not linear, as in the case of the mass (two bodies with the same velocity, one possessing twice the mass of the other, will also have twice the kinetic energy), but quadratic (two bodies of the same mass, one with twice the speed of the other, will have four times more kinetic energy).

Therefore, it is easily recognized that the use of vehicles (means of transport) for movement imply in additional energy expenditures (fuels), being both higher, the higher the masses and speeds of these vehicles. The lower limit is walking (pedestrian) when there are no additional masses set in motion. For the case of cyclists, there is already a need to move the bicycle; in the case of passenger vehicles, the cars and so on for each case. Just to get an idea of values, Table 1 shows the relative masses of some transport vehicles, considering people with 70 kg of mass.

Table 1: Relationship between Passenger Masses (70 kg each) and Vehicle

Middle	Average additional mass (kg)	Average number of passengers	Mass ratio
Pedestrian	0	1	0
Bicycles	10	1	0.1
Car ride	1,500	1	21.4
Car ride	1,500	4	5.4
Bus	20,000	40	7.1
Train (Subway/SP)	300,000	2.000	2.1
Aircraft (A380)	300,000	800	5.4

As can be seen in Table 1, the increase in additional mass can be compensated by the greater number of passengers, which reduces the specific expenditure of fuel and therefore of atmospheric emissions.

Regarding the speed of the means of transport, the concept of kinetic energy implies that the faster the displacement, the more the energy is consumed. In fact, with the occasional exception, the smallest time expenditure for everyday commutes (coming and going to the workplace) or exceptional ones (international travel, for example) are always desirable. Shorter time means higher speeds to travel the same distance, and the price to pay for this benefit is increased energy expenditure. This fact is also reflected in the cost of transportation, being the fastest (airplane), as a rule, is more expensive than the slower (trains).

It is also worth mentioning that the displacement of any mass on a surface (streets, roads, etc.) and/or submerged in a fluid (water and air) is subject to frictional forces, which tend to reduce movement, dissipating its kinetic energy. Thus, even if a constant velocity is maintained without variation of kinetic energy, there will be a need for some energy expenditure to compensate for these frictional losses. This expense will take place along the entire route, indicating that the longer the distance traveled, the higher the fuel consumption.

Therefore, the energy consumption will be different for each mean of transportation. Furthermore, any comparison between the different modes of transport also implies in the adoption of criteria depending on the distance travelled (per meter or kilometer) and the number of passengers transported (per person). In the case of energy consumption, these relationships can be seen in Figure 1.

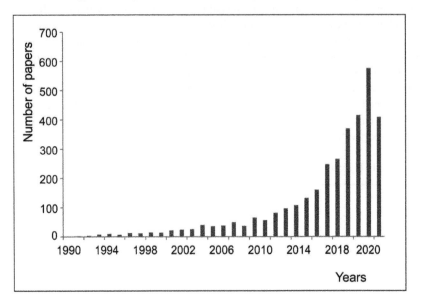

Figure 1: Specific Energy Consumption by Each Mean of Transportation
(Reprinted with Permission from Vadim Povkh and Joao. Pimentel. Ferreira in https://
pt.wikipedia.org/wiki/Eficiência_energética_nos_transportes, adapted)

As can be seen in the figure, there is a hierarchy in terms of fuel expenditure per passenger and per kilometer traveled, evidencing that from the usage of the planet's energy resources point of view, one should privilege those of lower consumption in relation to the others, whenever this is possible.

2. The Concept of Sustainability

Since it was conceived in the Brundtland Report in 1987, the term sustainability has undergone several redefinitions, expansions and criticisms but remains a concise and relatively simple way of understanding the need to maintain the natural resources and environmental conditions, in general, which are essential for life on the planet. Given its scope, its application in particular situations or systems imposes the need for a more specific definition in each case. Here, applied to mobility, it is understood as the displacement of people and goods with minimum expenditures of natural resources and the lowest possible environmental impacts, especially those that directly and indirectly affect people's health with an emphasis on emissions of pollutants and greenhouse gases in the atmosphere.

The natural resources used in mobility are mainly of two natures: materials for the manufacture of means of transport (vehicles) and energy resources (fuels and electricity), necessary for the movement of vehicles. In the first case, the reduction in the use of materials is strongly stimulated by competitiveness among manufacturers since higher quantity means higher cost and lower sales. This mass reduction also brings the benefit of reducing energy expenditure, as described in the previous item, increasing the efficiency of vehicles and again their competitiveness in the market. Even so, a significant amount of metals (iron, aluminum, copper, etc.), synthetic materials derived from petroleum (various plastics, rubbers, mineral oils) and other substances are used in the manufacture of each vehicle (Holden et al. 2020). The rationalization of this production, from the point of view of the sustainability of the process, goes through the analysis of the life cycle of vehicles, from reducing the use of materials to recycling all possible components, at the end of the useful life of each one. This aspect is not the object of this article, which focuses on the second case, that is the usage of energy resources in the propulsion of vehicles.

Analyzing the main means of transport, shown in Table 1 and Figure 1, it can be seen that all of them require an energy source, whether electrical or chemical, even though in electric vehicles electricity is stored equally in chemical form in the form of batteries. While in vehicles such as cars and buses the chemical energy is visibly stored in their tanks, in the case of pedestrians and the use of bicycles this is not so evident. In fact, it is common to find some texts defending that the consumption and/or emissions associated with pedestrians are null. This is not true since during walks humans consume an amount of chemical energy acquired during the process of feeding and digestion. To maintain its motor capacity, this energy must be recharged by ingesting more food. Despite their organic origin, these foods were grown, processed, stored and transported so there was an energy expenditure, including, in most cases today, the use of fossil fuels such as diesel and gasoline. Thus, it is important to associate the emission of greenhouse gases and pollutants by pedestrians and cyclists indirectly, such as electric vehicles, which emit nothing in their operation, but depending on how the electric energy they use is produced is responsible for the release of these products onto the atmosphere (in natural gas thermoelectric industries, for example).

Since the amount of energy used in each type of vehicle is associated with the corresponding emissions, both from a quantitative point of view (higher energy expenditure, more emissions) and qualitative (fossil fuels emit more greenhouse gases than the renewables ones), it is also true that in the case of emissions, correlations must be established according to the distance traveled and the number of people transported. Thus, Figure 2 shows a similar relationship to Figure 1 with respect to greenhouse gas emissions for each type of vehicle.

Therefore, in this case too, there is a hierarchy between the different means of transport, and those with lower emissions should be privileged to the detriment of those that contribute most to air pollution and climate change. In one case or the other, it appears that collective means are more aligned with the objective of promoting the sustainability of mobility, whether of people or material goods. This hierarchy is, in a simplified way, shown in Figure 3.

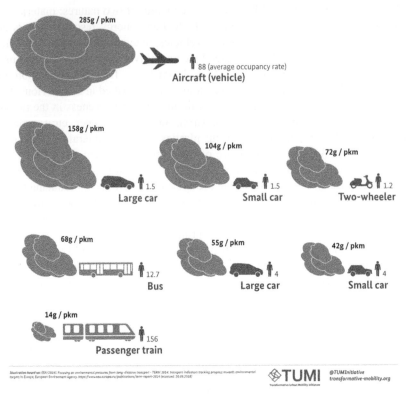

Figure 2: CO_2 Emissions by Type of Transport, Per Kilometer Traveled and Per Passenger Transported (Reprinted with permission from Transformative Urban Mobility Initiative (TUMI) in https://en.wikipedia.org/wiki/Sustainable_development, adapted)

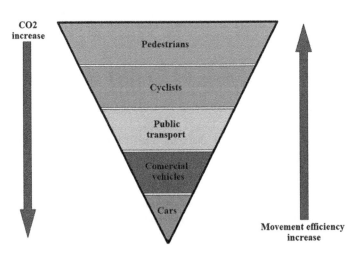

Figure 3: Hierarchy between Some Modes of Mobility, According to Energy Consumption and CO_2 Emissions

3. Individual and Collective Rights to Mobility

At the time this article was written, the world was going through the pandemic of the coronavirus SARS-CoV-2. In the context of public health measures adopted by all countries, a number of controversies arose, in support of or against certain measures. Some of them involved issues related to individual and collective rights, raising many ethical, political and economic considerations, some even about the police force. In a way, this type of discussion can also be found when it comes to public policy actions regarding mobility, especially in large urban centers in which these policies are essential.

Mobility is a necessity and a desire of the vast majority of people, and as such it constitutes an individual and collective right established in the Universal Declaration of Human Rights in article XIII, item one (UN 2020), being present in virtually all the constitutional norms of the countries; for example, in the Brazilian case, in Article 5, item XV, chapter I, Title II – Fundamental Rights and Guarantees, of the Constitution of the Federative Republic of Brazil (FSC 2020).

Although individual and collective (or social) rights are guaranteed by constitutional norms, these same normative instruments guarantee the so-called fundamental rights which are in higher authority than those above (in the Brazilian constitution, Title I, Fundamental Principles). Because they extend to the entire nation, these fundamental rights usually have a collective character, such as the guarantee of national development and the well-being of the society itself.

In the case of the pandemic, a controversy arose about the restriction of social meetings, avoiding the formation of agglomerations, which could result in a larger spread of the virus. In some countries, states and cities, this restriction has acquired a legal character, seeing as it is a product of government decrees or determinations, the non-compliance of such would result in fines and even arrests. Is this not a violation of people's constitutional rights to meetings and associations? This understanding led to conflicting situations, including physical aggression and police interventions.

When it comes to public policies in mobility, which seek to meet the collective, these can also be questioned for restricting individual freedom. This is the case, for example, of preventing traffic of vehicles on certain streets or avenues, which became exclusive to pedestrians. Would not that be a violation of individuals' rights to travel with their vehicles on these roads? Or, to also situate the issue of collective health, the implementation of the rotation of vehicles, seeking to reduce emissions of air pollutants, does not also hurt the constitutional right of free movement throughout the national territory?

The legal aspect of such questions, in democratic countries, is dealt with through laws, decrees and other instruments drafted and approved according to the reigning constitution in the country. From this point on, as long as they are classified as constitutional measures, they become mandatory and offenders are subjected to the respective penalties provided in each instrument. A person who does not agree with the established law must then question the constitutional terms that allowed such a law to be passed. In this case, within a democratic process, a change in the constitution must be sought, achieved by parliamentary or popular action. Utopically, the parliamentary majority that established such constitutional norms, as legitimate representatives of the majority of the society, would be considering the interests of the collective over individual interests. In practice, the usual is that the parliament represents a dominant elite, whose interests may or may not coincide with the interests of the majority of the population. Here is seen the controversial Principle of Supremacy of Public Interest Over Private Interest (Gabardo and Hachem 2010), an important part of Social Administrative Law, discussion of which is – beyond this article's scope.

As for the ethical aspect, the prevalence of collective interests over individual ones is usually recognized, provided that minority rights are respected. However, important issues arise when one unnecessarily extrapolates assumed collective interests, removing individual rights without justification. For example, in the case of urban mobility, there is no doubt about the loss of human life due to the emission of pollutants by motor vehicles. This fact motivates the adoption of public policies in favor of collective modes of mobility, reducing the emission of greenhouse gases into the atmosphere, as previously shown. On the other hand, the exaggeration of preventing people from using their individual vehicles would not be justified. Even in more serious cases of urban pollution, one can prevent the movement of part of the vehicles, as in plate restriction car rotations adopted in larger cities. But the ban on the movement of all vehicles, all the time, is configured as a denial of the citizens' rights.

Therefore, it can be observed that, despite the predominance (but not supremacy) of collective interest over individual interest in the case of conflict, the scenario should happen with minimal losses or restrictions on the citizens' freedom. This task is a major challenge to those in charge of decision making since in many cases there is no clear line delimiting this separation of interests. A good strategy is to adopt, whenever possible, solutions with less impact that can be evaluated progressively over time, in accordance with the obtained results. For example, in the case previously mentioned of the car rotations, one can initially establish the movement restrictions for specific plate endings on each day of the week, which may block plates with two or three specific digits according to the reduction (or lack of it) of the air pollution measured daily.

In any case, certain individual attitudes, even when within the scope of personal rights, cannot harm the collectivity, even when it takes place in an indirect or diffuse manner, like in the case of refusal to wear face masks during the pandemic. Even more, the point here is whether people have the ability to understand the problem, or if they act deliberately against established norms, motivated by beliefs, ideologies or simple actions that violate the law. The first case presents a more complex issue because the cognitive abilities of people in certain countries are compromised, whether by cultural, religious and/or, more fundamentally, by precarious and ancient educational structures.

To overcome the cognitive barrier, there is a clear need of reforming the educational structures, evidently; but when thinking about short term effects, it is possible to obtain results through advertising campaigns and other forms of dissemination. In the case of mobility policies, measures adopted require wide dissemination and explanation to society, in a way that it is able to cover even the lowest levels of education available. Similarly, the appeal to individual benefits is also more necessary in these precarious conditions since the ability to perceive collective gains suffers greater blockages. In cases where individuals are aware of how their actions may cause collective problems but still insist on maintaining their attitudes, there are many possible reasons, some quite complex that run away from the scope of this article.

An important aspect to be highlighted is that, in addition to the awareness campaigns and dissemination processes of the measures adopted, the so-called community awareness, there is always the need to establish legal punishments to offenders, otherwise the decisions made end up ignored and/or fall into disuse quickly, sometimes even at the moment the campaign ends. This aspect has been verified in numerous situations but with different results, which vary according to societies (culture of respect for laws, educational level, etc.), the ability to surveil, etc. The certainty and rigour of punishments have ensured the success of many initiatives. The lack of supervision and the certainty of impunity are guarantees that a specific public action will not work. In the current case of the SARS-CoV-2 pandemic, these aspects have been confirmed in several countries, where mask use measures and restrictions on agglomerations have been more or less successful, depending exactly on these factors.

In the case of mobility, the issues of sustainability and the impacts of this sector on the environment with damages to the health of communities fit into this type of problem. Public policies aimed at the well-being of the whole are often not understood by individuals, who disagree with rules and norms, disobeying and often hindering them, even going as far as preventing (or actively going against) the effectiveness of these policies.

An interesting issue brought up by the SARS-CoV-2 pandemic with impacts on mobility and its sustainability was a very large adherence to the 'home office' system. In fact, in view of the problems caused by mobility, if it is made possible to avoid displacements, the smaller the environmental impacts. Thus, to avoid the transportation of employees at peak times also means to reduce congestion, freeing up spaces in public transport and avoiding the use of many vehicles by a single passenger. However, it should be noted that the impacts of this model are not null since many home office workers make use of home deliveries, opting for individual meals (greater losses than on collective meals in restaurants), which often bring greater environmental impacts than in the case of working and eating in the company. But mainly because of the general effects on urban mobility, this model is environmentally advantageous, especially in larger urban agglomerations.

Another relevant aspect of the home office is that this system is restricted to some professional categories, many in the service sector, as well as to the social classes of medium to high purchasing power. For the lower-income classes, usually manual workers, who constitute in many countries the majority of the population, this modality does not apply. For them, leaving home is a necessity and the availability of economically accessible means of transport is essential.

In addition to the technical issues of the mode of use as well as to the traction mode, one aspect that impacts mobility in large urban centers is the organizational form adopted, that is the public mobility policies. The first aspect that draws attention is that the vast majority of cities were not planned for their respective fleets of vehicles. In fact, the significant increase in the number of urban vehicles occurred after World War II (1939-1945) with the resumption of economies worldwide. In general, even in relatively young countries, their urban structures were established long before the emergence of vehicle fleets. Only the cities that emerged from the middle of the last century were able to plan their road systems for a large number of vehicles, which in many cases ended up creating new mobility problems, mainly for the displacement of pedestrians and bicycles as was the case of the Brazilian capital, Brasilia, inaugurated in 1960.

Some cities, where urban mobility is considered more strategic, have promoted expropriations, removals of public spaces and other measures for the implementation of streets, avenues, parking lots, etc., facilitating the movement of its large fleets of vehicles. Others only promoted the organization of urban traffic, implementing lanes and/or exclusive areas of circulation for certain vehicles, without significantly altering the existing urban profile. The results of each strategy showed advantages and disadvantages, but few were consecrated since local, cultural, historical characteristics, etc., greatly influence any of the actions.

As can be seen, in addition to aspects directly involved in mobility, such as types of vehicles, fuel consumption, etc., and their sustainability, such as emissions of pollutants and greenhouse gases, other aspects greatly influence the search for solutions. The complexity of the theme requires that multiple aspects of each proposal or public policy be examined always having advantages and disadvantages in each one. Even lesser obvious alternatives, such as the replacement of petroleum products with biofuels, need their effects considered beyond the analysis on mobility, involving land-use policies (land-use issue, monocultures, etc.), competition with areas of food cultivation, the use of water for irrigation and many others that without proper planning and application of mitigating measures will only be changing the problem faced, perhaps even for the worse.

References

FSC. 2020. Brazilian Federal Supreme Court, Constitution of the Federative Republic of Brazil. http://www2.senado.leg.br/bdsf/handle/id/243334.

Gabardo, E. and D.W. Hachem. 2010. O suposto caráter autoritário da supremacia do interesse público e das origens do direito administrativo: uma crítica da crítica. *In*: Bascellar Filho, R.F. and Hachem, D.W. (Coord.). Direito administrativo e interesse público: estudos em homenagem ao professor Celso Antônio Bandeira de Mello. Fórum: 155-201.

Holden, E., D. Banister, S. Gössling, G. Gilpin and K. Linnerud. 2020. Grand narratives for sustainable mobility: a conceptual review. Energy Research & Social Science. 65: 101454.

Lelieveld, J., K. Klingmuller, A. Pozzer, R.T. Burnett, A. Haines and V. Ramanathan. 2019. Effects of fossil fuel and total anthropogenic emission removal on public health and climate. Proceedings of the National Academy of Sciences U.S.A. 2019. 116: 7192-7197.

Oxford. 2020. Oxford English and Spanish Dictionary. 2020. https://www.lexico.com/en/definition/mobility

UN. 2020. United Nations. Universal Declaration of Human Rights. https://www.ohchr.org/EN/UDHR/Documents/UDHR_Translations/eng.pdf.

Estimating Vehicular CO$_2$ Emissions on Highways Using a *"Bottom-up"* Method

Estevão Brasil Ruas Vernalha[1,2]* **and Sônia Regina da Cal Seixas** [1,3]

[1] Energy Systems Planning Postgraduate Program, Faculty of Mechanical Engineering
State University of Campinas, Brazil
[2] Centre for Sustainability and Cultural Studies, Centro Universitário UNIFAAT
(NESC/CEPE/UNIFAAT), Estrada Municipal Juca Sanches 1050, ZIP 12954-070,
Atibaia, SP, Brazil
[3] Center for Environmental Studies and Research, NEPAM/State University of Campinas,
UNICAMP, Rua dos Flamboyants, 155, Cidade Universitária, ZIP 13083-867,
Campinas, SP, Brazil

1. Introduction

One of the main categories of GHG emissions associated with human activity and climate change is classified as "production and use of energy and the transport sector" (Laurmann 1989). While the transport sector is in this case embedded in a larger category, other authors work directly with a larger number of more specific emission sources, e.g., Marcotullio et al. (2013), who suggest six main sources of GHG emission, whereby the transport sector is decoupled from the energy sector.

In this system, emission sources are distinguished according to agricultural, energy-related (electricity and heating), industrial and production-related, residential, transportation, and waste-related provenance. It should be noted here that the large-scale burning of biomass was not considered in the database used by Marcotullio et al. (2013) for the Emissions Database for Global Atmospheric Research (EDGAR).

This article only considers the transport sector and evaluates the methodological possibilities for estimation of vehicular CO$_2$ emissions on a major highway in Brazil – D. Pedro. I Highway (SP-65). These emissions are considered the main source for GHGs concentration increasing in the atmosphere and are thus associated with climate change.

In this context, the exportation corridor Campinas – São Sebastião, which was announced by the state government of São Paulo in 2005, consists of three highways:

*Corresponding author: estevao.gestao@gmail.com

D. Pedro I (SP-65), Carvalho Pinto (SP-70) and Tamoios (SP-99) (São Paulo 2005). With this announcement, the state government also implemented a number of investments in order to increase the product flow capacity.

According to the State Program for Public-Private Partnerships (PPP) in São Paulo, these investments should include the improvement of highway Dom Pedro I, the duplication of the Tamoios highway on the plateau stretch and the Serra stretch, the construction of the contour road Caraguatatuba – São Sebastião (condition to moving the port of São Sebastião), the construction of the urban contour roads within São Sebastião, and the expansion of the port of São Sebastião (São Paulo 2006). These investments were justified by growth forecasts from the Development Director Planning for Transport in São Paulo (PDDT-Vivo 2000/2020), based on a predicted increased annual demand for cargo transport in the order of 3.3% (Braga 2008).

Highway SP-65 is, especially in the region of Campinas, an integral part of the export corridor Campinas – São Sebastião, which is an important route for the international importation and exportation of products, as well as for goods distribution throughout the state of São Paulo interior (Brazil). From 2000 to 2010, vehicles flow on SP-65 almost doubled, which significantly enhanced the contribution to GHGs emissions accounted for the state of São Paulo.

In fact, traffic intensification on SP-65 can be seen in line with the state government of São Paulo strategy regarding the traffic capacity expansion in the export corridor Campinas – São Sebastião. It should be noted here that the annual vehicle numbers have increased every year, which is reflected by the recorded vehicles number at the Itatiba toll station (situated in SP-65): 1997 – 5,741,178 vehicles; 2000 – 5,926,778 vehicles; 2003 – 7,239,736 vehicles; 2006 – 8,426,400 vehicles; 2010 – 10,508,350 vehicles; 2013 – 11,948,275 vehicles (Dersa 2013, DER 2014a, b).

In contrast to the inherent economic benefits upon increasing product flow capacity of the export corridor, the increase of highway vehicles number is also associated with a correspondingly increasing of air pollutants and GHGs emissions, which have a local, regional and global impact. In the transport sector, GHGs (mainly CO_2) arise predominantly from burning fossil fuels and are correlated to climate change and global temperature rise (Rossetti 2002, Penteado 2008, Cerri et al. 2010). The impact of such high atmospheric concentrations of GHGs is manifold: they are related to the increased occurrence of droughts, intensified rainfalls and floodings, the increased occurrence of heat waves and raging forest fires, the sea level rise, and the damage of water resources, agriculture, wildlife, and natural ecosystems (Mcmichael et al. 2003, EPA 2009).

In this context, accurate estimation of GHG emissions—from both stationary and mobile sources—is of fundamental importance to improve understanding with regards to the global impact of these gases atmospheric concentration. The uncertainty regarding changes that GHG emissions have been suffering overtime should also be due to its structural character associated with changes in these emissions (Lesiv et al. 2014). Promoting the increasingly precise determination of GHG emissions by national registration agencies is crucial in order to identify all existing global GHG sources and sinks, and thus contribute to improved comparative studies regarding these emission data and already known concentrations of GHGs in the atmosphere.

We intend to contribute to this discussion by offering a method for the estimation of vehicular CO_2 emissions on a major highway in Brazil (SP-65). This method is based on a 'bottom-up approach', which was suggested as one of the two main methods by Intergovernmental Panel on Climate Change (IPCC) for conducting GHG emission estimates for mobile sources (IPCC 2006).

1. Results and Methodological Discussion

2.1. Principal Equation for the Calculation of Emission Estimates

According to Borsari (2009), the calculation of estimated vehicular CO_2 emissions by the *bottom-up* method can be accomplished using the following equation:

$$\text{Emission} = Fr \times Fe \times km_{av.} \qquad (1)$$

wherein Fr = number of vehicles; Fe = emission factor; $km_{av.}$ = average distance travelled per vehicle per year.

For this estimate, some authors (Bales et al. 2013) present a formula which initially seems different from equation 1, but is in essence identical:

$$E = Fe \times Iu \times Fr \qquad (2)$$

wherein E = mass of pollutant emitted for the period considered (e.g. g/yr); Fe = emission factor or mass of vehicle-emitted pollutant for a given distance (g/km or g/kWh); Iu = annual mileage of the vehicles (km / year); Fr = number of vehicles divided by vehicle type and year.

This formula, presented by the Environmental Sanitation Technology Company (CETESB) for the calculation of the exhaust emissions, is identical to that suggested by Bales et al. (2013) (CETESB 2013). Thus, the methodological discussion for estimating the vehicle-borne CO_2 emissions on highway SP-65 in this article contains information from three sources cited in relation to the primary equation for calculating emission estimates.

2.2. Emission Factors

Emission factors (vehicular pollutants mass for a defined distance) should be defined for vehicles that are differentiated according to the category, model, manufacture year, and type of used fuel. Data collected herein are related to emission factors that were developed by CETESB (2013).

Furthermore, vehicular CO_2 emission rates are affected by vehicles deterioration levels, as their efficiency decreases over time, i.e. vehicles consume more fuel per distance upon aging. Therefore, age-dependent deterioration factors should be considered which would adjust the emission factors that were originally calculated for new vehicles. The application of such deterioration factors would generate fixed emission factors according to the following equation (Esteves et al. 2004):

$$FE_{corr} = FE \times FD \qquad (3)$$

wherein FE_{corr} refers to the corrected emission factor after inclusion of the deterioration factor, FE refers to the new vehicle emission factor (measured before leaving the factory), and FD refers to the decay factor.

2.3. Profile of the Circulating Vehicle Fleet

As noted by the IPCC (2006), when a particular fleet is unknown with respect to the vehicle and fuel types, it can be estimated from national statistics. The document also states that the use of locally estimated data should reduce uncertainty, especially for estimates from using the bottom-up methodology. Accordingly, the number of registered vehicles by model, manufacture year and region can be obtained from national licensing records (IPCC 2006). For this article purpose, the profile of SP-65 fleet was considered identical to the state of São Paulo fleet profile. Data for smaller regions in which the highway is located, e.g. the metropolitan area of Vale do Paraíba, or São Paulo Macrometropolis were also obtained from documents developed by CETESB (2013).

For the fuel used, it is necessary to identify regional averages regarding the use of hydrated ethyl alcohol fuel (AEHC) and gasoline in so-called flex-fuel vehicles, i.e. vehicles that can use either of these two fuels. These averages vary mainly driven by the price difference between these fuels. Flex-fuel car owners decide at the petrol station which fuel to buy, and the vast majority usually chooses the one that presents the best cost-benefit, considering the fuel price in relation to its efficiency (Goldemberg et al. 2008).

2.4. Circulating Fleet Volume and Average Distance Traveled

In order to accurately determine the fleet volume and the average distance traveled per year, it is necessary to divide the highway into sections, considering the fluctuation behavior of the fleet along the route and the different volumes of vehicles that can be observed in the different sections.

The division into parts depends essentially on the highway points where the vehicles are counted, e.g. toll stations. For sections that count at both ends, it is reasonable to consider the average between the values obtained at both ends as a circulating fleet volume proxy. In order to refine the thus obtained estimate, points in these sections where the fleet volume may change should be identified, e.g. junctions and intersections with other traffic routes and the section should accordingly be subdivided further.

The circulating fleet in these subsections may therefore be weighted according to the capacity of each of these junctions and crossings to influence on the SP-65 vehicle flow. This influence may be assessed via an analysis of the circulating fleet of the highways that interact with SP-65. In sections that count only at one extremity, the thus obtained value must be used for the entire section.

3.   Results and Discussion

3.1.   Emission Factors

Table 1 presents some emission factors for new cars differentiated according to manufacture year and fuel type (CETESB 2013).

Table 1: Average Emission Factors for New Light Vehicles

Year	Fuel type	PROCONVE stage[b]	CO_2 (g/km)
2010	Gasoline C[a]	L5	208
	Flex-Gasol. C[a]		177
	Flex-AEHC		171
2011	Gasoline C[a]	L5	198
	Flex-Gasol. C[a]		178
	Flex-AEHC		170
2012	Gasoline C[a]	L5	195
	Flex-Gasol. C[a]		180
	Flex-AEHC		173

[a] Gasoline C: Commonly used for standard gasoline sold in Brazil, containing about 22% of anhydrous ethanol (EACA).

[b] PROCONVE = Programa de Controle da Poluição do Ar por Veículos Automotores, i.e. the control program for air pollution from motor vehicles, whereby L5 refers to the program stage 2010-2012.

Source: Emissões Veiculares no Estado de São Paulo 2012 (CETESB 2013)

As mentioned in the section "Deterioration Factors" of the report "Vehicle Emissions in the State of São Paulo 2012" (CETESB 2013), CETESB does not considers vehicle deterioration factors for their estimates of vehicular pollutants emissions in the case of missing data or validated studies.

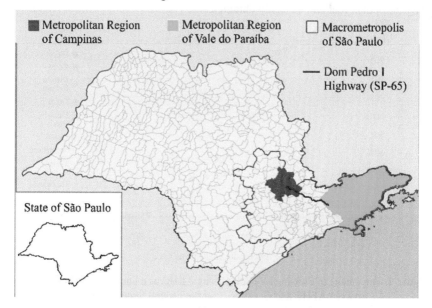

Figure 1: Localization of the SP-65 Highway within the State of São Paulo
Source: Produced by the Author – Data from São Paulo (2016)

3.2. Circulating Vehicle Fleet Profile

In order to determine the circulating fleet profile that travels on SP-65, regional average percentages should be established for each type of vehicle (model, manufacturing year, and fuel type) along the highway. Table 2 shows the state of São Paulo circulating fleet in 2012 (CETESB 2013). From these values, it is possible to determine the individual percentage values for each vehicle type, and values may subsequently be applied to each SP-65 section. In general, state-average values can be used, but if available, more specific regional averages should be used, e.g. metropolitan region of Campinas, metropolitan region of Vale do Paraíba or São Paulo Macrometropolis. Those values have also been reported by CETESB (2013).

3.3. State of São Paulo Circulating Fleet Estimate

Next, Figure 2 shows the percentage of AEHC fuel used in flex-fuel vehicles as a function of the price ratio between AEHC and gasoline C for each federal state in Brazil (Goldemberg et al. 2008). The power density of AEHC is about 70% of that of gasoline C. Thus, it is expected that when the price relationship AEHC/gasoline C is 70%, half the fleet operates on AEHC. The green dashed curve in Figure 2 represents this "ideal curve". However, each federal Brazilian state exhibits a unique use resistance toward AEHC, resulting in the Brazilian average (red curve), which is by 9% lower than the ideal curve. Accordingly, AEHC price has to be 61% of that of gasoline C for 50% of the flex-fuel fleet in the country to operate on AEHC. Only the behavior in the state of São Paulo coincides with the ideal curve.

Given that the price ratio between the average prices of the two fuel types is available for any particular year, the percentage of flex-fuel vehicles that used AEHC in any particular year can be easily obtained.

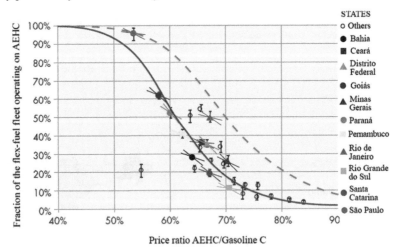

Figure 2: Percentage of Flex-Fuel Vehicles Using AEHC as a Function of the Price Ratio between AEHC and Gasoline C for the Federal States in Brazil
Source: Bioenergia no Estado de São Paulo: Situação atual, perspectivas, barreiras e propostas (Goldemberg et al. 2008).

Table 2: Estimation of Vehicle Fleet in the State of São Paulo in 2012

Category	Fuel	Sate of Sao Paulo	Average age	MR of Sao Paulo	County of Sao Paulo	MR of Campinas	MR of Vale do Paraiba	Macro metropolis
Car	Petro	4.73.008	14	2.313.673	1.501.375	317.901	198.256	3.257.967
	Etanol	406.215	23	174.707	116.773	29.647	16.720	259.515
	Flex	4.818.146	4	2.505.830	1.607.066	427.463	236.757	3.752.200
Light Duty	Petrol	686.0.51	10	397.459	274.667	53.082	31.053	554.414
	Etanol	40.873	22	16.156	10.745	2.890	1.702	25.018
	Flex	664.066	4	287.037	180.637	61.067	30.493	460.824
	Diesel	343.784	8	156.340	98.542	26.206	14.852	233.675
	Diesel	42.928	16	17.054	8.123	3.390	1.746	27.951
Semi-Light	Diesel	119.788	14	47.671	22.602	9.477	4.842	78.113
Trucks Light		73.814	16	29.646	14.114	5.816	2.992	48.382
Medium		101.147	8	41.324	19.506	8.233	3.982	67.190
Semi-Heavy		10.238	8	40.924	19.301	8.163	3.937	66.594
Heavy	Diesel	93.285	11	49.205	29.853	8.092	4.293	70.479
Urban		10.384	11	5.460	3.303	906	477	7.835
Road	Petrol	2.262.277	7	799.911	455.673	175.031	123.502	1.430.818
Motorcycles	Flex	348.766	2	80.146	41.403	30.379	21.255	185.124
Total		14.344.770	8	6.962.543	4.403.	1.167.769	696.861	10.526.099

*MR: Metropolitan region.
Source: Emissões Veiculares no Estado de São Paulo 2012 (CETESB 2013)

3.4.   Circulating Fleet Volume and Average Distance Traveled

SP-65 highway includes three toll stations: Itatiba (P1 – km 110), Atibaia (P2 – km 79.9) and Igaratá (P3 – km 26.6), where vehicles are counted in both directions.

Vehicle volumes recorded at the three toll stations in 2010 exhibit quite similar behavior for the circulating fleet in both directions. The toll station that recorded the most similar fleet volumes in 2010 in both directions was Itatiba (southbound: 5,262,205; northbound: 5,246,145) (DER 2014a).

An initial analysis revealed that the circulating fleet of vehicles varies along the highway. While the vehicle volumes measured at P1 and P2 were comparable, P3 registered a significantly lower number of vehicles (~ 50% relative to P1 and P2). The highest number of vehicles in 2010-2013 was recorded at P2 (DER 2014th). The observed differences between the vehicle numbers recorded at P1, P2 and P3 suggests that along the highway different volumes of vehicles travel in different sections. In order to estimate the CO_2 emissions for the circulating fleet that runs along SP-65 most accurately, the highway should be subdivided into the largest possible number of parts.

Considering the three toll stations where the vehicle fleet is counted (P1, P2 and P3), in addition to one point before the first toll station (beginning of the highway) and one point after the last toll station (end of the highway), a division of the highway into four main sections seems feasible in order to facilitate the circulating fleet analysis: section 1 – the **city of Campinas (P0)** to **P1**; section 2 – **P1** to **P2**; section 3 – **P2** to **P3**; section 4 – **P3** to the **city of Jacareí (P4)**.

The determination of the circulating fleet to be considered in each one of the four sections is simplified according to Figure 3. For Section 1, the fleet is the value measured at P1 (as the fleet is not counted at P0); for Section 2, the average of the values measured at P1 and P2 is used; for Section 3, the average of the values measured at P2 and P3 is used; finally, for Section 4, the values measured at P3 are used (no count at P4).

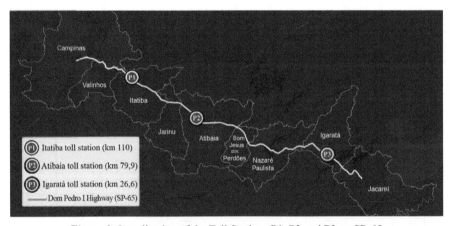

Figure 3: Localization of the Toll Stations P1, P2 and P3 on SP-65
Source: Prepared by the Author using ArcGis Webmap Tool – Data from Rota das Bandeiras (2015).

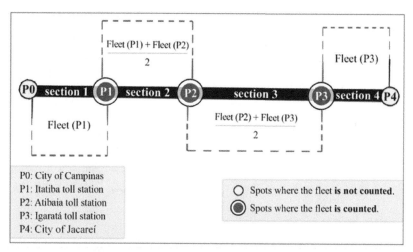

Figure 4: Vehicle Fleet Definition for the Four Subsections of SP-65
Source: Produced by the Author – Data from Rota das Bandeiras (2015)

However, upon closer inspection, each section exhibits characteristics that indicate the need for adjustments to the simplified definition shown in Figure 3. Therefore, we analyzed each of the four sections separately in more detail.

3.4.1. Section 1: P0-P1

The determination of the starting point for section 1 (P0) should directly impact the estimated vehicular CO_2 emissions on SP-65. Initially, setting the starting point (P0) right at the beginning of SP-65 highway (km 146), at its junction with Anhanguera highway, seems to be the most natural decision. In Figure 5, this point is marked by the beginning of the light blue line on the left side of the map.

Nevertheless, P0 was set at a more advanced point of SP-65 (128 km), predominantly on account of one particular reason: regardless of the location chosen for this starting point (P0), vehicles are not counted at this point. For Section 1, only the vehicles counted at P1 will be relevant and used for the entire section. If P0 was placed further toward the beginning of SP-65 (left side of the light blue line), there would be several possibilities of increase or decrease involving the number of vehicles circulating between P0 and P1 because of the intersections with highways SP-332, SP-340, SP-081, and SP-083. In that case, using the values recorded at P1 for the entire section would be associated with substantial discrepancies between model and reality.

In total, the section represented by the light blue line, which is not considered for the estimate, amounts to 18 kilometers. Thus, the distance between P0 (128 km) and P1 (110 km), which is defined as 'Section 1', is 18 km.

3.4.2. Section 2: P1-P2

In this section, the fleet is recorded at both ends (P1 and P2). Vehicle volumes recorded at these toll stations are relatively close. For example, in 2010, 10,508,395

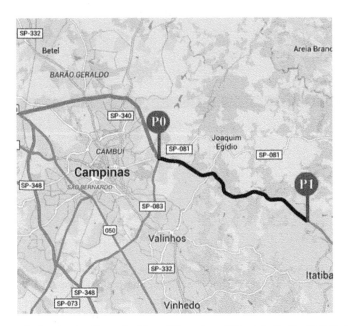

Figure 5: Section 1 – From the City of Campinas (P0) to the Itatiba Toll Station (P1)
Source: Produced by the Author using Google Maps

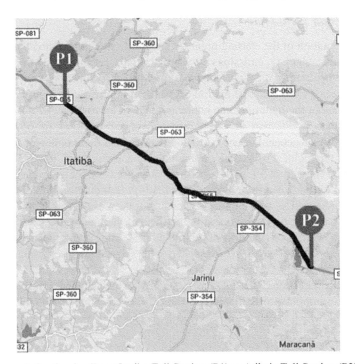

Figure 6: Section 2 – From Itatiba Toll Station (P1) to Atibaia Toll Station (P2)
Source: Produced by the Author using Google Maps

vehicles were registered at P1, while 12,183,700 were counted at P2 (DER 2014a). Similarly, close numbers were recorded in 2011, 2012 and 2013 (DER 2014b), suggesting that this close correlation is a characteristic feature of the vehicle fleet on this highway section.

Considering the small variation between the vehicle volumes recorded at both ends of this section, the circulating fleet to be considered for vehicular CO$_2$ emissions estimate in this section should simply be the average of values measured at P1 and P2.

The distance between P1 (km 110) and P2 (km 79.9) is 30.1 km.

3.4.3   Section 3: P2-P3

Among the four subsections defined in this study, this is the longest with a distance of 53.3 km between P2 (79.9 km) and P3 (26.6 km). A particularly interesting factor is the relatively high discrepancy between vehicle volumes recorded at each end of the section. In 2010, for example, 12,183,700 vehicles were registered at P2, while only 6,297,415 were recorded at P3 (DER 2014a), i.e., the number of vehicles counted at P3 was almost half of that at P2. Similar numbers were observed for 2011, 2012 and 2013 (DER 2014b), suggesting that this pattern is characteristic for the circulating fleet on this highway subsection.

The quite substantial difference between the vehicle numbers measured at P2 and P3 requires a more detailed analysis. As shown in Figure 7, three main points can be identified, where the fleet volume could change: **A** – the junction with the Fernão

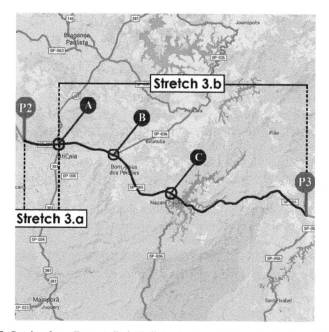

Figure 7: Section 3.a – From Atibaia Toll Station (P2) to Point A; Section 3.b – From Point
A to Igaratá Toll Station (P3)
Source: Produced by the Author using Google Maps

Dias highway (BR-381); **B** – the first junction with highway SP-036; and **C** – the second junction with the SP-036 highway.

An analysis of the fleet volume that travels on these three roads that access SP-65 in points A, B and C should reveal which point is predominantly responsible for the significant variations in this section circulating fleet. On the highway that joins SP-65 at point B, 2,177,955 vehicles were recorded in 2010, while 414,640 were counted on the highway joining at junction C (DER 2014b). In 2009, the volumes measured on highway BR-381 immediately before (13,651,000) and after (14,096,300) the junction with SP-65 (point A) are nearly seven times higher than those of the roads linked at points B and C (DNIT 2009).

Thus, the ability of point A (intersection with Fernão Dias highway) to interfere with SP-65 traffic volume is much higher than that of B or C. Accordingly, in order to describe the behavior of the circulating fleet more accurately, Section 3 should be divided into two more subsections:

Section 3.a: **P2-A** (intersection with the Fernão Dias highway).

Section 3.b: **A-P3**.

Considering the substantial influence of A on the volume of vehicles that travel on SP-65, the circulating fleet in Section 3 is defined as follows: for Section 3a, values measured at P2 are used, while for Section 3.b volumes measured at P3 shall be considered. This model is based on the assumption that the large discrepancy between values measured at P2 and P3 is almost entirely due to the influence of point A, and thus affords a more coherent value for Section 3 fleet than by using the average between P2 and P3, as has been done for Section 2.

Section 3a comprises a total of 5.9 kilometers from P2 (79.9 km) to Point A (74 km), while Section 3.b includes 47.4 km from Point A (74 km) to P3 (26.6 km).

3.4.4. Section 4: P3-P4

Similar to Section 1, a small part at the end of section 4 is discarded in order to avoid major distortions between the estimates made in this study and reality. Intuitively, it seems feasible to place P4 at the end of SP-65, i.e. at the right end of the light blue line in Figure 8. However, under this definition, the circulating fleet volume in Section 4 would be strongly influenced by two highways with extremely heavy traffic (Dutra highway – SP-060 – and Carvalho Pinto highway – SP-70). Vehicles in Section 4 are counted only at one end (P3), and this value has to be extrapolated for the entire section.

Therefore, P4 was set on the first intersection with one of these major highways (SP-060). The impact of these roads on the downstream traffic on SP-65 will thus not affect the values considered for Section 4.

The light blue section, which is not considered in the estimate, covers 5 km. The distance covered by Section 4 includes 21.6 km from P3 (26.6 km) to P4 (5 km).

4. Conclusion

The bottom-up methodology, presented by the IPCC (2006) as one major option to estimate vehicular GHG emissions, is based on three primary pieces of information:

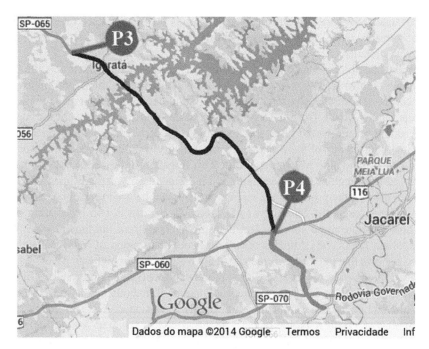

Figure 8: Section 4 – From Igaratá Toll Station (P3) to the City of Jacareí (P4)
Source: Produced by the Author using Google Maps.

the size of the examined fleet, the emission factor (which, in turn, depends on the vehicle type—category and model—its age and the used fuel) and the distance covered by the fleet. The analysis of the methodological possibilities for the vehicular CO_2 emissions estimation on a major highway via the bottom-up methodology should allow identifying the parameters that could increase the estimate accuracy.

In order to obtain accurate values for the size of the fleet and the distance traveled by it, subdividing the highway into as many parts as possible is essential. As the roads contain a flux of extremely dynamic traffic, including several intersections with other roads, where vehicle numbers may change, vehicle fleets of different composition are expected for different sections. In order to account for this fractionation, vehicle fleets have to be recorded at different points on the highway. As outlined in Section 3.3 of this article, highway SP-65 may be divided into four main parts, which are separated by the three toll stations (P1, P2 and P3) where vehicle numbers are registered. For the sections that count the fleet only on one of their ends (Sections 1 and 4), the thus obtained values have to be used for the entire section. For the sections that count the fleet on both ends (Sections 2 and 3), the average of the recorded values from both ends allow a more accurate estimate.

As detailed information on the circulating fleet profile (vehicle category, model and age, and used fuel) is usually not measured at the toll stations (except for the vehicle category – private or commercial), recorded average values for the study area should be used. In this context, it should be noted that the estimate accuracy improves with the location specificity that the average refers to. Thus, the use of federal states averages afford more accurate results than national averages, and the use of averages from specific areas where the highway is included should provide even better results.

In the specific case analyzed in this article, Figure 1 demonstrates that the use of average values for the state of São Paulo, which contains the SP-65 highway, provides more accurate estimates for the circulating fleet profile than using national average values. If the Brazilian average would be used to generate the estimate, the resulting profile would take trends from other federal states into account, thus reducing the accuracy of the estimate. Figure 1 also shows that SP-65 is inserted in the so-called São Paulo Macrometropolis. This fragment of the federal state of São Paulo and its dynamics have been analyzed individually by research institutes; the resulting data on the profile of the circulating fleet can be used to generate an even more accurate estimate. The highway is moreover located in two other areas that may also generate more specific data than the average from the federal state: the metropolitan region of Campinas and the metropolitan region of Vale do Paraíba. Figure 1 shows that the former is located at the western end of highway SP-65, while the latter is located at its far east. Compared to using only numbers from one of the two regions, the use of an average value combination from both regions may produce an improved fleet profile estimate that effectively travels on SP-65.

Similar to the fleet profile, special attention is also required for the fuel type used. For example, in the case of flex-fuel vehicles, the use of AEHC or gasoline C relies on fuel prices. The graph in Figure 2 shows how each state response to fuel price ratio differs. At a price ratio of 0.7 between AEHC and gasoline C (balance of autonomy: 70%), 50% of flex-fuel vehicles should be expected to operate on AEHC. The dashed green curve represents the 'ideal curve' consistent with this notion. However, the national average use of AEHC shows a deviation of ~ 9% in relation to this ideal curve – the continuous curve, representing the Brazilian average, is situated well below the ideal curve. Accordingly, only when AEHC price is as low as 61% of that of gasoline C (instead of 70%) than 50% of the flex-fuel vehicles in the country opt for AEHC biofuel (Goldemberg et al. 2008).

However, calculations for the state of São Paulo revealed that in this specific case, the ideal zero resistance curve depicts the use of AEHC in flex-fuel vehicles more accurately (Goldemberg et al. 2008). Thus, this curve should be used to determine the fuel choice for flex-fuel vehicles on SP-65. In 2011, for example, AEHC average price in São Paulo was R\$ 1.865/liter, while gasoline C average price was R\$ 2.642/liter (ANP 2013). Thus, the price ratio AEHC/gasoline C was 70.59%, which according to the ideal curve by Goldemberg et al. (2008) corresponds to ~ 50% of the flex-fuel vehicles using AEHC.

The emission factors also improve with locally estimated indexes, which provide a greater degree of reliability for the established methodology. For example, CO_2 emission factors of gasoline-powered vehicles in Brazil substantially differ from those in other countries due to the presence of significant proportions of EACA in gasoline sold in Brazil. CO_2 emitted upon burning the biofuel fraction of this mixture is characterized as biogenic carbon combustion. CO_2 emissions that occur from the combustion of biogenic material (taken from the place where it grew) should not be accounted for in the GHG emission inventories. This recommendation is based on the assumption that these CO_2 emissions are counterbalanced by the CO_2 that was absorbed by this material before its harvest (fixation via photosynthesis). Nevertheless, these emissions should be reported for the sake of completion (Gazzoni 2002, Macedo et al. 2004).

In the interest of a prudent analysis, it is moreover recommended to account for the emissions occurring during the entire biofuel life cycle, including so-called 'upstream' or 'indirect' emissions, which relate to the production phase, processing, storage, transport, and the fuel distribution. Thus, it is possible to estimate the volume of CO_2 and other emitted fossil-origin GHGs throughout the biofuel life cycle, not only regarding the direct emissions associated with its combustion. It is also prudent to investigate and measure possible changes occurrences in land use for the production of biofuels. For example, in the event of forest clearance for sugarcane cultivation, the benefits of biofuel carbon sequestration and storage are compromised considering the previously forested area clearance. This way, it is possible to generate a carbon debt, considering the total emissions – direct and indirect (Borsari 2009, Macedo et al. 2004).

Finally, the vehicle age is also important in order to verify the necessity to apply a deterioration factor to the emission factor that is designated for new vehicles. Such a deterioration factor should be applied when the vehicle age and use substantially affect its efficiency. In that case, more fuel will be consumed per km, which increases the emitted CO_2 volume per kilometer and consequently also the emission factor.

Thus, a more precise understanding regarding the features that characterize different scenarios associated with global GHG emissions depends on the understanding of the conditions and factors that are responsible for the emissions in these scenarios. In each specific situation, a detailed investigation into concrete alternatives that may be capable of significantly reducing these emissions should be desirable. For the transport sector, the analysis of a major highway such as SP-65 in Brazil may help to quantify GHG emission volume from this source and to find strategies for the reduction of these GHG emissions. One such strategy may be the use of biofuels, e.g. AEHC and EACA for passenger vehicles or biodiesel for commercial and cargo vehicles.

In the near future, we propose the application of the methodology discussed in this article to separately evaluate biofuels performance in the context of regional passenger transport as well as the regional cargo transport in Brazil, especially for vehicles that transport heavy goods, e.g. on SP-65.

Acknowledgements

The authors acknowledge the Sao Paulo Research Foundation financial support (FAPESP) process number 2013/17173-5. To the Coordination for the Improvement of Higher Education Personnel – CAPES/Brazil, for granting the Master's scholarship to the first author, and part of this article comes from his master's dissertation (reference 35), and the National Council for Scientific and Technological Development – CNPq/Brazil, for granting of the Research Productivity Scholarship to the second author.

References

ANP. 2013. Agência Nacional do Petróleo, Gás Natural e Biocombustíveis. Anuário estatístico brasileiro do petróleo, gás natural e biocombustíveis 2013. MME, Rio de Janeiro, Brazil.

Bales, M.P., C. Dias and S.R. Silva. 2013. Estimativa de emissões de poluentes e GEE em frotas: Aplicação Prática. CETESB. 19° Congresso Brasileiro de Transporte e Trânsito. Brasília, Brazil.

Borsari, V. 2009. Caracterização das emissões de gases do efeito estufa por veículos automotores leves no Estado de SP. Masters Dissertation, University of São Paulo, São Paulo, Brazil.

Braga, V. 2008. Logística, planejamento territorial dos transportes e o projeto dos Centros Logísticos Integrados no Estado de São Paulo. e-premissas, Revista de estudos estratégicos. Campinas. 03: 68-92.

Cerri, C.E.P. and A.J. Melfi. 2010. Energia, Ambiente e Aquecimento Global - Curso de Especialização em Gerenciamento Ambiental. São Paulo University, Esalq, Piracicaba, Brazil.

CETESB. 2013. Companhia de Tecnologia de Saneamento Ambiental. Emissões Veiculares no Estado de São Paulo 2012. Série Relatórios. São Paulo, Brazil.

DER. 2014a. Departamento de Estradas de Rodagem do Estado de São Paulo. SICSP - Solicitação de Informação [mensagem pessoal]. Serviço Estadual de Informações ao Cidadão. Mensagem recebida por estevao.gestao@gmail.com em 15 janeiro 2014a.

DER. 2014b. Departamento de Estradas de Rodagem do Estado de São Paulo. Volume diário médio das rodovias (VDM). São Paulo, Brazil.

DERSA. 2013. Desenvolvimento rodoviário S/A. Relatórios de Volume de Tráfego. São Paulo, Brazil.

DNIT. 2014. Departamento Nacional de Infraestrutura de Transportes. Estimativa do volume médio diário anual – VMD – 2009. Brazil.

EPA. 2009. United States Environmental Protection Agency. EPA Finds Greenhouse Gases Pose Threat to Public Health, Welfare. Washington, USA.

Esteves, G.R.T., P.D. Araújo, E.P. Silva and S.R.C.S. Barbosa. 2004. Análise dos Impactos da Contaminação do Ar: o caso da cidade de São Paulo. II Encontro da ANPPAS. Indaiatuba, São Paulo, Brazil.

Gazzoni, D.L. 2012. Balanço de emissões de CO_2 por biocombustíveis no Brasil: histórico e perspectivas. Embrapa soja. Documentos. Londrina, Brazil.

Goldemberg, J., F. Nigro and S. Coelho. 2008. Bioenergia no Estado de São Paulo: Situação atual, perspectivas, barreiras e propostas. Biblioteca da Imprensa Oficial do Estado de São Paulo. São Paulo, Brazil.

IPCC. 2006. Intergovernmental Panel on Climate Change. Guidelines for national greenhouse gas inventories. Volume 2 – Energy. Chapter 3 – Mobile Combustion.

Laurmann, J.A. 1989. Emissions control and reduction. Climatic Change. 15: 271-298.

Lesiv, M., A. Bun and M. Jonas. 2014. Analysis of change in relative uncertainty in GHG emissions from stationary sources for the EU 15. Climatic Change. 124: 505-518.

Macedo, I.C., M.R.L.V. Leal and J.E.A.R.S. Silva. 2004. Balanço das emissões de Gases do Efeito Estufa na produção e no uso do etanol no Brasil. Secretaria do Meio Ambiente, Governo do Estado de São Paulo. São Paulo, Brazil.

Marcotullio, P.J., A. Sarsynski, J. Albrecht, N. Schulz and J. Garcia. 2013. The geography of global urban greenhouse gas emissions: an exploratory analysis. Climatic Change. 121: 621-634.

Mcmichael, A.J., D.H. Campbell-Lendrum, C.F. Corvalán, K.L. Ebi, A.K. Githeko, J.D. Scheraga and A. Woodward. 2003. Change and human health: risks and responses. World Health Organization. Geneva, Switzerland.

Penteado, H. 2008. Ecoeconomia – Uma nova abordagem. 2nd edn. Lazuli, São Paulo, Brazil.

São Paulo. 2005. Estado anuncia corredor de exportação. Portal do Governo do Estado de São Paulo. São Paulo, Brazil.

Rossetti, J.P. 2002. Introdução à economia. 19th edn. Atlas, São Paulo, Brazil.

Rota das Bandeiras. 2015. Pedágio. Odebrecht TransPort. São Paulo, Brazil.

São Paulo. 2016. Subsecretaria de Comunicação. São Paulo: aspectos territoriais. Biblioteca Virtual. São Paulo, Brazil.

Urban Collective Mobility: Global Challenges Toward Sustainability

Daniela Godoy Falco*, Leonardo Mattoso Sacilotto and Carla Kazue Nakao Cavaliero

State University of Campinas, Mechanical Engineering School, Mendeleyev Street,
200, Barão Geraldo, ZIP CODE 13083-860, Campinas, SP

1. Introduction

Urban collective mobility is expected to continuously develop as a result of the expansion of public transportation services oriented toward human rights[1] and consequently toward sustainable development[2]. In this sense, public transport fleets[3] are expected to increase, in line with the Sustainable Development Goals (SDGs)[4], boosting urban accessibility without sacrificing future generation's needs. By 2050, public transport might account for 35% of the world's urban passenger transport, 2.4 times more than in 2015 (International Transport Forum 2019).

The literature identifies some challenges that hinder the development of public transportation toward sustainability. Lu et al. established five priorities for sustainable development: (i) designing metrics; (ii) establish monitoring mechanisms; (iii) assess progress; (iv) improve the infrastructure; and (v) standardize and verify

---

[1] "Human rights are norms that aspire to protect all people everywhere from severe political, legal, and social abuses" (Nickel 2019) without discrimination of any kind.

[2] In 1987, the Brundtland Report first introduced the concept of sustainable development, including the principle of intergenerational equity (current resources management should not reduce adequate living standards for future generations) and integrating the environmental and social dimensions into the economic one, thus achieving the idea of sustainable development, although the definition of environmental and social sustainability was not yet clear (World Commission on Environment and Development 1987).

[3] Bus, coach, rail and metro.

[4] Global guidelines which include 17 goals and 169 targets agreed in order to foster sustainable development. It was adopted by all United Nations Member States in 2015 as an universal call for action to end poverty, protect the planet and ensure that all people enjoy peace and prosperity by 2030 (United Nations 2015, United Nations Development Programme 2020b).

---

*Corresponding author: falco@fem.unicamp.br

data (Lu et al. 2015). According to the United Nations Development Program (UNDP), the challenges to achieving 2030 Agenda and the SDGs have further increased due to the Coronavirus pandemic (COVID-19). In 2020, global human development—a combination of education, health and living standards—might fall this for the first time since 1990, when measurements began (United Nations Development Programme 2020a). For SUM4ALL, the path for sustainable mobility should address the following challenges: the acknowledgment of the urgent systemic changes in mobility, addressing technologies, policies and its interactions among other sectors in collaboration with stakeholders; the monitoring and evaluation of the performance to achieve the SDGs and the policy objectives that are encompassed by sustainable mobility (universal access, efficiency, security and green mobility); the definition of a path for sustainable mobility that is based on local circumstances (Sustainable Mobility for All (SuM4All) 2019). In addition, it has been pointed out the four common challenges faced by countries when developing and implementing sustainable urban mobility plans: (i) stakeholder participation; (ii) institutional cooperation; (iii) identify effective measures; and (iv) monitor and evaluate the process of preparing the plan and its measures (CH4LLENGE 2013).

However, the literature does not acknowledge the whole value chain when referring to the challenges of sustainable public transportation. Considering the entire urban public transport value chain makes it possible to analyze the related challenges through a systematic rather than a coordination approach. This is similar logic to life cycle thinking, which goes beyond the traditional focus on production and use of a product or service and considers its entire life cycle (Life Cycle Initiative 2020). It should be remarked that some challenges and impacts, below mentioned, are not exclusively concerned or associated to the transport sector alone; as an example, diesel is used as a fuel in the transport sector but can also be used to produce electricity, so the impacts cannot be traced back solely to the transport sector. The life-cycle thinking assesses a distributed allocation of impacts among different products or services in different areas, but the connections and overlaps should be also noticed.

The chapter discusses challenges toward sustainability that emerge from the demand for urban collective mobility and its value chain. Many aspects and direct and indirect impacts of the urban collective mobility value chain are presented and distributed amid the SDG, whether delimited by the energy production sector, i.e., oil and gas[5], biofuels[6]

---

[5] Diesel is the most widely used fuel in public transport systems; natural gas is expected to be the second largest fuel source by 2035. The oil and gas sector can provide access to affordable energy, decent employment, business development and tax revenue. However, this sector has historically contributed to some of the challenges that the SDGs aims to address, such as climate change and environmental degradation, displacement of populations, armed conflicts, tax evasion and corruption, health problems and human rights violations (UNDP, IFC, IPIECA, Columbia Center on Sustainable Investment 2017).

[6] The SDGs do not make direct reference to biomass, bioenergy or biofuels. However, efforts regarding the 'bioeconomy' are expected to escalate globally, considering its renewable nature (Fritsche et al. 2018).

and electricity[7]; by vehicles production sectors, i.e. batteries[8], and by public transportation services sectors. The goals and challenges are assigned below in five 'Ps'[9] that shaped SDG.

2. People

The first 'P' focuses on people and addresses the eradication of poverty and hunger in all its forms and dimensions. It aims to ensure all people have equal means to fulfill their potential with dignity and without discrimination and in a healthy environment. It also aspires the empowerment of all vulnerable people. Achieving sustainable development is not possible if only half of humanity have opportunities and their human rights respected (United Nations 2015).

2.1. No Poverty (SDG 1)

The SDG proposes the eradication of poverty in all its forms everywhere (United Nations 2015). Poverty maintains a close relationship with the transport sector, considering, as an example, affordability and the biofuel industry labor related to harvesting crops. Supply-side poverty is typically rural, composed mostly of youth (Roser and Esteban 2013) and dependent on agriculture for subsistence or income generation (Starkey and Hine 2014). Demand-side poverty is urban. The lack of universal access to transportation services affects vulnerable groups, in particular, low-income populations who live largely in the suburbs, limiting these populations basic access to services and opportunities (Booth et al. 2000).

Considering this goal, the more challenging target for the public transport value chain might be the 1.4. This target proposes that by 2030, everyone, particularly the poor and the vulnerable, might have access to basic services (United Nations 2015).

---

[7] The electricity sector has a highly varied generation capacity, according to its geographic region. However, balancing electricity demand and supply will become a global major challenge, particularly concerning the variable energy generation (based on renewable energy) and the electrification of various sectors (IEA 2020b). The main goals related to the electricity sector include climate action (SDG 13), universal access (SDG 7) and the reduction of energy-related air pollution and the impacts associated with public health (SDG 3) (IEA 2020a).

[8] Battery manufacture is a highly greenhouse gases-intensive activity and can be responsible for up to one third of the emissions in a vehicle life cycle. This drawback can be reduced by some actions, such as increasing energy density of the batteries, expanding manufacturing facilities, increasing productivity and energy efficiency using energy sources with low carbon content in some steps of cell's manufacture and assembly of packaging, ensuring proper battery disposal management and promoting second life automotive batteries at more attractive costs than new batteries (IEA 2020b).

[9] All the 17 SDG can be categorized into five 'Ps', which summarize the whole States commitment to 2030 Agenda: People, Prosperity, Planet, Peace and Partnerships. People category is composed by the first six goals. Prosperity category combines Goals 7, 8, 9, 10, 11 and 12. Planet category involves Goals 13, 14 and 15. Peace category is represented by the sixteenth goal and finally partnership category is represented by the seventeenth and last SDG.

Poverty is a multidimensional challenge and is intertwined with all other Agenda 2030 goals. Worldwide, poor people suffer most from hunger (SDG 2), from poorer health conditions (SDG 3), have much less access to education (SDG 4), frequently have no light at night (SDG 7) (Roser and Esteban 2013) and especially in developing countries, they are more subject to non-decent work (SDG 8) (ILO 2016). This complex relations also shed light to the link between poverty and infrastructure (SDG 9) as poor people are more often exposed and compelled to low productivity, high time-consuming and energy-consuming tasks (SDG 7), which are mostly done by women (SDG 5). Poor people lack capital assets in general—natural (SDG 6, SDG 14 and SDG 15), social (SDG 16 and SDG 17), human (SDG 4 and SDG 8), physical (SDG 9 and SDG 11) and financial capitals (SDG 10)—and do not have the means to cumulate them. Lack of income and insufficient consumption (SDG 12) are affected by institutions (SDG 16) and structure. Likewise, poor social participation is associated with lacking social (SDG 16 and SDG 17) and human capital (SDGs 4 and 8). This scenario increases social, natural and economic vulnerability (SDG 10, SDG 13, SDG 16 and SDG 17). Inadequate transport services (SDG 11) limits poor people's options and capabilities to cope with vulnerabilities and violence (SDG 16). As a trade-off, expanding transport infrastructure into isolated (or commonly segregated) communities is also associated with greater exposure of those communities to risks, e.g., people exposure to new diseases (SDG 3) (Booth et al. 2000).

2.2. Zero Hunger (SDG 2)

SDG 2 is related to food security, nutrition and sustainable agriculture (United Nations 2015), which can be associated to the production of raw materials in the value chain of the bioenergy[10] sector, although it also interacts with oil and gas and electricity sectors. Considering the value chain of fossil fuels, the onshore exploration and production of oil and gas, which mainly occurs in rural areas, can affect agriculture that competes for the use of land and water, causing changes in ecosystems that can affect local food production. From a financial perspective, both the bioenergy sector and the oil and gas sector have their price fluctuations affecting food production (UNDP, IFC, IPIECA, Columbia Center on Sustainable Investment 2017).

This SDG has eight targets (United Nations 2015), four of which are suitable for the bioenergy, oil and gas and electricity sectors. Targets 2.3., 2.4., 2.5. and 2.C broadly address sustainable food production, in addition to specific challenges faced by agriculture, involving key points, such as productivity, labor compensation, land use, agricultural practices to promote resilience, genetic diversity and price.

The challenges concerning this SDG are related to more than half of other goals. Agricultural development is an essential part of the poverty (SDG 1) reduction agenda. Malnutrition becomes both the cause and the consequence of low income (Roser and Esteban 2013), not to mention the fact that low production, productivity and income can also impact poor communities in rural areas and increase their exposure to adverse shocks. Health (SDG 3) cannot be achieved without nutrition and agricultural practices may affect health directly, through its soil, water and

---

[10] It refers to the use of biological commodity for energy purposes (World Bioenergy Association 2019).

biodiversity and ecosystems management[11] and indirectly through changes in incomes (SDG 1) (International Council for Science 2017). Another social aspect comes from the failure to recognize women (SDG 5) as the main farmers or producers in many parts of the world (World Bank, Food and Agriculture Organization, International Fund for Agricultural Development 2009) and fail to highlight the challenge of achieving gender equality (SDG 10) in the food and biofuels productions. The need to increase the remuneration of rural labour (SDG 8), typically less than urban (ILO 2016), and access to land also show the social impacts experienced in the agriculture sector. Regarding the environmental linkages, productivity and practices that promote resilience and genetic diversity are specific challenges for agro-energy monocultures. Bioenergy production also affects the availability and quality of water (SDG 6). Agriculture currently accounts for about 70% of all freshwater consumption in the world (World Bank 2020) and can carry pollution to water bodies after irrigation due to the intensive use of pesticides and chemical fertilizers (Mateo-Sagasta et al. 2017). These chemical products, in addition to diesel used by machines, are consumed at all stages of food and bioenergy production and are therefore affected by the dynamics of fossil fuel prices (UNDP, IFC, IPIECA, Columbia Center on Sustainable Investment 2017). Agriculture is the biggest contributor to the loss of biodiversity (SDG 15) due to the conversion of natural ecosystems into farms, emission of pollutants and contaminants (Dudley and Alexander 2017). Regarding climate change, it is known that food production accounts for a quarter of the world's greenhouse gas emissions (Ritchie 2019). Large-scale commercial agriculture—primarily soybean and oil palm, some of the main biodiesel[12] feed-stocks—accounted for 40% of tropical deforestation between 2000 and 2010 (FAO and UNEP 2020). Competition over the land and water can result in trade-offs between energy (SDG 7) and agriculture (International Council for Science 2017).

2.3. Good Health and Well-Being (SDG 3)

This SDG aims to ensure healthy lives and the promotion of well-being (United Nations 2015) and therefore relates to fuel combustion processes and with transport services. It is related to health risks for workers and local communities, mostly in the agriculture and transportation sectors, which are labour-intensive. A pressing matter refers to illness and death from chemicals and pollution, concerning all the transport-related sectors. There are also specifics correlations between this SDG and the transport service to promote accessibility to universal health coverage and mainly with the decreasing of death and injuries from road traffic accidents. An amount of 5 million people dies every year to air pollution (Ritchie 2017). Globally, 40% to 50%

---

[11] The public transport services are public spaces where communicable diseases can spread, what should raise awareness to the health care and prevention. If biodiversity and ecosystems are not properly managed, increasing agricultural productivity, for example, could harm health through increasing pathogen habitats. The emergence of the COVID-19 disease reveals this situation and has a special correlation with the Target 3.3., related to the fight against communicable diseases.

[12] Liquid fuel produced from vegetable oil and animal fats (World Bioenergy Association 2019).

of traffic deaths occur in urban areas and the proportion of fatalities, which is already high, is increasing in low- and middle-income countries, where 90% of these deaths occur (Sustainable Mobility for All or SuM4All 2019).

In this sense, the target 3.6. aims to halve the number of global death and injuries from road traffic accidents. As mentioned by the World Health Organization, the challenges facing the transport sector for road safety are strengthening the road safety management capacity, improving the safety of road infrastructure and wider transport networks, continue to develop vehicle safety, improve the behaviour of road users and response to traffic collisions (World Health Organization 2011). Other challenges emerge from the targets 3.8.[13] and 3.9., each one regarding, respectively, accessibility to health facilities and concerns about levels of chemical contamination and pollution, mostly air pollution. These issues pose a significant health risk and cause millions of premature deaths worldwide every year (World Bank, IHME 2016). Accessibility to health care and facilities is also crucial to reduce maternal mortality and distance to care is known to influence the access to health services (World Health Organization 2018, Hanson et al. 2017).

In this light, the SDG is related to SDG 8 and SDG 11. Activities in all transport-related sectors can result in health risks for workers (SDG 8) and local communities (UNDP, IFC, IPIECA, Columbia Center on Sustainable Investment 2017). The SDG is also related to challenges in achieving sustainable and affordable transport (SDG 11): safe transport is one that promotes efficient access to basic services such as healthcare.

2.4. Quality Education (SDG 4)

The SDG aims to achieve inclusive and equitable quality education and promote lifelong learning opportunities for all. The transport services are implicated in this goal, mainly connected to the accessibility to schools, universities, and other facilities. All transport-related sectors should contribute to education for sustainable development.

The main challenges concerning transport and education refers to targets 4.1., 4.2., 4.3. and 4.5. that focus on accessibility to education facilities in relation to the transport services. Education for sustainable development, on the other hand, imposes a cross-cut challenge among the transport-related sectors.

The fourth SDG can be related to almost all other goals. Just to point out some relations, sustainable and accessible transport (SDG 11) can contribute to access formal education and promote equality and equitable educational opportunities for all (SDG 10). The challenges arise even more considering the accessibility of students (SDG 5) and people with disabilities to educational facilities. Human resources (SDG 8) are crucial for a high-quality transport system. Education sector can promote innovation on this subject (UNEP, UN-Habitat, SLoCaT 2015).

---

[13] i.e. including specific targets such as 3.2 – end all preventable deaths under 5 years of age; 3.4 – reduce mortality from non-communicable diseases and promote mental health; 3.5 – prevent and treat substance abuse; 3.7 – universal access to sexual and reproductive care, family planning and education.

2.5.   Gender Equality (SDG 5)

The goal aims to achieve gender equality and empowerment to all women and girls. Concerning the eradication of all discrimination, violence and exploitation and the promotion of participation in decision-making and leadership all transport-related sectors are implied.

The challenges have implications for targets 5.1., 5.2., 5.5., 5.6., 5.A. and 5.C. Gender-based harassment is an issue in public transport services. On the other hand, adequate access to transport services is imperative to ensure women accessibility to health care and health facilities, ensuring women and maternal rights.

SDG 5 is also related to other SDGs. Gender equality can help achieve the targets related to sustainable and increased food production and nutrition and can enhance the role of women in agriculture (International Council for Science 2017). It is essential to integrate women in the transport-related sectors from research and planning to decision-making and policymaking, as women basic mobility need are mostly associated with work and family care. The female representation must also be present in the sectors' workforce, which is a challenge since traditionally (except the agriculture sector) they are fields dominated by men. Public transport, when accessible, increases opportunities for access to health services (SDG 3), educational facilities (SDG 4), jobs (SDG 8) and social activities. When providing services, it is important for women to have safe, reliable and sustainable transport available (SDG 9 and SDG 11) as women who travel by public transport or on foot are particularly vulnerable in terms of crime, sexual harassment or theft (UNEP, UN-Habitat, SLoCaT 2015).

2.6.   Clean Water and Sanitation (SDG 6)

The SDG 6 aims to ensure availability and sustainable management of water and sanitation for all. All transport-related sector is committed within the goal as every sector consumes water in their processes but most crucial subjects should be addressed by the agriculture, which utilizes 70% of all water consumed in the world (World Bank 2020). Moreover, all sectors should guarantee to their workforce adequate work conditions, including drinking water. Integrated management of water resources should be also pursued to plan efficiency consumption and promote the protection of ecosystems.

All sectors should engage on guaranteeing safe drinking water and access to sanitation and hygiene (targets 6.1. and 6.2.) to their workforce, ensuring adequate work conditions. Furthermore, challenges concerning water-use efficiency and the protection of basins and other water-related ecosystems cross-cut all sectors (targets 6.3., 6.4., and 6.6.). In the manufacturing phase, it is estimated that industry withdraws 19% of the world's freshwater resources; the water demand in industry and energy sectors is expected to grow to 24% by 2050. This can increase scarcity in water-stressed areas because less water will be available in the hydrological cycle (Koncagül et al. 2020).

There are correspondences between SDG 6 and many other SDG. From the environmental point of view of the transport sector (SDG 9 and SDG 11), emissions and residues can result in water contamination by metals and minerals impacting water quality and availability. A series of diffuse emissions of substances generated by transport can reach soils, drainage systems or surface waters via water and air, some of them can partially reach groundwater. As a result, aquatic ecosystems are contaminated by bioaccumulation[14] of heavy metals, which through biomagnification[15] processes increases concentration as food chain trophic levels advance. The increase in traditional agricultural production can also result in the run-off of nutrients and pollution to water bodies (International Council for Science 2017). Wastewater discharge and fertilizer-rich run-off often cause eutrophication os water bodies, raising methane emissions from lakes and reservoirs; it is estimated that these levels will increase by 30-90% through 2100 (Koncagül et al. 2020). Petrol stations can also represent sources of groundwater pollution (UNEP, UN-Habitat, SLoCaT 2015). Regarding the SDG 15, biodiversity loss and ecosystem degradation are in a grand part induced by water depletion and pollution and at the same time, the correlation reduces ecosystem resilience and makes societies more vulnerable. Eutrophication is known as a major issue concerning available water supplies worldwide (Koncagül et al. 2020).

3. Prosperity

All people should be able to enjoy prosperous lives and all the economic, social and technological progress in harmony with nature. All governments should build strong, sustainable, dynamic, innovative, inclusive and people-centered economic foundations that promote full employment for the youth and bolster women's economic empowerment. Governments should ensure decent work for all, therefore eradicating forced labor, human trafficking and child labor in all its forms and support a healthy and a well-educated workforce. In order to accomplish this ambition, income inequality should be addressed by policies that promote increased productive capacities; productivity; financial inclusion; sustainable agriculture; pastoral and fisheries development; sustainable industrial development; universal access to affordable, reliable, sustainable and modern energy services; sustainable transport systems and resilient infrastructure. Sustainable urban management is crucial to people's quality of life. City and human settlement planning should consider community cohesion and personal security, besides stimulating innovation and supporting employment. From an environmental perspective, reducing negative impacts of urban activities must result in environmentally adequate management with waste reduction and recycling, safe use of chemical products and more efficient use of water and energy (United Nations 2015).

3.1. Affordable and Clean Energy (SDG 7)

The objective of SDG 7 is to ensure affordable, reliable, sustainable and modern energy for all. All transport-related sectors must focus on increasing the use-efficiency

---

[14] Accumulation of a contaminant in an organism, due to its uptake from the abiotic environment and from the diet. Thus, top carnivores may be at greater risk from heavy metals (Ali and Khan 2019).

[15] Increase in concentration of contaminant along trophic levels in a food chain (Ali and Khan 2019).

and foster the adoption of renewable sources of energy. The metal sector has major obstacles as it is very energy-intensive. In public transport fleets, internal combustion engines are the main technology present, being less efficient than electric ones and powered mainly by non-renewable fuels, such as diesel and natural gas. Although electric mobility is an alternative, average global electricity is mostly non-renewable. Therefore, a quick electrification process can be a great challenge due to the need to increase the production of renewable electricity and the reliable management of the electricity grid.

Energy is a central subject to transport sectors. Both ensuring universal access to modern energy and increasing global percentage of renewable energy (targets 7.1. and 7.2.) are the leading challenges imposed on the sectors. Energy efficiency (target 7.3.) is an aim to be pursued. In this light, it is estimated that global energy demand will increase by more than 25%, but the demand could double if no efficiency improvement is made (Koncagül et al. 2020). Aside from that, renewable energy transport options are mainly represented by biomass-based fuels (bioethanol, biodiesel, etc.) and electricity; the former consisted in a 3% share in the transport sector in 2017; the latter, 1.1%. Electricity can be originated from renewable and non-renewable energy, but about 75% of the electricity sector has non-renewable origins (World Bioenergy Association 2019).

The seventh SDG is related to other SDG. Development of renewable energy industries has the potential to reduce poverty in some of the poorest States (SDG 1). Energy supports agriculture (SDG 2), which also plays an important role in meeting the energy target, mainly through biofuels. Correlated with SDG 3 and 13, it is known that combustion of biomass-based energy like biodiesel produces fewer air pollutants than petroleum diesel fuel, although it can produce slightly higher emissions of nitrogen oxide which is also a greenhouse gas (EIA 2019). Some energy production steps, such as thermal cooling and resource extraction, require substantial amounts of water that can impact aquatic ecosystems depending on temperature and chemical components of residual water. The deployment of renewable energies and energy-efficient technologies can foster innovation (SDG 9) and reinforce industrial local and national employment objectives (SDG 8). In 2018, 11 million people were employed in the renewable energy industry worldwide; it is estimated that the bioenergy sector alone employed 3.2 million people along the complete value chain, accounting for one-third of the global renewable energy employment and second-largest employer among renewable energy technologies after solar power (World Bioenergy Association 2019). Nevertheless, the expansion of renewable energies can increase the pressure on water resources. Another (potential) trade-off is the competition between biomass for energy and crops for food. Decarbonizing energy systems (SDG 13) can also have trade-offs, as promoting renewable energy and increasing energy efficiency can cause energy price rise and, thus, hinder universal access to modern energy supplies. The decarbonization of energy systems (SDG 13) through greater use of renewable energies and energy efficiency can also restrict economic growth in some countries (International Council for Science 2017). However, the combustion of biomass-based energy like biodiesel produces fewer air pollutants than petroleum diesel fuel, although it can produce slightly higher emissions of nitrogen oxide which is also a greenhouse gas (EIA 2019).

3.2.   Decent Work and Economic Growth (SDG 8)

The SDG 8 aims to promote sustained, inclusive and sustainable economic growth, full productive employment and decent work for all, involving all transport-related sectors. In the production's perspective, it encompasses increasing efficiency, diversifying and innovating. Employment should be "human-centred" (ILO 2019), ensured the decent work and an equitable and safe work environment. In the next decades, there are concerns that driverless vehicles may reduce the demand for drivers in the road transport sector (ILO 2018).

Most challenging targets refer to productive processes and the need to diversify, innovate and upgrade toward economic productivity, that also encompasses improving resource efficiency (targets 8.2. and 8.4). Targets 8.5., 8.6., 8.7. and 8.8. impose work-related challenges concerning, in general, full employment with decent work and equality, the protection of labor rights and the promotion of safe working environments.

The SDG 8 can be also associated with other SDGs. There's relation with SDG 4, 7 and 9 as low-quality education can exclude workers without the requisite skills from enjoying the benefits of economic growth, besides discouraging investment in new technologies (ILO 2018). The eighth SDG is related to SDG 1, 2, 3, 5, 10 and 16. Investments in transport can create economic opportunities by providing employment (SDG 8) (Booth et al. 2000). When this economic opportunity is sustainable and equitable (SDG 10), through access to decent work, it also enables housing (SDG 1), food (SDG 2), health, well-being and medical assistance (SDG 3), and education (SDG 4), which in turn contribute to greater productivity and income generation (SDG 1) (International Council for Science 2017). The creation of sufficient jobs that ensure decent work, especially for women and the youth, is considered a pressing challenge, taking into account the potential for job losses due to automation and robotics, common in the automotive industry, in the decades to come (ILO 2018). Finally, the resource efficiency target is related to SDG 12. Business practices and electronic waste are highlighted as two of the most critical challenges facing the electronics industry (ILO 2018).

3.3.   Industry, Innovation and Infrastructure (SDG 9)

As this SDG aims to build resilient infrastructure, promote inclusive and sustainable industrialization and foster innovation, all transport-related sectors are implied.

Targets 9.1., 9.4., 9.5. and 9.A emerge as the major challenges regarding the transport sector. Upgrading infrastructure for sustainability and facilitating it are key issues to be addressed, mostly in developing countries.

The SDG 9 is correlated with other goals, namely, SDG 2 and 11, relating to social inclusion; SDG 7, regarding economic development; and SDG 6, 13 and 15, concerning environmental sustainability. Reliable transportation can meet security and emergency response needs. Serious disruptions to transport infrastructure can have significant impacts on the city's ability to prepare for and recover from a disaster (SDG 13). The transport infrastructure must be efficient (SDG 11), equitable (SDG 10) and resilient. A wide range of innovative technologies is generated in the transport sector. Some actions and technological routes can improve the operational efficiency of the entire transport sector and reduce energy consumption for all forms of motorized transport (UNEP, UN-Habitat, SLoCaT 2015). Today's reality delivers a mixed picture for the future of collective mobility (e.g. alternatives to bus technologies carried by natural gas, biofuels, electricity and hydrogen, using fuel or hybrid), driven by tightening regulation to meet fleet CO_2 targets and rising customer demand. There are expectations in regard to future technologies and business models to promote autonomous driving, connectivity, electrification, and shared mobility (ACES), but there is also a need for making the business crisis resistant. The challenge of making these technologies and business models more profitable than internal-combustion-engine (ICE) vehicles remains, mostly due to currently available battery technology, economies of scale, non-native design, and cooperation (SDG 17) lack across the value chain (Möller et al. 2019).

3.4. Reduced Inequalities (SDG 10)

SDG 10 intends to reduce inequality within and among countries. All transport-related sectors are engaged in making efforts to reduce discrimination and inequality, besides guaranteeing equal opportunities for all. The goal becomes even more critical in the post-COVID-19 scenario, as income inequality is known to rise in bad economic times (OECD 2015).

Related to transport sectors, the most pressing targets are 10.2. and 10.3., regarding the promotion and supportive action to universal social, economic and political inclusion, that is also ensured by equal opportunities and no discrimination. In this sense, access to urban resources should be equally granted for all people as well as equal opportunities in work and school.

There are interlinkage between SDG 10 and other SDGs. Poverty is a known cause for inequality, from which emerges the correlation with SDG 1. Good quality education and gender issues are addressed by SDG 4 and SDG 5, also concerning the goal for achieving equality. Urban resources access is focused on SDGs 9 and 11, as SDG 12 also calls for action toward responsible consumption and a sustainable lifestyle. SDG 10 is closely related to SDG 8, considering that inequality also results from unsustainable economic growth and inadequacy of work conditions. Likewise, inequality is bolstered by corruption and discriminatory laws and policies (SDG 16).

3.5. Sustainable Cities and Communities (SDG 11)

Promotion of cities and human settlement inclusive, safe, resilient and sustainable is the main proposal of this goal and in this light, it concerns all transport-related sectors. In the transport services, equity analysis can be difficult: it can be considered many types of equity and various ways to categorize people, the numerous impacts and ways of measuring them (Geurs et al. 2016).

The most noticeable challenges are addressed by targets 11.2., 11.4., 11.6. and 11.7. Target 11.2. aims to ensure affordable and sustainable transport systems, shedding light on the necessity of availability to all and green, efficient and safe transport systems (Sustainable Mobility for All or SuM4All 2019). These should foster the economic growth of all sectors, mostly in the labor-intensive sectors, as the accessibility is increased and the travel obstacles decreased. Target 11.4. refers to the protection of the world's cultural and natural heritage. Acid rain, a byproduct of the burning of sulfur and nitrates that are present in fossil fuel, has a corrosive effect on buildings. Target 11.6. aims to reduce the environmental impact of cities, an objective that cross-cuts all transport-related sectors. Transport services are responsible for a considerable share of particulate material emissions and there are issues derived from fuels and oils leaking and tires discarded inappropriately. Lastly, there is the challenge to provide access to safe and inclusive green public spaces, which include public transport (target 11.7.).

The SDG 11 can be associated with SDGs 1, 3, 4 and 5, mainly regarding the accessibility to all people and urban resources, without discrimination and ensuring safe transport to all women. Cities have an influence on health (SDG 3), supporting mental health, accessing health services, reducing non-communicable diseases and limiting environmental impacts (International Council for Science 2017). A correlation with SDG 6 raises awareness about fuels and oils drained to water resources and the treatment of industrial process water. SDGs 8 and 10 are also related to eleventh SDG, as accessibility is generally linked to economic efficiency and equity (Geurs et al. 2016). There is also an association with SDG 16, concerning the corruption risks and the fact that transport fees can reinforce discriminatory policies.

3.6. Responsible Consumption and Production (SDG 12)

As SDG 12 aims to ensure sustainable consumption and production patterns, it is expected to regard all transport-related sectors.

The most challenging targets concerning SDG 12 and the transport sectors refer to the sustainable use of natural resources (12.2.), encompassing water and renewable energy; the increase of efficiency and reducing of waste (12.3.), notably in the biofuels value chain, and the promotion of sustainable lifestyles, regarding the public transport as an activity-oriented to sustainable development (12.8.). Concerning renewable energy and the bioenergy sector, bioenergy is heavily associated with bioethanol and biodiesel, with the USA and Brazil as major producers, accounting for 87% of bioethanol global production. Moreover, biodiesel is largely produced in South America and Europe with a respective share in the global production of 37% and 44% (World Bioenergy Association 2019).

The twelfth SDG can be associated with other SDGs, such as SDG 1, sustainable agriculture; SDG 6, sustainable use of water and protection of ecosystems; SDG 7, sustainable production and use of energy; SDG 10, as the sustainable resources use can foster equality; and SDG 11, development and transition to a sustainable transport system. Besides, it is related to SDG 13, as research and sustainable lifestyle can reduce carbon dioxide emissions; SDGs 14 and 15, adequate waste disposal and SDG 17, as sustainable production and consumption cannot be achieved without partnerships. The innovation (SDG 9) of technologies and processes in public transport is crucial to seek more sustainable forms of production and service consumption (UNEP, UN-Habitat, SLoCaT 2015). Defining guidelines applied to local needs and financing these innovations can be a great challenge. Vehicles used in public transport are largely made from metals, which are obtained from mining and recycling processes. Mining sector can contribute to impacts on lands (SDG 15) and natural resources (SDG 6), carbon emissions (SDG 13) mainly because it is energy-intensive (SDG 7), poverty (SDG 1), gender-based violence (SDG 5), inequality (SDG 10), armed conflicts and corruption (SDG 16), health problems (SDG 3), among others (UNDP, World Economic Forum; Columbia Center on Sustainable Investment 2016). However, some of these metallic minerals are increasingly being recycled in the end-of-life stage for a circular economy, reducing pressure on primary production. This phenomenon results from the great demand for limited natural resources that depend on the geological disposition, technological feasibility, social, environmental, political and economic issues (UNEP 2020).

4. Planet

As social and economic development depends on the sustainable management of our planet's natural resources, the 'P' encompasses the goals that address the threats posed by climate change and environmental degradation. Further international cooperation is required to reduce and mitigate the emissions of greenhouse gases and to adapt communities and economy to the impacts of climate change. The cooperation should also act in the conservation and sustainable use of resources such as freshwater, oceans and seas, as well as forests, mountains and drylands; besides the protection of biodiversity, ecosystems and wildlife (United Nations 2015).

4.1. Climate Action (SDG 13)

The SDG 13 calls for urgent actions to combat climate change and its impacts, therefore, concerning all transport-related sectors. Green House Gas (GHG) emissions need to be drastically reduced, especially with the use of fossil fuels. Technological approaches such as improving energy efficiency, renewable energies and carbon capture and storage (CCS) also have a role to play.

The main challenges defying the transport sectors are related to targets 13.1. and 13.3. The former concerns the necessary strengthening of resilience and adaptive capacity to climate-related to disasters and side effects, which should impact mostly agriculture, shedding light on the importance of biofuels cultivation and adoption. The latter refers to building knowledge and capacity to meet climate change: every sector is responsible, but mainly those who are responsible for the greatest emissions of carbon dioxide, like fossil fuels based, electricity, beyond others that are energy-intensive. In addition, knowledge concerning climate change should be scientific but equally accessible for all, especially for more vulnerable human groups to climate change effects. Combustion processes are most responsible for greenhouse gas emissions. Regarding the energy sector, biogas represents an alternative that can contribute to abate greenhouse gas emissions; nonetheless, its production is also associated with undesired emissions of methane and nitrous oxide, mainly during incomplete combustion of biogas and other diffusive emissions related to biomass[16]

---

[16] Derived from plants, animals or algae (World Bioenergy Association 2019).

storage and management and both gases are linked to greenhouse gas effect (Paolini et al. 2018).

Climate change has wide-ranging impacts on the security of agriculture, food (SDG 2) and water (SDG 6) through extreme weather events, besides being considered a "poverty multiplier" (SDG 1) that could force one hundred million people into extreme poverty by 2030. It is expected that climate change will aggravate water quality degradation, not only by inducing the blooming of harmful algae in warmer water temperatures, intoxicating drinking water in water bodies worldwide but also increasing the frequency of flood and droughts that further the risks of water pollution and pathogenic contamination and pollutant concentration (Koncagül et al. 2020). Climate change is already causing mass movements of people (SDG 10) and even significant impacts on human health (SDG 3), through the spread of disease and the effects of heat stress, affecting the ability to work outside (SDG 8) (International Council for Science 2017). About two-thirds of the world's greenhouse gas emissions is originated from energy production, therefore the correlation with SDG 7 (Koncagül et al. 2020). Finally, concerning the seventeenth SDG, the correlation shed light to partnerships involving knowledge and technology transfer among States to propagate actions to mitigate climate change effects.

4.2. Life Below Water (SDG 14)

The conservation and sustainable use of the oceans, seas and marine resources is the focus of this SDG. It is naturally linked with SDG 6, as all the freshwater (and the human wastes) eventually meet the seas. In this sense, the most implicated sector within the goal is the fossil fuels sector.

The major challenges regard the reduction of marine pollution and ocean acidification (14.1. and 14.3., the latter concerning carbon dioxide emissions), the protection and restoration of ecosystems (14.2.) and the conservation of coastal and marine areas. Offshore petroleum drilling poses a significant challenge, especially deep and ultra-deepwater drilling, which can damage marine habitats due to accidents, pollution or spills (UNDP, IFC, IPIECA, Columbia Center on Sustainable Investment 2017).

SGD 14 is associated with other SDGs as well. Productive and resilient oceans and coasts are noted for ensuring food security (SDG 2), a critical enabler of poverty alleviation (SDG 1), especially for coastal communities; it also supports employment and economic growth (SDG 8), as more than three billion people are dependent on marine resources for their livelihood (UNDP, IFC, IPIECA, Columbia Center on Sustainable Investment 2017). Responsible and sustainable consumption (SDG 12) defies ending overfishing and sustainably managing marine and coastal ecosystems. Coasts are attractive for urban development and therefore achieving the fourteenth goal reinforces the sustainability of urban planning and resilient coastal human settlement (SDG 11). Without sustainable urban planning, conflicts can occur where coastal conservation limits the options for housing, infrastructure or transportation. Coastal ecosystems and oceans are affected by climate change (SDG 13) as the conservation of these ecosystems act as sinks for blue carbon.

4.3. Life on Land (SDG 15)

The SDG 15 aims to protect, restore and promote sustainable use and management of terrestrial ecosystems, combat desertification and to halt land degradation and biodiversity loss.

The main challenges of transport-related sectors within the goal are related to targets 15.1., 15.3., 15.5. and 15.A. The targets encompass the conservation and restoration of the terrestrial and freshwater ecosystem (15.1.), the eradication of desertification as the restoration of degraded land (15.3) and the protection of biodiversity and natural habitats (15.5). As a supportive action, all parts must increase financial resources to achieve the targets, namely, to conserve and to ensure sustainable use of ecosystems and biodiversity.

The SDG 15 is closely related to SDG 2, being somewhat a precondition for its achievement. In fact, agriculture is known as one key driver impacting the ecosystems; unsustainable agricultural practices can result in deforestation and land degradation, threatening food security. The resilience of human food systems and their capacity to adapt to future change depends on biodiversity. The correlation between SDGs 15 and 12 sheds light to the challenge of improving production efficiency; today, globally, crops produce more efficiently than before, considering the same area of land (World Bioenergy Association 2019). Human health and well-being (SDG 3) are also closely associated with forests due to the medicinal use of plants and the relationship of forests with cultural activities and physical and mental human health. Habitat loss due to changes in the forest area increases human health exposure to greater risks of contact with new infectious zoonotic diseases. Besides, forests can also mitigate climate change (SDG 13) by capturing and storing carbon mainly at their growing phase (FAO and UNEP 2020).

5. Peace

The goals within this 'P' aims to promote peaceful, fair and inclusive societies that are free from violence, as sustainable development and peace should be addressed altogether. Institutions should be inclusive, transparent and accountable, based on the respect for human rights and in the rule of law. In this sense, intercultural tolerance, respect of global citizenship and shared responsibility must be encouraged. There are factors such as injustice, inequality, corruption and poor governance, among others, that feed violence (United Nations 2015).

5.1. Peace, Justice, and Strong Institutions (SDG 16)

As a cross-cut goal that concerns all sectors, included transport-related ones, SDG 16 intends to promote peaceful and inclusive societies for sustainable development, to provide access to justice for all and to build effective, accountable and inclusive institutions at all levels.

Considering the goal, reducing violence (target 16.1.) represents the major challenge concerning the transport sectors, both in the work environment of the transport value chain and in the public space of public transport. In addition, target

16.2. also poses an important challenge concerning the protection of children from abuse and violence, especially in the agriculture sector, where there is a large concentration of poor children and youth. Besides, corruption should be broadly addressed within the transport-related sectors (target 16.5.).

The SDG is linked to other social-based SDGs, mainly SDG 1 and 2 regarding the children's safety and well-being, especially in the agriculture sector. There is also a correlation with SDG 5 and the protection of women in public transports, making them safe from harassment. Other SDGs are also included and can also be derived from the SDG 16 since corruption can promote less availability of resources to meet environmental and vulnerable group demands. Countries rich in natural resources tend to suffer from weak institutions and high levels of corruption, compromising economic growth (SDG 8), inequality reduction (SDG 10) and poverty alleviation (SDG 1) (UNDP, IFC, IPIECA, Columbia Center on Sustainable Investment 2017).

6. Partnerships

This 'P' calls for action toward the implementation of the aligned goals through a Global Partnership for Sustainable Development, based on a spirit of global solidarity, focused mainly on the needs of the poorest and most vulnerable and with participation from all governments, stakeholders and peoples. The mobilizing of all actors and all available resources is important due to the scale and ambitions of the Agenda – it encompasses the mobilization of financial resources and the training and transfer of technologies to developing governments on favourable terms (United Nations 2015).

6.1   Partnerships for the Goals (SDG 17)

The SDG 17 aims to strengthen the means of implementation and revitalize the global partnership for sustainable development.

The main targets associated with the transport-related sectors are targets 17.16. and 17.17. While the first aims to improve the global partnership for sustainable development, the second seeks to encourage effective partnerships.

The implementation of universal urban access, directly in SDG 9 and SDG 11, faces some challenges. Adequate data, monitoring and evaluation and the production of information and knowledge are very weak elements in many cities when it comes to dimensions of sustainable transport. The challenge is to apply the knowledge available to provide policy guidance, technical support and financial assistance, redirect the allocation of hidden funds, policies, investments and grants to that interest and learn from these best practices. The growing global demand for mobility from sustainable transport (SDG 11) points out to the need for more efficient use of resources (SDG 7 and SDG 12) and a reduction in emissions (SDG 13) of the public transport life cycle. The transportation of people can influence and promote changes in areas of the entire economy and society (SDG 8), as its technologies and practices condition personal behavior and social habits. Regarding the innovation toward sustainable and demand-oriented public-transport, there is a tension between the long-term benefit of innovation and the short-term cost of disruptive innovation, in addition to the fact that no single model is available to move from disruptive to productive innovation (SDG 9). While the productive aspects of innovation are proven, broader systems need to adapt to integrate innovation. Transportation has always faced complex patterns due to the multimodal and interactive nature of its sub-sectors. Existing mechanisms, institutions, arrangements and standards provide a starting point for improvement. Governments need to cooperate to establish knowledge, increase investments, reinforce work that is already delivering results and address challenges and opportunities provided by further innovations (Sustainable Mobility for All or SuM4All 2019)

7. Final Considerations

The lifecycle-based approach in focus here, intends a holistic, although not exhaustive, analysis of the public transport-related issues, from the manufacturing and energy production phases up to the use of transport services itself. As the literature review revealed, this analysis had not yet addressed the whole value chain regarding public transportation, encompassing all transport-related sectors. Therefore, the perspective hitherto adopted is significant because it enables a systematic approach of the challenges concerned, relating them with the respective sustainable development goal, and thus favouring a comprehensive view of the hurdles ahead.

In this sense, the interactions here assigned show the complex and intricate challenges associated with the sustainable use of public transport services worldwide. Regarding the SDG, it was shown that all five "Ps" (People, Peace, Prosperity, Planet and Partnership) are implicated within the matter. The analysis of each SDG shed light to some of the most pressing challenges in the public transport sector, exposing the entanglement in the value chain relations and with the stakeholders implied. Urban mobility is a fundamental concern in modern cities, one that facilitates and implements access to other essential services. On the other hand, public transportation is one key issue to address and accomplish urban sustainable mobility and can also implicate in substantial environmental impacts. Besides the interconnections, there are trade-offs which must also be noticed.

The achievement of the sustainable development goals is still permeated by challenges and apparently, public transportation has a long journey to this accomplishment. The cross-cut correlations here exemplified depict the complexity of the challenges ahead but it is expected that this perspective can also assist well-informed decision-making and facilitate consistent policies that consider possible interactions, synergies and eventual trade-offs.

Acknowledgements

The authors would like to thank CNPq and CPFL (PD-00063-3043/2018) for funding this research.

References

Ali, Hazrat and Ezzat Khan. 2019. Trophic transfer, bioaccumulation, and biomagnification of non-essential hazardous heavy metals and metalloids in food chains/webs—concepts and implications for wildlife and human health. Human and Ecological Risk Assessment: An International Journal. 25(6): 1353-1376. https://doi.org/10.1080/10807039.2018.146 9398.

Booth, David, Lucia Hanmer and Elizabeth Lovell. 2000. Poverty and Transport: a report prepared for the World Bank in collaboration with DFID. Final report and Poverty and Transport toolkit. London: ODI.

CH4LLENGE. 2013. RUPPRECHT CONSULT. CH4LLENGE – Addressing Key Challenges of Sustainable Urban Mobility Planning. Publishable final report. 2016. http://www.sump-challenges.eu/sites/www.sump-challenges.eu/files/01_ch4llenge_final_report.pdf.

Church, Clare and Laurin Wuennenberg. 2019. Sustainability and Second Life: The Case for Cobalt and Lithium Recycling. Manitoba. https://www.iisd.org/sites/default/files/publications/sustainability-second-life-cobalt-lithium-recycling.pdf%0A.

Dudley, Nigel and Sasha Alexander. 2017. Agriculture and biodiversity: a review. Biodiversity. 18(2-3): 45-49. https://doi.org/10.1080/14888386.2017.1351892.

EIA. 2019. Biofuels Explained: Biomass-Based Diesel and the Environment. U.S. Energy Information and Administration. 2019. https://www.eia.gov/energyexplained/biofuels/biodiesel-and-the-environment.php.

FAO and UNEP. 2020. The State of the World's Forests 2020: Forests, Biodiversity and People. Rome.

Fritsche, Uwe R., Ulrike Eppler, Horst Fehrenbach and Jürgen Giegrich. 2018. Linkages between the Sustainable Development Goals (SDGs) and the GBEP Sustainability Indicators for Bioenergy (GSI). Darmstadt, Berlin, Heidelberg.

Geurs, Karst T., Roberto Patuelli and Tomaz Ponze Dentinho. 2016. Accessibility, Equity and Efficiency: Challenges for Transport and Public Services. Edited by Karst T Geurs, Roberto Patuelli and Tomaz Ponze Dentinho. Cheltenham and Northampton: Edward Elgar Publishing.

Hanson, Claudia, Sabine Gabrysch, Godfrey Mbaruku, Jonathan Cox, Elibariki Mkumbo, Fatuma Manzi, Joanna Schellenberg and Carine Ronsmans. 2017. Access to maternal health services: geographical inequalities. United Republic of Tanzania. Bulletin of the World Health Organization. 95: 810-820. https://doi.org/http://dx.doi.org/10.2471/BLT.17.194126.

IEA. 2020a. Clean Energy Innovation. Paris. https://www.iea.org/reports/clean-energy-innovation.

IEA. 2020b. Global EV Outlook 2020. Paris. https://www.iea.org/reports/global-ev-outlook-2020.

ILO. 2016. World Employment Social Outlook 2016: Transforming Jobs to End Poverty. Geneva.

ILO. 2018. Back to the Future: Challenges and Opportunities for the Future of Work Addressed in ILO Sectoral Meetings since 2010. Geneva.

ILO. 2019. Time for Act for SDG 8. Geneva.

International Council for Science. 2017. A Guide to SDG Interactions: From Science to Implementation. https://council.science/publications/a-guide-to-sdg-interactions-from-science-to-implementation/.

International Transport Forum. 2019. ITF Transport Outlook 2019. https://doi.org/https://doi.org/https://doi.org/10.1787/transp_outlook-en-2019-en.

Koncagül, Engin, Michael Tran and Richard Connor. 2020. The United Nations World Water Development Report 2020: Water and Climate Change, Facts and Figures.

Life Cycle Initiative. 2020. What Is Life Cycle Thinking? Life Cycle Initiative. 2020. https://www.lifecycleinitiative.org/starting-life-cycle-thinking/what-is-life-cycle-thinking/.

Lu, Yonglong, Nebojsa Nakicenovic, Martin Visbeck and Anne-Sophie Stevance. 2015. Policy: Five Priorities for the UN Sustainable Development Goals. Nature. 2015. https://www.nature.com/news/policy-five-priorities-for-the-un-sustainable-development-goals-1.17352.

Mateo-Sagasta, Javier, Sara Marjani Zadeh and Hugh Turral. 2017. Water Pollution from Agriculture: A Global Review. Rome and Colombo.

Möller, Timo, Asutosh Padhi, Dickon Pinner and Andreas Tschiesner. 2019. The Future of Mobility is at Our Doorstep. McKinsey & Company. 2019. https://www.mckinsey.com/industries/automotive-and-assembly/our-insights/the-future-of-mobility-is-at-our-doorstep.

Nickel, James. 2019. Human Rights. Stanford Encyclopedia of Philosophy Archive - Summer 2019 Edition. 2019. Stanford Encyclopedia of Philosophy Archive%0ASummer 2019 Edition.

OECD. 2015. In It Together: Why Less Inequality Benefits All. Paris: OECD Publishing.

Paolini, Valerio, Francesco Petracchini, Marco Segreto, Laura Tomassetti, Nour Naja and Angelo Cecinato. 2018. Environmental impact of biogas: a short review of current knowledge. Journal of Environmental Science and Health, Part A. 53(10): 899-906. https://doi.org/10.1080/10934529.2018.1459076.

Ritchie, Hannah. 2017. Air Pollution. OurWorldInData.Org. 2017. https://ourworldindata.org/air-pollution.

Ritchie, Hannah. 2019. Food Production is Responsible for One-Quarter of the World's Greenhouse Gas Emissions. Our World in Data. 2019. https://ourworldindata.org/food-ghg-emissions.

Roser, Max and Ortiz-Ospina Esteban. 2013. Global Extreme Poverty. Published Online at OurWorldInData.Org. 2013. https://ourworldindata.org/extreme-poverty.

Starkey, Paul and John Hine. 2014. Poverty and Sustainable Transport: How Transport Affects Poor People with Policy Implications for Poverty Reduction. London: ODI.

Sustainable Mobility for All (SuM4All). 2019. Global Roadmap of Action Toward Sustainable Mobility. Washington DC: License: Creative Commons Attribution CC BY 3.0.

UNDP; IFC; IPIECA; Columbia Center on Sustainable Investment. 2017. Mapping the Oil and Gas Industry to the Sustainable Development Goals: An Atlas.

UNDP; World Economic Forum; Columbia Center on Sustainable Investment. 2016. Mapping Mining to the Sustainable Empowered Lives Resilient Nations Development Goals: An Atlas.

UNEP; UN-Habitat; SLoCaT. 2015. Analysis of the Transport Relevance of Each of the 17 SDGs.

UNEP. 2020. Mineral Resource Governance in the 21st Century: Gearing Extractive Industries towards Sustainable Development.

United Nations. 2015. Transforming Our World: The 2030 Agenda for Sustainable Development.

United Nations Development Programme. 2020a. COVID-19 and the SDGs. 2020. https://feature.undp.org/covid-19-and-the-sdgs/?utm_source=web&utm_medium=sdgs&utm_campaign=covid19-sdgs.

United Nations Development Programme. 2020b. What Are the Sustainable Development Goals? 2020. https://www.undp.org/content/undp/en/home/sustainable-development-goals.html.

World Bank; Food and Agriculture Organization; International Fund for Agricultural Development. 2009. Gender in Agriculture Sourcebook. Washington, DC.

World Bank; IHME. 2016. The Cost of Air Pollution. Seattle.

World Bank. 2020. Water in Agriculture. World Bank. 2020. https://www.worldbank.org/en/topic/water-in-agriculture#2.

World Bioenergy Association. 2019. Global Bioenergy Statistics 2019. https://worldbioenergy.org/uploads/191129 WBA GBS 2019_HQ.pdf.

World Commission on Environment and Development. 1987. Report of the World Commission on Environment and Development: Our Common Future Acronyms and Note on Terminology Chairman's Foreword.

World Health Organization. 2011. Global Plan for the Decade of Action for Road Safety 2011-2020.

World Health Organization. 2018. Discussion Paper: Developing Indicators for Voluntary Global Performance Targets for Road Safety Risk Factors and Service Delivery Mechanisms.

Aiming for Sustainable Urban Transport in South America – The Cases of Brazil, Argentina and Chile

Alyson da Luz Pereira Rodrigues

[1] Ph.D. Candidate, Postgraduate Program Energy Systems Planning, Faculty of Mechanical Engineering, State University of Campinas, São Paulo, Brazil

1. Introduction

South America experienced accelerated urbanization and dynamic economic growth since the past century. These conditions led to an increase in industrial activities, motorization rates, and consumption of fossil fuels. This evolution of urbanization and increased demand for urban transport have led to serious environmental and social consequences.

Examples of environmental consequences can be seen in the megacities of South America (i.e., Buenos Aires, Santiago and São Paulo), where vehicular emissions of carbon monoxide (CO) and nitrogen oxide (NO_x) cause serious impact on public health. According to current inventories, traffic emissions are responsible for 1/3 to 2/3 of inhalable particles and the vast majority of ozone precursors.

Specific issues related to transport and mobility include the inadequate provision of public transport, congestion, noise pollution, accidents, lack of access to the city, and risks of social exclusion. The Covid-19 pandemic may have exacerbated these problems further. Reduced demand for mobility in urban regions of South America imply lower revenues and lead to the reduced provision of public transport services, particularly affecting the poorer segments of the population.

Consequently, governmental transport policy actions are required not only to resume services but also to review viable and sustainable alternatives to improve mobility services. Transport planning is required in close integration with social, economic and environmental policies. Only then can sustainable urban development be achieved.

*E-mail: a230391@dac.unicamp.br

Acknowledgments: Profa Dra Semida Silveira for corrections and recommendations.

This chapter discusses the challenges and possible solutions for urban mobility in Argentina, Brazil, and Chile (Figure 1A). These three South American countries are represented in the Sustainable Cities Index-2018 with the cities of Buenos Aires (81°) São Paulo (78°), Santiago (77°), respectively (Figure 1B). This index addresses the sustainability of cities on three pillars (social, environmental, and economic) with an indicative rating of 100 for the largest cities in the world.

Initially, a general description of the current situation of urban transport in each country is provided, considering the transport structures, energy consumption and GHG emissions, and the context of national urban mobility policies. Subsequently, solutions are proposed based on the UN recommendations and the Solutions Project (European Urban Mobility Observatory – ELTIS). Regarding the applicability and potential of the solutions for the countries analyzed, local and national conditions are considered, such as socioeconomic aspects, political context, and legal structures.

The information gathered here may assist decision-makers as they identify and categorize the profile of current urban mobility, develop guidelines to transition to low carbon transport systems, and define projects to catalyze change.

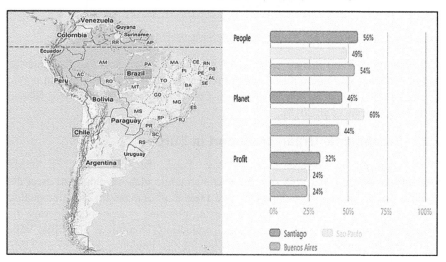

Figure 1: (A) Countries Investigated; (B) Sustainable Cities Index (SCI 2018)

2. A Vision Toward Sustainable Urban Transport

There are several definitions of sustainable urban transport, derived from the general concept of sustainability, but these concepts can be summarized in the main objective, which is to meet current needs without compromising the ability of future generations to meet their own needs (United Nations 1987). Figure 2 summarizes a possible scale of priorities when it is intended to project sustainable urban transport.

To reach sustainable urban transport, planning scholars increasingly advocate the use of sustainable mobility planning (SUMP) as a way to tackle the urban mobility crisis (Pinhate et al. 2020). This is defined as a concept/structure developed to support local authorities in exploring new strategies to promote the reduction

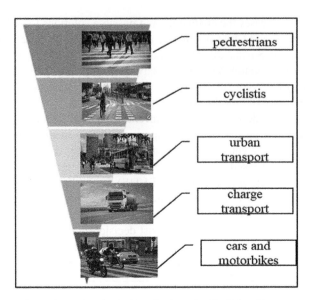

Figure 2: Scale of Priorities for Sustainable Urban Transport

in the use of private vehicles associated with transport measures that facilitate the population's access to activities through greater use of bicycles, hiking, and public transportation (Banister 2008, Hans 2016, Cavalcanti et al. 2017, Mozos-Blanco et al. 2018, Dorina et al. 2018).

2.1. Sustainable Urban Transport in Europe

Many countries in Europe adopt SUMP. From the perspective of SUMP implementation planning, this is defined as a strategic plan developed to meet the mobility needs of people, companies, and their environment for a better quality of life (Joo et al. 2012). Based on the existing planning practices considering the principles of integration, participation, and evaluation (Dominic et al. 2016). Table 1 presents the basic planning characteristics for the implementation of a SUMP (CIVITAS 2020).

On that continent, a world reference in the implementation of sustainable urban mobility plans, a total of 1,000 SUMPs were identified. The main contributors are the countries in which the adoption of a SUMP is mandatory by law with 37% of cities report having a plan qualified as SUMP. Compared to the 2013 situation, the total number of SUMPs adopted has increased from 800 to 1,000 with important contributions from Romania, Slovenia, and Sweden. The number of SUMPs in preparation has also increased significantly from 160 to 350. Given such advances in the adoption of SUMPs, experienced and pioneering cities have a role to share their experience with other start-up cities (ELTS 2014, CIVITAS 2020).

However, there are several barriers to the implementation, design, and implementation of SUMPs at national and regional levels in European countries, such as inter-administrative cooperation between different levels (city, regional, national); lack of national support and an adequate regulatory structure; absence

Table 1: Basic Planning Characteristics for the Implementation of a SUMP

Features	Explanation
Vision	• Plan for short-term implementation including schedule, budget, allocation of responsibilities, and required resources.
Participation	• The participatory approach that involves citizens and other key actors in the implementation process.
Balanced Development	• Cost-benefit actions that include technical service and infrastructure measures.
Integration and Cooperation	• Cooperation between departments at the local level, to ensure consistency with the policies of the related sectors: transport, land use, social services, health, energy and education.
Evaluation	• Analysis of the current situation and the establishment of a baseline for measurement.Identification of specific performance objectives that are realistic in the light of current urban areas,Definition of measurable goals, based on realistic assessments and that identify specific indicators for measuring implementation progress.
Monitoring	• Regular monitoring of progress towards the plan's goals and their achievement is regularly evaluated based on a table of indicators.
External Cost	• Cost-benefit analysis of all modes of transport.

Source: CIVITAS (2020)

of political incentives; inability to prioritize the implementation of measures with available resources that are aligned with the SUMP concept; lack of data and poor culture of evaluating and monitoring activities (Diez et al. 2018).

2.2. Sustainable Urban Transport in South America

This continent is characterized as the most urbanized developing region on the planet. However, motorization levels in South America and Mexico are still low compared to developed countries. Also, public policy measures and instruments are needed to reduce mobility needs and make urban transport more sustainable (Knowles 2020).

Besides, there are no performance indicators for the transport sector available in South America that can be used for comprehensive documentation and comparison of results with other continents. However, some initiatives seek to make this information available such as the Latin American Observatory of Urban Transport (OMU CAF) (Dalkmann and Sakamoto 2011).

Furthermore, there are different levels of progress in the generation of sustainable transport policies by the countries of South America. However, there are already initiatives to change this reality, such as the agreement reached during the Forum of Sustainable Transport in Latin America in Bogotá. The Bogotá Declaration in 2011, the latest regional initiative, emphasizes the need for reliable information from the transportation sector on the continent. For this, the development of technical capacity and coordination mechanisms in institutions is crucial (Hidalgo, Huizenga 2013).

Therefore, in South America, there is still no continental policy for sustainable urban transport, as is the case in Europe. Mobility measures are based on national policies according to each country's infrastructure and resources. Some national governments are trying to shift the prevailing paradigms toward a sustainable urban transport system. Thus, the following question arises: how realistic are national policies measures to reduce urban transport's greenhouse gases whilst maintaining the dynamics of individual mobility?

3. Urban Transport in Brazil

Brazil is the country with the largest population in South America with an annual rate of population growth in urban areas corresponding to 0.9% (CEPAL 2020). In this country, there was an intense territorial development between 1960 and 1970, which generated investment needs in the road system. However, this process was fast and caused several problems, such as insufficient road infrastructure and ineffective transport systems (Gomide and Galindo 2013).

3.1. Transport Structure

The Brazilian vehicle fleet grew 1.9% in 2018 with a greater concentration in the states of São Paulo (30.4%); Minas Gerais (11.9%); Paraná, (7.7%); Rio de Janeiro (7.6%); and the Rio Grande do Sul (6.8%). Aggregated, these five states represented 64.5% of all vehicles that circulate in the country. This is mainly supported by flex vehicles with 67.1% and gasoline (22.2%) (SINDIPEÇAS 2019).

Regarding the modal division, it was found that most trips made in the country in 2017 were: on foot and by bicycle (43%), individual motorized transport (29%), and public transport (28%). Specifically, public transport is mainly composed of buses (24%) (ANTP 2018).

In this setting, there are national initiatives for the development of clean transport through federal initiatives to promote the use of cane ethanol in the road transport sector (Marx et al 2015). Also, the use of electric buses is already a reality for the cities of São Paulo, Campinas, and Rio de Janeiro. These buses are part of the country's already diverse fleet, including buses powered by biodiesel, ethanol, and diesel, as well as electric trolleybuses (Estafane et al. 2015).

3.2. Impacts on Energy Consumption and GHG Emissions

In Brazil, the transportation sector accounts for approximately 32.7% of energy consumption. Much of this energy used comes from diesel oil (45%), gasoline (28%) (EPE 2019). Also, the diesel fleet in Brazil is a significant source of particles (CETESBE 2018).

In São Paulo, gasoline vehicles represent approximately half of the total mobile emissions, followed by diesel engines, which correspond to 25% of the total CO emissions. NO_x emissions are dominated by diesel engines that represent 83% of the total, followed by gasoline engines that contribute 12% of the total (CETESBE 2018).

However, Figure 3 shows the significant participation of biofuels in the Brazilian mix, since ethanol produced in the country is consumed mainly in the transport sector. One of the reasons for this high domestic demand for ethanol is the fuel mixing mandate, which currently mixes 27% of anhydrous ethanol (by volume) in gasoline (E27) (Junior et al. 2019).

Thus, Brazil's commitment to reducing GHG emissions is directly related to the promotion of renewable energy sources that correspond to more than 85% of the national energy matrix. This is reflected in the transportation sector, where Brazil has been a leader in clean transportation since the 1970s when it launched 'ProAlcool', a federal initiative to promote the use of cane ethanol for transportation (Mendes et al. 2020). The first phase of the program required oil suppliers to mix 10% anhydrous ethanol with oil; the second phase, starting in 1979, supported the production of hydrated ethanol as an alternative to gasoline (Mączyńskaa et al. 2019).

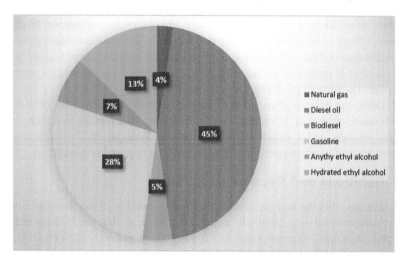

Figure 3: Fuels Used in the Brazilian Transport Sector. *Source*: EPE (2019)

Besides, the Brazilian government announced in its Nationally Determined Contribution (NDC) to the Paris Agreement the intention to further increase the supply and consumption of biofuels through the recently approved National Biofuels Policy, also known as RenovaBio, as specified by Law 13.576 / 2017 (Brazil 2017).

Also, Brazil has made important progress in combating air pollution in recent years. The CONAMA (National Council for the Environment) passed new resolutions to control vehicle emissions, causing significant reductions in the next decade. It also established new standards and instruments for the management of air quality by the states. This set of measures effectively seeks to protect the environment and population health (CONAMA 2019).

3.3. National Policy on Urban Transport

In 2012, the Brazilian federal government approved Law No. 12.587, known as the National Urban Mobility Law. This law requires cities with more than 20,000 inhabitants to develop urban mobility plans and determines the municipalities the task of planning and executing the urban mobility policy based on accessibility and integration between different modes of transport (Ministry of Cities 2015).

In general, urban mobility issues have been treated in Brazil as a matter of providing transport services. The results of this process can be described by providing infrastructure for road transport, mainly by opening highways and expressways, prioritizing individual transport over collective transport, considering non-motorized modes, and completely separating urban planning and transport (Silva et al. 2008).

However, surveys carried out by the Ministry of Cities (2017) with approximately 1.500 cities, showed that only 5.9% of eligible cities had an urban mobility plan in place, 26.7% of which were in the process of developing the plan. Therefore, it is noted that there are barriers to the implementation of Urban Mobility Plans (PMU).

Among the main barriers for Brazilian cities, the following stand out: difficulties in finding mobility solutions that adequately meet the needs of all interested parties, budgetary restrictions to implement actions toward more sustainable modes of transport; effective involvement of the population in the decision-making process; insufficient data collection for the preparation of urban mobility plans and the lack of qualified labor in municipal agencies to design and implement the actions contained in the plans (Bezerra et al. 2020).

Besides, the establishment of the National Urban Mobility Plan (PNMU) in Brazil is relatively recent. Therefore, there is still no specific tool for evaluating urban mobility projects under development (Mello and Portugal 2017). Consequently, plans are generally not carried out or do not meet the required requirements.

3.4.   Solutions and Recommendations

It was observed that the actions needed to improve mobility in the Brazilian context must be comprehensive and realistic, as they require local, state, and federal political interventions. Table 2 presents some of the sustainable transport solutions that have

Table 2: Applicable Solutions for Brazilian Cities

Axes of intervention	Measures	Mitigation actions
Public Transport	Development of low emission transport	Strategies for incentives to use biofuels
	Prioritizing investments in public transport	Implementation of electric buses
	Investments in other means of mobility	Bike-sharing and public bicycles
Transport Infrastructure	Changes to the current infrastructure	Dedicated bus lanes; intermodal interchanges; pedestrianizing city centers and streets.
City Logistics	New measures to regulate urban transport	Vehicle and operation restrictions on time, weight and size; low-emission zones
Integrated Planning/ Sumps	Stakeholder participation and citizen engagement	SUMP monitoring and evaluation
Mobility Management	Improvement of traffic and mobility management	Regulations for improving parking, traffic, and multimodal travel planning management

been identified as applicable to Brazilian cities, based on the UN recommendations and the Solutions project.

Brazil is the largest producer of sugarcane ethanol in the world and a pioneer in the use of ethanol as motor fuel (Novato et al. 2017). Investments in biofuels stand out as one of the main urban mobility solutions, as they contribute to the problem of dependence on fossil fuels and climate change (Afionis et al. 2020). Besides, the fact that this country is a world leader in the generation of renewable energy contributes to the implementation of electric buses as the main public transport solution (Sampaio et al. 2007).

For actions related to city logistics, urban planning activities are necessary to establish strategies that support land use planning and the organization of logistical flows. An example case is a city of Salvador-Bahia, which has streets with low traffic capacity, absence of regulated loading and unloading spaces, and this situation is worse in the stretches where traffic flows are shared with public transport (Viana et al. 2019).

Concerning the proposals for the adequacy of the integrated planning/sumps transport infrastructure, mobility management, and political interventions are necessary for adoption as goals in the existing mobility plans. Mainly for cities that intend to implement electric buses, and autonomous vehicles as it requires medium and long term infrastructure planning (Bakker et al. 2018, Carey et al. 2019).

An example is the city of Curitiba, known worldwide as the model city of public transport, as it has built a public transport system at a reasonable cost that integrates physically and operationally different bus lines in a single network. Also, bicycle and pedestrian areas were built as an integral part of the road network and public transport system (Lu 2020).

4. Urban Transport in Argentina

Argentina is the third-largest economy in South America and has an urbanization rate of over 90%, being one of the most urbanized countries in the world with an annual rate of population growth in urban areas corresponding to 0.9% (CEPAL 2020). In this country, urban mobility faces a high concentration of population in several urban areas, where the Metropolitan Area of Buenos Aires (AMBA) stands out, concentrating more than a third of the country's population (Anapolsky 2012).

4.1. Transport Structure

In Argentina, the fleet of vehicles in circulation at the end of 2018 was 13,950,048, representing an increase of 4.9% over the previous year. This comprises 85.2% of automobiles, 11.2% of light commercial vehicles, and 3.6% of heavy commercial vehicles, including trucks and buses. Additionally, 47% of the fleet in circulation is concentrated in the province of Buenos Aires. Also, the country has an increasing motorization rate of 3.15 inhabitants per vehicle (AFAC 2019).

Regarding vehicle technologies, currently, CNG vehicles represent 13.6% of the total fleet. Besides, hybrid vehicles (gasoline/electric) still did not add up to 1.000 units in the current fleet and they are starting to increase very slowly. Of the vehicles incorporated into the circulation fleet from 2008 to 2018, 81.3% went to gasoline and the remaining 18.7% to diesel (AFAC 2019).

Concerning urban passenger mobility, the number of trips per year reached almost 9 billion in 2012 with a strong share of buses (38%), cars (37%), railways (5%), taxi (4%), and subway (2%). The number of passengers on automotive intercity transport services in 2015 increased by 37.2 million with a total route of 708.5 million kilometers. About long-distance cargo, the preponderance of the road mode stands out (92.7%), followed by the railroad (3.7%) and the fluvial-maritime in ships and barges (3.6%) (PANTyCC 2017).

4.2. Impacts on Energy Consumption and GHG Emissions

According to Argentina's National Energy Balance (BEN), the transport sector is the largest consumer of primary energy with almost 31% of the total energy consumed in 2017. This energy consumption is fundamentally based on the consumption of oil products, compressed natural gas, and biofuels as shown in Figure 4.

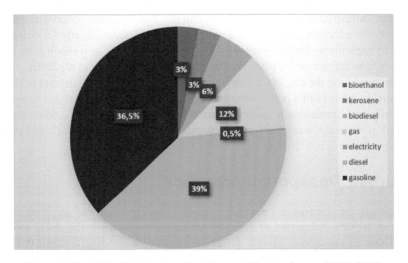

Figure 4: Fuels Used in the Argentine Transport Sector. *Source*: BEN (2019)

Consequently, the consumption of these fuels correlates with the greenhouse gas emissions generated. In Argentina, the transport sector was responsible for the emission of 50.2 MCO_{2eq} in 2016, which represents 13.8% of the country's total GHG emissions that year (AFAC 2019). The main responsible for such emissions is road transport, which represents more than 90% of GHG emissions in this sector.

Given this context, Argentina was the first country in South America to present a review of its national contribution to make it more ambitious. The absolute objective assumed is "not to exceed the net emission of 483 million tons of carbon dioxide equivalent ($MtCO_{2\ eq}$) in the year 2030". Therefore, this objective is reflected in initiatives aimed at the transport sector (Noto 2020).

To establish a state climate change policy, in 2016, the National Climate Change Office proposed the elaboration of the National Plan for Mitigation and Adaptation to climate change. To that end, the National Action Plan on Transport and Climate Change was developed, which consists of a set of initiatives that Argentina plans to contribute to reduce GHG emissions in this sector and adapt to the effects of climate change in the transport sector, according to the commitments made before the United Nations Framework Convention on Climate Change (PANTyCC 2017).

4.3. National Policy on Urban Transport

Several initiatives for urban mobility have been adopted in this country. The first of these was the creation of Argentina's National Transport Plan, which has the following strategic objectives: modernizing transport infrastructure through federal initiatives and prioritizing sustainable modes of transport for non-motorized individuals (Sptycop 2018).

Also, the federal government of Argentina passed Law No. 13.857 in 2019, known as the Sustainable Mobility Policy Law. The objective of this law is to establish a legal, economic, and public structure that progressively promotes the massive use of sustainable mobility alternatives in public transport across the country (Noto 2020).

Besides, for monitoring, the National Strategic Plan for Sustainable Urban Mobility was created. This is updated every 5 years with accessibility to the public and seeks to improve the quality of residents by prioritizing public transport. The developed plan is based on the Metrobús, EcoBici, and the Pedestrian Priority Plan (PANTyCC 2017).

However, in urban agglomerations in Argentina, urban transport plans in many cities are not coordinated with broader urban plans in cities. Long-term planning for public works is fragmented at various government levels. Limited responsibilities and capacity of municipalities often prevent the integration of transport systems with land use planning in urban areas (Muller 2017).

The autonomous city of Buenos Aires (CABA) has developed a Sustainable Mobility Plan that is in the process of implementing several programs trying to address the different problems identified that make mobility. The main barriers are related to long periods of information gathering that delay decision making, ineffective citizen participation processes, and politicians' lack of interest (PANTyCC 2017).

4.4. Solutions and Recommendations

Table 3 presents some of the solutions that have been identified as applicable to Argentine cities, based on the UN recommendations and the Solutions project.

Argentina has great potential for implementing electric mobility combined with an electrical matrix with a high penetration of renewable energy that will lead to GHG reductions, as this country together with Chile lead the world exploration of lithium which facilitates the manufacture of storage batteries used in electric buses. According to Benchmark Mineral Intelligence (BMI 2019), lithium demand projections are mostly for electric vehicles (88%).

The rapid transit bus (BRT) systems began to be implemented in the 1970s in Curitiba (Brazil) and are now present in 165 cities around the world. In Argentina, the first BRT system was implemented in Buenos Aires (capital), in May 2011, these have capital costs lower than 30 to 40% than the costs of subways (Valeria et al.

Table 3: Applicable Solutions for Argentine Cities

Axes of intervention	*Measures*	*Mitigation actions*
Public Transport	Development of low emission transport	Energy efficiency labeling in vehicles; electric and hybrid vehicles
	Prioritizing investments in public transport	Implementation of electric buses and new bus rapid transit (BRT)
	Use of indicators	Accessibility and proximity indicators
Transport Infrastructure	Changes to the current infrastructure	Dedicated bus lanes; safe cycling infrastructure; Intermodal interchanges
City Logistics	New measures to regulate urban transport	Vehicle and operation restrictions on time; urban consolidation centers
Integrated Planning/ Sumps	Stakeholder participation	SUMP monitoring and evaluation
Mobility Management	Improvement of traffic and mobility management	Regulations for improving parking, traffic management, multimodal travel planning

2020). Besides, benefits can be gained from decreased motorization, road safety, crime rates, travel time, greenhouse gases, and an increase in land value (Hensher 2020). Therefore, similar systems are recommended in different locations in the country's capital and several provinces (states).

About the development of indicators, this can be an important tool to inform political debates about changes in land use and Argentine transport (Vasconcellos 2018, Boisjolya et al. 2020), as accessibility indicators can support making decisions to improve the quality of public transport systems, helping people to reach their destinations (Berry et al. 2016).

5. Urban Transport in Chile

Chile is the sixth most populous country in South America with an annual population growth rate in urban areas corresponding to 0.8% (CEPAL 2020). In February 2010, this country was crossed by a natural disaster of the unexpected magnitude that directly affected the inhabitants' sense of survival. This historic event raises the need to recognize an emergency planning scenario for unconventional mobility (Tejeda 2012, Sagaris 2018).

5.1. Transport Structure

Chile has a circulating vehicle fleet that corresponds to 33,729,982 in 2018, including motorcycles (5%), cars (70%), vans (18%), buses (1%) and trucks (5%), which represents an increase of 2.61% over the previous year. Most of the vehicle fleet in Chile is located in the metropolitan regions that correspond to 41% of the country's total fleet (INE 2019). In 2016, the motorization rate in Chile was 3.9 people per vehicle-based on a park of 4,400,224 light and medium vehicles and a population of

17,373,831 inhabitants. The rate increased by 5%, compared to the previous year and 41% in the last 10 years (ANAC 2016).

In Santiago, 28% of trips are made privately and by car, 29% use public transport and 34.5% on foot. Of the total use of public transport, 52% are considered exclusively on buses (Aymeric and François 2017). Faced with this reality, in 2007, the city of Santiago, Chile, implemented a public transport plan (Transantiago), which included a new urban bus transport system and fare integration with the metro (Batarce et al. 2018). Consequently, in 2017, the Chilean capital won the International Sustainable Transportation award for major improvements in public space, cycling, and public transport (IDTP 2017).

Notwithstanding, Chile's mobility during the COVID-19 pandemic shows a 44% decrease in trips in Santiago, with the metro (55%), free ride (51%), and buses (45%) showing the greatest reduction. Modes with the lowest reduction are motorcycles (28%), automobiles (34%), and hiking (39%) (Astroza et al. 2020).

5.2. Impacts on Energy Consumption and GHG Emissions

Chile imports 74% of all fuels so it depends on energy from other countries. The 57% of these imported fuels are used in transportation (Ministry of Energy 2015). Therefore, 99.9% of the vehicle fleet in Chile is dependent on gasoline or diesel (INE 2014) as shown in Figure 5.

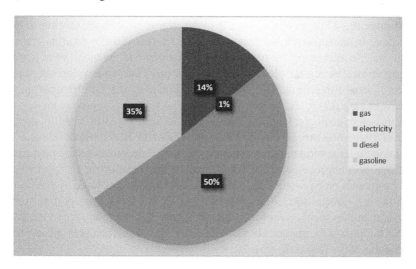

Figure 5: Fuels Used in the Chilean Transport Sector. *Source*: CNE (2019)

Also, Chile's transport sector is responsible for 22% of the country's GHG emissions with growth projections of 40% until 2020. In Santiago, where more than 5 million Chileans reside, it faces deteriorating air quality. In June 2015, the government temporarily closed 900 industries and avoided approximately 40% of the 1.7 million cars circulating in the city (SECTRA 2015). About particulate matter emissions in the last 30 years in Santiago, it was observed a reduction between 67% and 72% (Jorquera 2020).

In 2012, Chilean authorities nationally proposed Appropriate Mitigation Action Plans (NAMA) under the UN Framework Convention on Climate Change, incorporating electric vehicles in the country's emission reduction strategy. Chile's electric mobility readiness plan aims to increase the market of 70,000 electric cars by 2020 and promote the circulation of low and zero-emission vehicles in urban areas (Jorquera 2020).

5.3. National Policy on Urban Transport

The national urban mobility policy in Chile is based on a strategic development objective, whose main areas of work guarantee the mobility of people by improving infrastructure while considering socio-economic aspects (MTT 2014). The country has adopted investment plans for highways, railways, port, and airport constructions to meet long-term needs. As for social development, efforts aim to improve access to public transport and the efficiency of urban transport (Jorquera 2020).

Public transport subsidy law 20,378 was approved by the government in 2009. This law introduced a permanent and transitory subsidy for public transport in the country (INE 2014). In 2013, almost 4,200 public transport services (buses and trains) in the country received subsidies; more than 3,000 regional buses were able to offer a reduced fare; 631 free school buses were funded to allow 42,000 children from low-income families to attend school; about 2,260 regional buses were replaced and approximately 6,300 new stopping stations were built (MTT 2014).

Besides, the 2025 master plan was also developed in 2014, which aims to guide investments in management and strategic infrastructure to develop an urban transport system capable of meeting the needs of the population and the mobility of goods in the long term. The master plan for 2025 offers proposals to develop coordinated projects to establish an efficient, equitable, sustainable, and safe transport system (Aymeric et al. 2017).

Concerning ecological mobility initiatives, the zero-emission mobility program launched by the Ministry of the Environment (Chilectra 2012) was developed. This program includes several initiatives such as the PM 2.5 decontamination plan and the public transport improvement plan, which aims to promote electric transport, placing greater emphasis on environmental issues and health impacts caused by pollution.

However, government efforts are focusing on the public transport sector that provides subsidies for the renewal of electric taxis. The purpose of the incentive is to bring the purchase price of electric vehicles to the same competition level with conventional ICVs (Aymeric et al. 2017).

5.4. Solutions and Recommendations

Table 4 presents some of the sustainable urban mobility solutions that have been identified as applicable to Chilean cities, based on the UN recommendations and the Solutions project.

It was noted that Chile also has great potential for electric mobility, as it is the world leader in lithium production with 33% of global production in 2013 (USGS 2015) and lithium reserve estimates of 31% of world reserves (Fox-Davies Capital 2013). Also, there is government support through credit and tax incentives, and the potential for generating electricity from renewable sources, especially solar energy for recharging electric vehicles (Munoz et al. 2010). However, the widespread adoption of electric vehicles will require short, medium, and long-term incentives from public and private entities to link electric mobility to renewable energy generation initiatives.

Table 4: Applicable Solutions for Chilean Cities

Axes of intervention	Measures	Mitigation actions
Public Transport	Development of low emission transport	Bus fleet renewal; electric and hybrid vehicles
	Prioritizing investments in public transport	Implementation of electric buses
	Intelligent Transport System (ITS) for public transport	Intelligent accident reduction systems
Transport Infrastructure	Changes to the current infrastructure	Dedicated bus lanes; pedestrianizing city centers and streets
City Logistics	Regulatory measures for urban transport	Vehicle and operation restrictions on time; urban consolidation centers
Integrated Planning/ Sumps	Stakeholder participation	SUMP monitoring and evaluation
Mobility Management	Improvement of traffic and mobility management	Regulations for improving parking, traffic management, and multimodal travel planning

Concerning the adequacy of the existing infrastructure, only bus routes with ideal setbacks, short lanes only for buses without setbacks when approaching traffic lights, optimization of spacing and location, improvement of the project, installation of high capacity stops are presented as ideal solutions for the Chilean context. However, it will require a thorough analysis of detailed engineering studies (Basso et al. 2020).

In Santiago, Chile there are already vehicle restrictions at certain times, such measures were applied for the use of vehicles with catalytic converters between 7:30 am and 9 am. The pm on days was declared as environmental 'pre-emergencies' due to high levels of air pollution. Consequently, the flow of passengers to the subway increased by about 3% (Grange 2011). Given these results, the time restrictions solutions for private vehicles can be extended to other cities in the country.

Besides, the Santiago transport system already applies to monitor indicators that have undergone several changes since the launch of the current public transport system, Transantiago in 2007 (Basso et al. 2020). However, none of these indicators included accessibility to opportunities, which can have a considerable impact on the efficiency of public transport (Frez et al. 2019). Therefore, the proposed accessibility and proximity indicators can be a viable solution for the Chilean context.

6. Conclusions

Life in cities requires appropriate policies for using public space and the design of infrastructure and services, including urban transport. Therefore, the daily functioning of a city involves the coordination of urban transport systems through a complex network of interrelations, taking into account local, regional, and national socio-environmental problems.

In this context, it was observed that the advances in urban transport vary widely since the nature of transport is that there is no 'single solution'. The adequacy of a solution depends on many attributes that include the built environment, population culture, conditions economic, national policies, etc. It was also noted that each country analyzed has different potentials that directly interfere with mobility solutions, such as Brazil, which is a world reference in the generation of biofuels and Chile and Argentina, which are global leaders in lithium production.

Therefore, this chapter provides evidence for important political interventions and actions that can improve urban transport in the investigated countries of South America. Such exposed actions can be extended to other countries on this continent, as these highlight sustainability as a solution target for systems urban transport.

Acknowledgments

The authors gratefully acknowledge the contributions of Coordination of Improvement of High-Level Personnel-Brazil (CAPES).

References

AFAC – Asociación de Fabricas Argentinas de Componentes. 2019. Flota circulante de vehículos Argentina 2018. Final Report, 2019. Argentina.

Afionis, S. and L.C. Stringer. 2020. Fuelling friendships or driving divergence? Legitimacy, coherence, and negotiation in Brazilian perceptions of European and American biofuels governance. Energy Research & Social Science. 67: 101487-101499.

Anapolsky, S.P. 2012. Desafios de la gestión y la planificación del transporte urbano y la movilidad em ciudades argentinas. Transporte y Territorio. 7: 57-75.

ANTP – Agência Nacional de Transporte Público. 2018. Final Report, 2018. SIMOB, São Paulo, Brazil.

Astroza, S., T. Alejandro, H. Ricardo, A. Ca. Juan, G. Angelo, M. Marcela, F. Macarena and T. Valentina. 2020. Mobility changes, teleworking, and remote communication during the COVID-19 pandemic in Chile. Transport Findings. 12: 1-8.

Aymeric, Girard and Simon François. 2017. Case study for Chile: the electric vehicle penetration in chile. Electric Vehicles: Prospects and Challenges. 245-285.

Aymeric, G. and S. Francois. 2017. Case study for Chile: the electric. pp. 246-285. *In*: Elsevier [ed.]. Vehicle Penetration in Chile Electric Vehicles: Prospects and Challenges. Kristiansand, Norway.

Banister, D. 2008. The sustainable mobility paradigm. Transport Policy. 15: 73-87.

Basso, F., J. Frez, L. Martinez, R. Pezoa and M. Varas. 2020. Accessibility to opportunities based on public transport gps-monitored data: the case of Santiago, Chile. Travel Behaviour and Society. 21: 140-153.

Batarcea, M. and P. Galileab. 2018. Cost and fare estimation for the bus transit system of Santiago. Transport Policy. 64: 92-101.

BEN – Balance Energético Nacional. 2019. Balance Energético Nacional de la República Argentina. Final Report, 2019. Secretaría de Gobierno de Energía, Argentina.

Berry, A., Y. Jouffe, N. Coulombel and C. Guivarch. 2016. Investigating fuel poverty in the transport sector: toward a composite indicator of vulnerability. Energy Research & Social Science. 18: 7-20.

Bezerra, B.S., A.L. Santos and D.V. Delmonico. 2020. Unfolding barriers for urban mobility plan in small and medium municipalities – a case study in Brazil. Transportation Research Part A: Policy and Practice. 132: 808-822.

BMI – Benchmark Mineral Intelligence. 2019. Lithium: challenges and opportunities of the EV revolution. Final Report, 2019. Chile.

Boisjolya, G., B. Serra and G.T. Oliveira. 2020. Accessibility measurements in São Paulo, Rio de Janeiro, Curitiba and Recife, Brazil. Journal of Transport Geography. 82: 102551-102564.

Bakker, S. and R. Konings. 2018. The transition to zero-emission buses in public transport – the need for institutional innovation. Transportation Research Part D: Transport and Environment. 64: 204-215.

Brasil, 2017. Lei n. 13576/2017. Disposições sobre a Política Nacional de Biocombustíveis (RenovaBio) e dá outras providências- Regras sobre a política nacional de biocombustíveis (Renovabio) e outras populações- Presidência da República, Brasília, Brasil.

Carey, C., S. John, L. Crystal and A. David. 2019. Governance of future urban mobility: a research agenda. Urban Policy and Research. 37: 393-404.

CONAMA – Conselho Nacional do Meio Ambiente. 2019. Resolução CONAMA nº 493, de 24 de junho de 2019. Relatory PROMOT M5,2019. Brasília, Brazil.

Cavalcanti, C.O., M. Limont, M. Dziedzic and V. Fernandes. 2017. Sustainability of urban mobility projects in the Curitiba Metropolitan Region. Land Use Polyce, 60: 395-402.

CEPAL. 2020. Comisión Económica para América Latina y el Caribe. Bases de Datos y Publicaciones Estadísticas. Final Report, 2019. Santiago, Chile.

CNE – Comision Nacional de Energia. 2019. Anuario Estadístico de Energia 2018. Final Report, 2019. Santiago, Chile.

CETESB, 2018. Companhia Ambiental do Estado de São Paulo. Relatório de Emissões veiculares do Estado de São Paulo. Final Report, 2019. São Paulo, Brazil.

CIVITAS. 2020. Cleaner and better transport in cities. The status of SUMPS in EU member states. European Platform on Sustainable Urban Mobility Plan, Final Report, 2020. Freiburg, Germany.

Dalkmann, H. and K. Sakamoto. 2011. Investing in energy and resource efficiency. Green economy report UNEP.

D'elia, V.V., Grand, M.C. and León, S. 2020. Bus rapid transit and property values in Buenos Aires: combined spatial hedonic pricing and propensity score techniques. Research in Transportation Economics. 80: 100814-100834.

Diez, J.M., M.E. Lopez-Lambas and M. Rojo. 2018. Garcia-Martinez Methodology for assessing the cost effectiveness of sustainable urban mobility plans (SUMPs): the case of the city of Burgos. Journal of Transport Geography. 68: 22-30.

Dominic, S. 2016. Identifying key research themes for sustainable urban mobility. International Journal of Sustainable Transportation. 10: 1-8.

Dorina, P. and S. Dominic. 2018. Policy design for sustainable urban transport in the global South. Policy Design and Practice. 2: 90-102.

ELTIS – The Urban Mobility Observatory. 2014. The poly-SUMP methodology: how to develop a sustainable urban mobility plan for a polycentric region. Final Report, 2014. European Platform on Sustainable Urban Mobility Plans, Spain.

EPE – Empresa de Pesquisa Energética. 2018. Brazilian Energy Balance: full historical series. Chapter 3, energy consumption by sector 1970–2018. Final Report, 2019.Rio de Janeiro, Brazil.

Estefania, M. and V. Lisa. 2015. The World Bank. Green Your Bus Ride: Clean Buses in Latin America.

Frez, N. Baloian, J.A. Pino, G. Zurita and F. Basso. 2019. Planning of urban public transportation networks in a smart city. Journal of Universal Computer Science. 25: 946-966.

Gomide, A. And E. Galindo. 2013. A mobilidade urbana: uma agenda inconclusa ou o retorno daquilo que não foi. Estudos Avançados. 79: 27-39.

Grange, L. and R. Troncoso. 2011. Impacts of vehicle restrictions on urban transport flows: the case of Santiago, Chile. Transport Policy. 18: 862-869.

Hans, N. 2016. Sustainable Urban Transport. Transport Reviews. 36: 682-682.

IDTP – Institute for Transportation & Development Policy. 2017. Sustainable Transport Award. Final Report, 2017. Bogotá, Colombia.

Hensher, A.D. 2020. Sustainable bus systems: moving towards a value for money and network-based approach and away from blind commitment. pp. 183-190. *In*: Elsevier [ed.]. Bus Transport: Demand, Economics, Contracting, and Policy. Amsterdam, Netherlands.

Hidalgo, D. and C. Huizenga. 2013. Implementation of sustainable urban transport in Latin America. Research in Transportation Economics. 40: 66-77.

INE – Instituto Nacional de Estadística. 2014. Parque de Vehículos en circulacion. Final Report, 2014. Santiago, Chile.

Joo, Y.G. and L. Seungil. 2012. An appraisal of the urban scheme for sustainable urban transport. International Journal of Urban Sciences. 16: 261-278.

Jorquera, H. 2020. Ambient particulate matter in Santiago, Chile: 1989–2018: A tale of two size fractions. Journal of Environmental Management. 258: 110035-110050.

Junior, M.A.U., C.A.V. Soterroni and A.R.M.F. Halog. 2019. Exploring future scenarios of ethanol demand in Brazil and their land-use implications. Energy Policy. 132: 110958-110972.

Knowles, R.D. 2020. Book review. Journal of Transport Geography. 83: 102587-102593.

Lu, H. 2020. Eco-Cities and Green Transport. Elsevier, Curitiba, Brazil.

Mączyńskaa, J. and P. Hanna. 2019. Production and use of biofuels for transport in Poland and Brazil – The case of bioethanol. Fuel. 241: 989-996.

Marx, R., A.M. Mello, M. Zilbovicius and F.F. Lara. 2015. Spatial contexts and firm strategies: applying the multilevel perspective to sustainable urban mobility transitions in Brazil. Journal of Cleaner Production. 108: 1092-1104.

Mello, A. and L. Portugal. 2017. Um procedimento baseado na acessibilidade para a concepção de Planos Estratégicos de Mobilidade Urbana: o caso do Brasil. Eure. 43: 99-125.

Mendes, D. and T. Denny. 2020. Competitive renewables as the key to energy transition— RenovaBio: the Brazilian biofuel regulation. pp. *In*: 223-242. Elsevier [ed.]. The Regulation and Policy of Latin American Energy Transitions. Elsevier, Amsterdam.

Ministry of Cities. 2015. Guia PlanMob: Caderno de referência para elaboração de Plano de Mobilidade Urbana. Secretaria Nacional de Transporte e Mobilidade Urbana. Final Report, 2015. Brasília, Brazil.

Mozos-Blanco, M. and L. Ángel. 2018. The way to sustainable mobility. A comparative analysis of sustainable mobility plans in Spain. Transport Policy. 72: 45-54.

MTT – Ministerio de Transportes y Telecomunicaciones. 2014. Maestro de Transporte 2025 Santiago. Ministry Final Report, 2014. Santiago, Chile.

Muller, A.E.G. 2017. Transporte urbano e interurbano en la Argentina: aportes desde la investigación. Academic Press, Buenos Aires.

Noto, G. 2020. System Dynamics for Performance Management & Governance. Springer, Switzerland.

Novato, M. and M.I. Lacerda. 2017. RenovaBio – towards a new national biofuel policy and a truly sustainable world. Innovative Energy and Research. 6: 125-138.

Pinhate, T.B., M. Parsons, K. Fisher, R.P. Crease and R. Baars. 2020. A crack in the automobility regime? Exploring the transition of São Paulo to sustainable urban mobility. Cities. 107: 102914-102930.

PANTyCC – Plan de Accion Nacional de Transporte y Cambio Climatico. 2017. Ministerio de Ambiente y Desarrollo Sustentable, Final Report, 2017. Argentina.

Sagaris, L. 2018. Citizen participation for sustainable transport: lessons for change from Santiago and Temuco, Chile. Research in Transportation Economics. 69: 402-410.

Sampaio, M.R., P.L. Rosa and A.M. D'Agosto. 2007. Ethanol–electric propulsion as a sustainable technological alternative for urban buses in Brazil. Renewable and Sustainable Energy Reviews. 11: 1514-1529.

SCI – Sustainable Cities Index. 2018. Citizen Centric Cities. Final Report, 2018. Arcadis, Los Angeles, USA.

SECTRA. 2015. Secretaría de Planificación y Transporte. Actualizacion y recoleccion de informacion del sistema de transporte urbano. Final Report, 2015. Universidad Alberto Hurtado. Santiago, Chile.

Silva, R.N.A. and S.M. Costa. 2008. Multiple views of sustainable urban mobility: the case of Brazil Transport Policy. 15: 350-360.

SINDIPEÇAS – Sindicato Nacional da Indústria de Componentes para Veículos Automotores. 2019. Relatório de Frota Circulante, 2018. São Paulo, Brazil.

SPTyCOP. 2018. Argentina Urbana Plan EstratégicoTerritorial. Final Report, 2018. Ministerio del Interior, Obras Públicas y Vivienda, Buenos Aires, Argentina.

Tejeda, D.J. 2012. Movilidad sustentable em chile: oportunidades, experiencias locales y referencias globales. Urbano. 25: 29-37.

United Nations. 1987. Report of the World Commission on Environment and Development, 1987. General assembly resolution 42/187.

UNO – United Nations Organization. 2016. Mobilizing sustainable transport for development. Final Report, 2016. High-level Advisory Group on Sustainable Transport, New York, USA.

Vasconcellos, E.A. 2018. Urban transport policies in Brazil: the creation of a discriminatory mobility system. Journal of Transport Geography. 67: 85-91.

Viana, M.S. and P.M. Delgado. 2019. City logistics in historic centers: multi-criteria evaluation in GIS for city of Salvador (Bahia – Brazil). Case Studies on Transport Policy. 7: 772-780.

Final Considerations

Sônia Regina da Cal Seixas and João Luiz de Moraes Hoefel

The great challenge undertaken during this year to bring to the reader this book, *Environmental Sustainability – Sustainable Development Goals and Human Rights*, was only possible because we could count on the generosity of fellow authors and collaborators of this project, who gladly accepted our invitation. They were willing to dedicate their precious time in a tough year for everyone and offer us extremely stimulating chapters with excellent approaches within their specific areas of expertise.

This project directly involved 31 authors from four different countries, that is Brazil, Mozambique, the United States of America and Wales/UK. We want to thank all of them and their institutions. In addition to the completion of this project, their work and dedication allowed them to create an international collaboration network around the main themes that generated this book on sustainable environment and human rights.

We hope that we can continue to work in our collaboration network and contribute to a fundamental reflection, not only for contemporary science but also for the construction of a more sustainable planet where the values of solidarity, empathy, and generosity can effectively build the dignity of collective life.

Index

Printed and bound by CPI Group (UK) Ltd, Croydon, CR0 4YY

17/10/2024

01775709-0007